Thermoluminescence and Thermoluminescent Dosimetry

Volume II

Editor

Yigal S. Horowitz, Ph.D.

Associate Professor of Physics
Head
Radiation Physics Laboratory
Ben Gurion University of the Negev
Beersheva, Israel

CRC Press is an imprint of the
Taylor & Francis Group, an **informa** business

general introduction to TL and TL dosimetry as well as to the common quantities used in radiation dosimetry. The definitions follow the recommendations of the 1980 International Committee on Radiation Units. The second and third chapters deal with models of TL trapping and recombination centers and TL kinetics and serve to outline the complexity of the TL mechanisms as well as to present an up-to date discussion of our current understanding of these subjects. The fourth chapter is useful in that it comprehensively lists and discusses the important TL and dosimetric characteristics of the commonly used TLD materials.

The second volume (four chapters) is a hybrid dealing with both TL characteristics (Chapter 4 is a comprehensive discussion of phototransferred TL) and TL dosimetry. In the first chapter, that most important of dosimetric quantities, the TL dose response, is comprehensively reviewed with special emphasis on TL models for supralinearity and sensitization. The second chapter discusses the use of TLDs in various radiation fields (X-rays, beta rays, neutrons, heavy charged particles) and a review and evaluation of general cavity theory is also included. The third chapter introduces the concepts of track structure theory and shows how these can be successfully applied to our understanding of heavy charged particle and neutron TL response.

The third volume (three chapters) is the most applied in nature. The first chapter on TL instrumentation includes many important technical details and is primarily intended to help the TL dosimetrist avoid many of the common pitfalls and errors often encountered in TLD. The second and third chapters treat the clinical and archaeological-geological applications of TLD. These will be especially useful for current workers in the field seeking updated reviews on their specialties.

Yigal S. Horowitz
May 1983

THE EDITOR

Yigal S. Horowitz, Ph.D., is an Associate Professor of Physics and Head of the Radiation Physics Laboratory in the Physics Department of the Ben Gurion University of the Negev, Israel.

Dr. Horowitz received the B.Sc., M.Sc., and Ph.D. degrees in Physics from McGill University in 1961, 1965, and 1968, respectively. He carried out postdoctoral research in experimental nuclear physics at the University of Toronto Linac in 1969 and the Weizmann Institute of Science Tandem-Van de Graaf in 1970. He also carried out medical physics research at the Massachusetts General Hospital and the Massachusetts Institute of Technology in 1971. He joined the Ben Gurion University of the Negev in October 1971 serving as a senior lecturer to 1979 and Head of the Radiation Physics Laboratory from 1975. It was in 1980 that he assumed his present position as Associate Professor of Physics.

Dr. Horowitz is a member of the Canadian Association of Physicists, the Israeli Physical Society, the Israeli Nuclear Society, and the Israeli Association of Medical Physicists. He has been a member of the American Physical Society and the American Nuclear Society and has been the recipient of many research grants from the International Atomic Energy Agency and the United States — Israel Bi-National Science Foundation.

Dr. Horowitz is the author of more than 70 research papers and several review articles in the areas of thermoluminescence and thermoluminescent dosimetry. His current major research interests are in the areas of radiation induced thermoluminescence, radiation damage, track structure theory applied to thermoluminescence, thermoluminescence dosimetry in exotic radiation fields, radiation transport, general cavity theory, and medical radiation physics.

This volume is respectfully dedicated to Professor John Cameron for his vast contributions to the field of thermoluminescence and thermoluminescent dosimetry.

ACKNOWLEDGMENTS

I would like to thank Mrs. Helena Paskal of the Ben Gurion University graphic arts department for her heroic efforts in the careful preparation of the over 100 figures used in my own contribution to these volumes. I am also grateful to Professor Shapira, Dean of the Faculty of Natural Sciences, for financial support in the preparation of the manuscript, and to Professor Reuben Tiberger, Chairperson of the Physics Department for his support and for use of the departmental and university facilities.

CONTRIBUTOR

Vinod K. Jain, Ph.D.
Scientific Officer
Power Projects Safety Section
Health Physics Division
Bhabha Atomic Research Center
Bombay, India

THERMOLUMINESCENCE AND THERMOLUMINESCENT DOSIMETRY

Volume I

Volume II

Volume III

TABLE OF CONTENTS

Chapter 1

TL DOSE RESPONSE

Yigal S. Horowitz

TABLE OF CONTENTS

I. TL DOSE RESPONSE: DEFINITION

The TL signal is a function of dose, F(D), usually showing a linear, then supralinear, then sublinear behavior with increasing dose. We define the TL dose response

$$f(D) = \{F(D)/D\}/\{F(D_\ell)/D_\ell\}$$

where $F(D_\ell)$ is measured at low dose, D_ℓ, i.e., somewhere in the linear region of F(D). An ideal TL detector for many dosimetric purposes would satisfy the criterion f(D) = 1 (i.e., show linear response) into the dose range of MGy. This would not only make calibration as a function of dose unnecessary, but would also ensure that the TL yield per Gy would be independent of ionization density and in the framework of track structure theory would lead to universal relative TL responses of unity ($\eta = 1$) for high density ionizing heavy charged particles or low energy X-rays. Unfortunately, f(D) = 1 only up to a few Gy in LiF and most other TL materials. Above a few Gy f(D) is greater than unity (supralinearity) reaching values as high as 10 or more at approximately 100 Gy. The phenomenon of supralinearity in the radiation response of solid state systems is by no means unique,[1] however, the dependence of TL supralinearity on ionization density is apparently unique to TL mechanisms. At higher doses f(D) decreases due to recombination, saturation, and/or radiation damage and reaches values substantially less than unity in the 10^3 to 10^5 Gy dose region. A typical TL dose response curve for LiF (Harshaw®) measured with ^{60}Co gamma radiation and lower energy X-rays is illustrated in Figure 1.

A. Dependence of f(D) on Impurity Composition in LiF-TLD

There is extensive evidence in the literature illustrating the very quixotic dependence of supralinearity on impurity composition in LiF. Extreme examples are given by Jasinska et al.[2] who have described a LiF:Cu,Ag phosphor with f(D) = 1 up to 10^3 Gy (this TLD is also ten times more sensitive than LiF:Mg,Ti, but has not been exploited commercially because of the greatly enhanced fading relative to LiF:Mg,Ti) and DeWerd and Stoebe[3] and Vora and Stoebe[4] who have illustrated that the incorporation of OH^- radicals into the LiF crystal structure can reduce $f(D)_{max}$ from approximately three to unity with a concurrent loss in the overall TL efficiency. Stoebe and Watanabe[5] have reviewed the effects of OH^- incorporation on the TL properties of LiF. At the other extreme, Burgkhardt et al.[6] have described the properties of a LiF:Mg,Na phosphor with $f(D)_{max} = 6$ at approximately 10^3 Gy. Other groups[7-11] have studied the supralinearity of LiF as a function of Mg,Ti concentration. The following general picture has emerged although the results definitely vary in detail from author to author.

1. The onset of supralinearity is affected by Ti concentration (1 to 3 Gy in samples with 2 to 10 ppm Ti) and 10 to 50 Gy when the Ti concentration exceeds approximately 20 ppm.
2. The maximum value of f(D) varies between 2 and 4, however, Niewiadomski[11] using 35 to 40 kV_p X-rays reports little variation of $f(D)_{max}$ with Ti concentration in contradiction to the other groups who used ^{60}Co or ^{137}Cs gamma radiation.
3. The dose at which maximum supralinearity occurs is similarly correlated with Ti concentration. At low Ti concentration (2 to 10 ppm) this dose is 50 to 200 Gy, whereas at higher concentration (>20 ppm) the dose increases to 500 Gy. An extreme example is given by Vana et al.,[10] doping "pure" LiF with 107 ppm Ti produced a phosphor with $f(D)_{max} = 3$ at 2×10^3 Gy. Figure 2 shows the dependence of f(D) on Mg and Ti concentration.

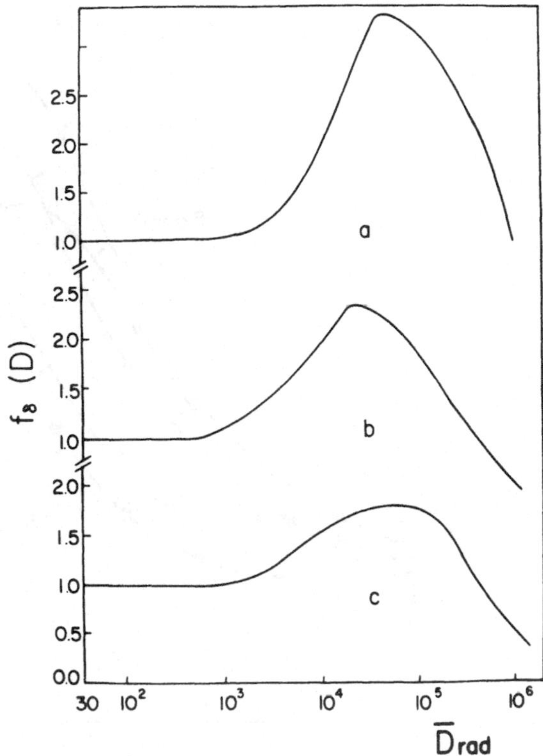

FIGURE 1. TL dose response of air-annealed LiF-TLDs (Harshaw®). (a) ^{60}Co gamma rays. (b) 50 kV_p X-rays. (c) 20 kV_p X-rays.

A question of considerable practical importance is the degree of universality of f(D) in various batches of TLDs purchased from the same supplier (e.g., Harshaw). Various groups have reported data for TLD-700 irradiated by ^{60}Co gamma rays.[6-8,12-14]

Although $f(D)_{max}$ always appears to occur at ~500 Gy, the onset of supralinearity varies from 1 to 10 Gy and $f(D)_{max}$ varies between 2.5 and 4. Similar variations are reported for TLD-100.[4,10,15]

Unfortunately insufficient experimental details are included with respect to annealing conditions, glow curve analysis, etc. to even guess at the contribution of these factors. Piesch et al.[16] and Burgkhardt et al.[6] report definite variations in f(D) between TLD-600 and TLD-700 (Figure 3) and even greater variations between two different batches of LiF-7 (Teledyne®) probably taken under identical experimental conditions.

B. Supralinearity as a Function of Ionization Density

1. Electrons and Gamma Rays

Supralinearity in LiF-TLD is a function of electron and gamma-ray energy.[13-15,17-23] Several general features emerge.

1. The maximum superlinearity of ~3.5 in TLD-100 is observed with ^{60}Co gamma rays with only very slight increase or no increase of f(D) with increasing electron energy[15,20] (Figure 4).
2. The maximum supralinearity begins to decrease with decreasing electron energy below approximately 275 keV reaching $f(D)_{max}$ = 1.2 for 5 keV electrons. Figure 1 shows

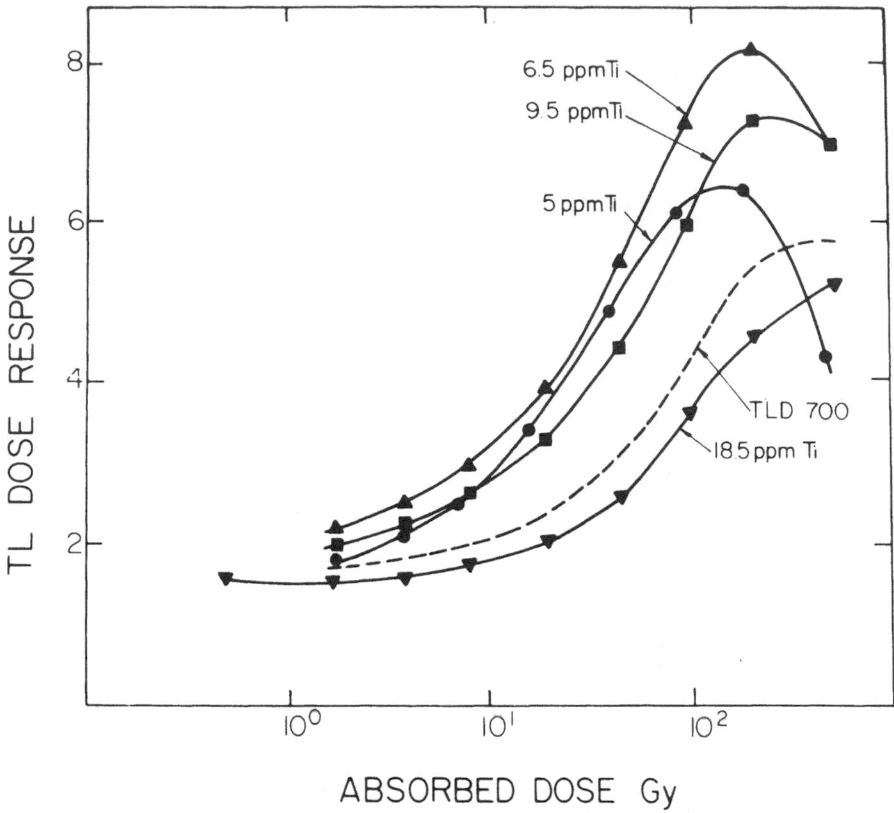

FIGURE 2. TL dose response of various forms of LiF as a function of Ti concentration using ^{60}Co gamma rays. (Adapted from Rossiter, M. J., Rees-Evans, D. B., Ellis, S. C., and Griffiths, J. M., *J. Phys. D*, 4, 1245, 1971.)

the dose-TL response of air-annealed LiF-TLDs for ^{60}Co gamma rays, 50 kV_p X-rays, and 20 kV_p X-rays.

The onset of supralinearity is also energy dependent. Rossiter[13] reported 120.9 ± 1.1 TL per Gy (95%) confidence limit) at 1.877 Gy in TLD-700 and 125.3 ± 0.4 TL per Gy at 3.749 Gy for ^{60}Co, whereas for 220 kV_p X-rays no supralinearity was observed between 1 and 4 Gy. Lasky and Moran[22,23] claim no electron-induced supralinearity below 5 keV for the integral of peaks 3 to 5, however, this claim appears to be based on data mainly limited to the sublinear high dose region. The one curve extended to relatively low dose at 200 eV is definitely supralinear! Barber et al.[21] report supralinearity for 4, 2, and 1 keV electrons with $f(D)_{max}$ increasing with decreasing electron energy, $f(D)_{max}$ = 1.4 and 2 for 4 and 1 keV electrons, respectively. Barber et al. measured the "net" TL probably implying the integral of peaks 4 to 8 if they followed standard annealing procedures. The increasing supralinearity with decreasing electron energy might therefore be due to the greater supralinearity of peaks 6 to 8 coupled with an opposite dependence on ionization density as that observed with peaks 3 to 5. Also, Barber et al. used air-annealing and their values of $\eta_{e\gamma}$ for 1.0 and 2.0 keV electrons equal to 0.2 indicate surface contamination (probably via OH^- ions) so that their results may not be of general significance. The increasing supralinearity is, however, not consistent with OH^- contamination.[5] Supralinearity also decreases with increasing ionization density in BeO[24,25] and $Li_2B_4O_7$.[14] No other detailed studies of

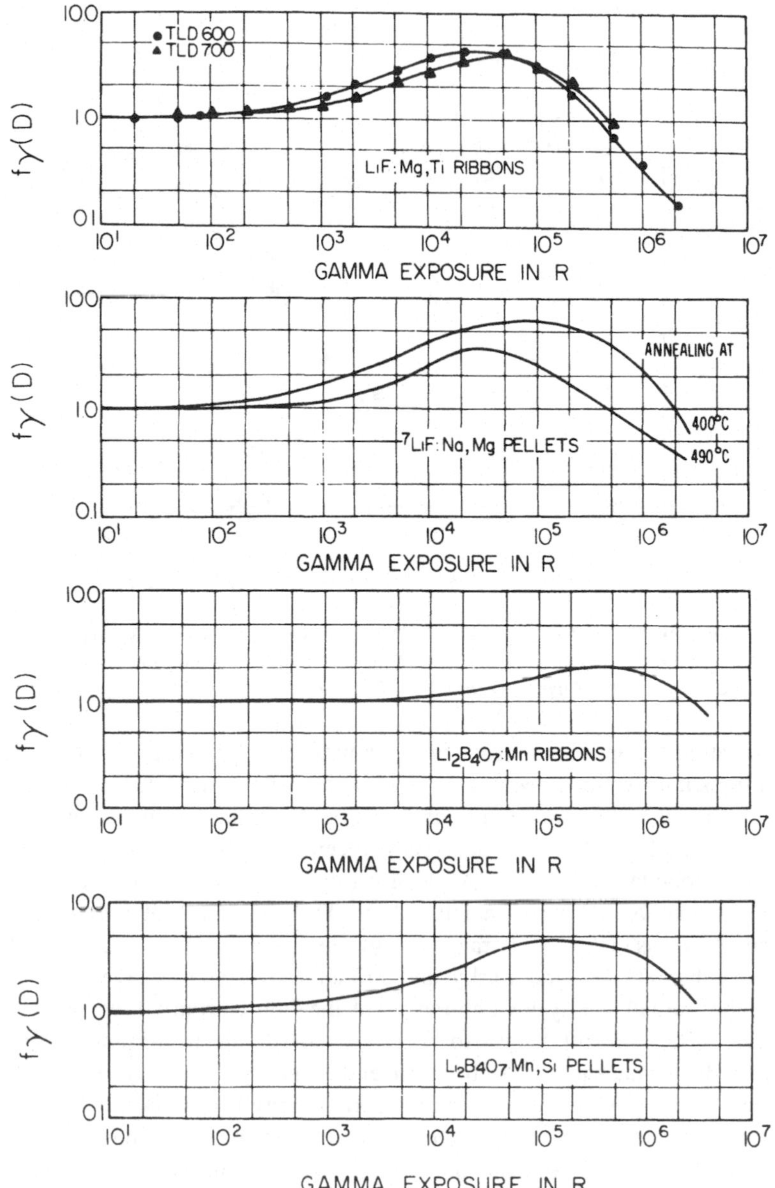

FIGURE 3. TL dose response of various TLDs. Note the effect of T_1 on the supralinearity of ^{7}LiF:Na,Mg pellets. (Adapted from Burgkhardt, B., Piesch, E., and Singh, D., *Nucl. Instrum. Methods*, 148, 613, 1978.)

the dependence of supralinearity on ionization density have been reported for the other supralinear TL materials except in CaF_2:Dy (Harshaw) where the glow curve structure is reported strongly dependent on radiation quality,[26] implying varying dependence of the supralinearity from glow peak to glow peak on X-ray energy.

2. Neutrons and Heavy Charged Particles

In the framework of track structure theory (TST) heavy charged particle (HCP) or neutron-induced supralinearity can arise only from the supralinear behavior of the secondary electrons liberated by the HCP via a track intersection between the HCP tracks. As such, the exper-

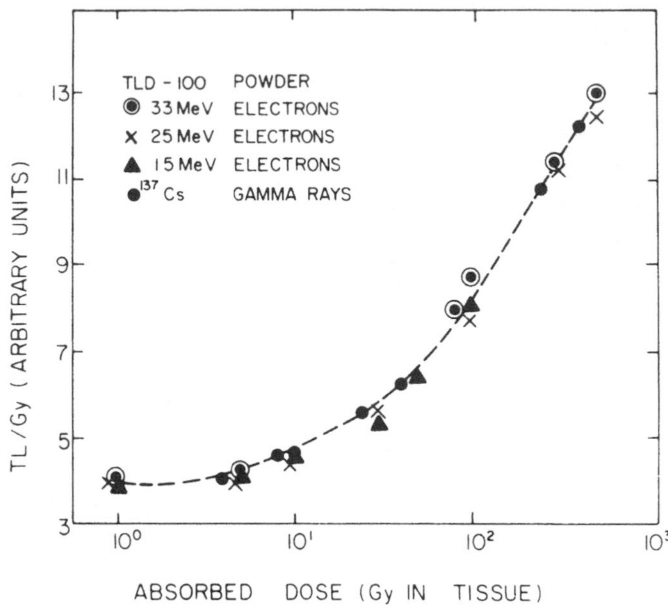

FIGURE 4. TL dose response for TLD-100 powder irradiated by ^{137}Cs gamma rays, 15, 25, and 33-MeV electrons. (Adapted from Suntharalingam, N. and Cameron, J. R., *Phys. Med. Biol.*, 14, 397, 1969.)

imental observation of thermal neutron- or alpha-induced supralinearity is (in TST) definite evidence for track intersection or equivalently track interaction in the nomenclature of Attix.[27] Since most of the HCP TL-induced research has been via thermal neutrons or low energy alpha particles (~1 MeV/AMU) our analysis will center on these radiations. The maximum electron energy liberated by a 4 to 5-MeV alpha particle or the alpha-triton pair liberated in the $^6Li(n,\alpha)\tau$ reaction is approximately 2 keV. Therefore, it is of critical importance to know whether electrons in the energy range of 0 to 2 keV exhibit supralinearity. Unfortunately the work of Barber et al.[21] and Lasky and Moran[22,23] is inconclusive. Experiments are underway in the Radiation Physics Laboratory of the Ben Gurion University of the Negev on this subject using ultrasoft X-rays to study low energy electron-induced supralinearity. Preliminary studies with tritium electrons (average energy 7 keV) yielded $f(D)_{max} = 2$. Nonexistence of low energy electron-induced supralinearity below about 2 keV would imply that thermal neutron and low energy alpha-induced supralinearity cannot be understood in the framework of conventional track interaction TST. An alternate explanation of HCP-induced supralinearity is via a track interaction that involves new traps or luminescent centers created via enhanced production of lattice displacements (relative to low energy electrons) due to atom-atom collisions. It is difficult to eliminate this possibility, but the recent work of Horowitz and Kalef-Ezra[28] and Kalef-Ezra and Horowitz[14] supports the contention that HCP-induced displacements do not play a role in the efficiency of the TL mechanism.

Thermal neutron supralinearity of peaks 4 and 5 in TLD-600 has been observed by Piesch et al.[16,29] and Nash and Johnson[30] (Figure 5). It is interesting to note that contrary to the supralinear behavior, the decrease in f(D) above 3×10^5 R is very similar for thermal neutrons and gamma rays indicating that saturation is somehow more closely related to the *average macroscopic dose* absorbed by the crystal irrespective of the details of the individual particle microscopic dose distributions. Nash and Johnson using peak height measurements observed $f(D)_{max} \simeq 1.2$ at 10^2 Gy. When the glow curve was integrated to 300°C, Nash and Johnson observed $f(D)_{max} = 1.4$ at 500 Gy indicating more pronounced neutron-induced

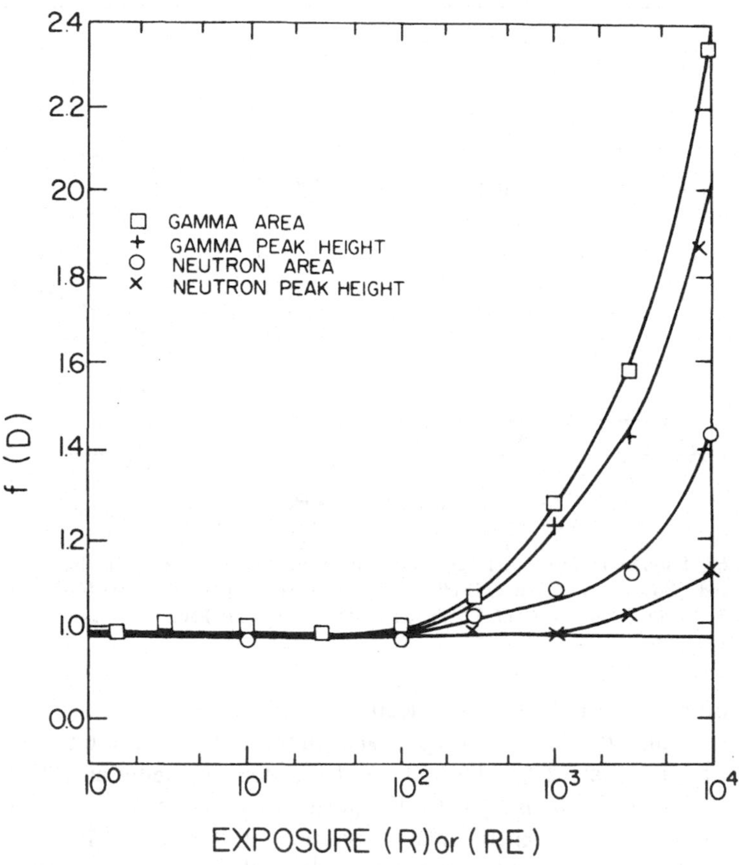

FIGURE 5. Thermal neutron-induced supralinearity for TLD-600. Peak height is for peak 5; area is the total area integrated to T_8 = 300°C. (Adapted from Nash, A. E. and Johnson, T. L., Proc. 5th Int. Conf. Luminescence Dosimetry, Sao Paulo, Physikalisches Institut, Giessen, 1977, 393.)

supralinearity of the high temperature peaks. Piesch et al.[16,29] used T_8 = 240°C so that part of the supralinearity observed by Piesch et al. might be due to the enhanced supralinearity of the high temperature peak components below 240°C. Indeed, Jain and Ganguly,[31] Jain,[32] and Busuoli et al.[33] observed peak 7 (255°C) supralinear from 1 Gy with $f(D)_{max}$ = 1.5 at 20 Gy (the maximum dose studied). Various other authors have also observed thermal neutron-induced supralinearity.[34-38] In the case of Wingate et al.[37] the supralinearity may have been due to the 4 to 20% γ component in the thermal neutron radiation field. Mason[35] observed very strong thermal neutron-induced supralinearity in Conrad type 7 LiF. Onset of supralinearity was at 0.1 Gy (~5 Gy for ^{60}Co gamma rays) with $f(D)_{max}$ = 1.4 at 10 Gy. To the best of our knowledge this is the only report of thermal neutron-induced supralinearity with a lower dose threshold than ^{60}Co-induced supralinearity. However, Mason[35] also observed in the same experiment n − γ nonadditivity which was probably due to contamination of the TLDs used in his study. Let us look at the question of the thermal neutron-induced supralinearity of peaks 4 and 5 in greater detail. Nash's[30] results of f(D) = 1.18 at 10^4 R equivalents were obtained for the height of peak 5. Our own studies[39] of the overlap of peak 5 with the higher temperature peaks indicate that the maximum contribution of the low temperature tail of peak 6 to the peak height of peak 5 is approximately 5% at the highest doses. Nash et al.[30] measured the ratio of the 250 to 190°C peak heights

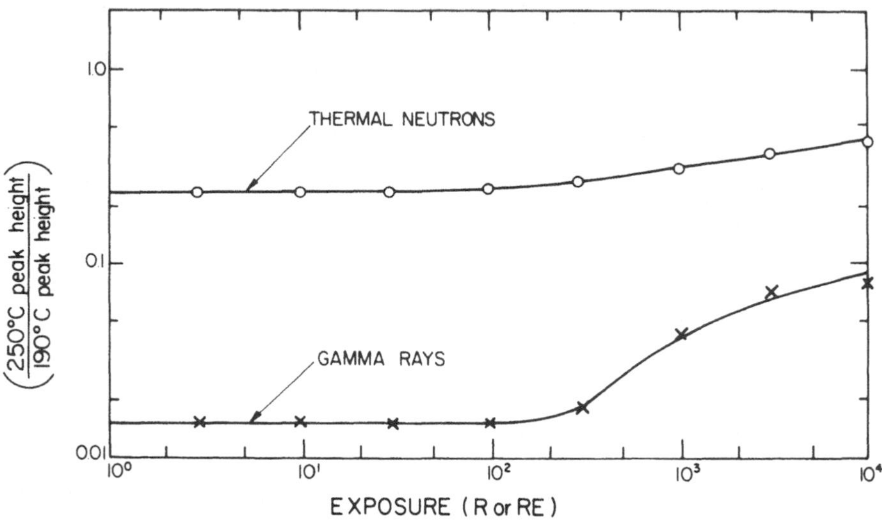

FIGURE 6. Ratio of the 250°C peak height to the 190°C peak height vs. exposure for both thermal neutrons and ^{60}Co gamma rays in TLD-600 for $T_7' = 5°C\ sec^{-1}$. (Adapted from Nash, A. E. and Johnson, T. L., Proc. 5th Int. Conf. Luminescence Dosimetry, Sao Paulo, Physikalisches Institut, Giessen, 1977, 393.)

as changing from 0.3 to 0.45 over the neutron supralinear dose region (Figure 6). This reduces the maximum effective percentage change in the overlap (peak 6 tail to peak height of peak 5) to approximately 2.5%. For a 2.5% change to bring about an 18% effect would require f(D) = 7 for peaks 6 and 7 at 10^4 R equivalents which is approximately a factor 5 greater than that indicated by experimental observations. In the case of Piesch et al.[16,29] the ratio of the peak heights increased even more slowly so that the above argument maintains its validity. We emphasize this point because of the possibly very important significance of the observation of thermal neutron-induced supralinearity to track interaction track structure applicability to TL mechanisms. It should be further noted that the thermal neutron dose distribution in TLD-600 is very nonuniform because of the very strong exponential absorption. This implies a smearing of the true thermal neutron-induced supralinearity that would tend to artificially depress the experimentally observed supralinearity.

In the case of low energy alpha particles, Jahnert[40] used 3.7 MeV alphas to irradiate TLD-700 and observed f(D) = 1.3 at 10^3 Gy (maximum dose studied). Unfortunately Jahnert made no attempt at differential glow peak analysis so that the observed supralinearity is for the integral glow to 370°C. Aitken et al.[41] used 4 MeV alphas to irradiate TLD-100 and reported peak 5 (230°C) linear to 3×10^3 Gy followed by saturation, whereas the high temperature peaks (290°C) showed $f(D)_{max} \simeq 1.5$ at 3×10^3 Gy with onset of supralinearity at ~100 Gy. Jain and Ganguly[31] report only peak 8 (280°C) supralinear [$f(D)_{max} = 2$ at 10^4 Gy], with the lower temperature peaks all linear to 10^4 Gy. Apparently peaks 4 and 5 are not supralinear for alpha-induced TL, but are slightly supralinear for thermal neutron-induced TL. But only Jain and Ganguly[31] studied the same material with both alphas and thermal neutrons so that this conclusion may be premature. The maximum electron energy given up by the 2.7-MeV triton is slightly less than the 2 keV which can be liberated by a 4-MeV alpha particle so that it is difficult but not impossible for the secondary electron theory of track structure to explain this anomaly. The explanation must come from the different geometry of the overlapping track dose distributions in the two cases. Monte Carlo calculations on this problem are currently underway at the Ben Gurion University (BGU). Another alternative is that the supralinearity is somehow suppressed for surface irradiation (alphas) compared to bulk irradiation (thermal neutrons). Suntharalingam and Cameron[20] reported no supralinearity with 4.4-MeV deuterons to 10^3 Gy. Wingate et al.[37] and Tochilin et al.[24] report supralinearity for very high energy HCP (745-MeV protons and 930-MeV alpha particles, respectively) in LiF, $Li_2B_4O_7$, and BeO similar to ^{60}Co-induced supralinearity. In the framework of TST this latter resemblance is easily understood because of the similarity in the secondary electron spectrum created by these radiation fields. Finally, Ayyangar et al.[38] and Lakshmanan et al.[42] report thermal neutron-induced supralinearity in $Li_2B_4O_7$ with $f(D)_{max}$ = 1.5 and 1.25, respectively, at $\sim 5 \times 10^3$ Gy.

Kitahara et al.[43] using home-grown LiF (two glow peaks at 150 and 335°C) report the first peak linear and the second peak slightly supralinear to thermal neutrons and Mehta and Sengupta[44] observe no alpha-induced supralinearity in Al_2O_3.

C. Supralinearity of Individual Glow Peaks

Several groups have studied the TL dose response of the individual glow peaks in LiF.[9,31,45-50] A general observation is that increasing glow peak temperature results in increasing supralinearity, although there are exceptions, e.g., peak 1.[9] Podgorsak et al.[45] showed that the maximum supralinearity of peaks −4 (−128°C), −1 (−6°C), 0 (14°C), 2 (102°C), and 5 (186°C) varies linearly as the glow peak temperature cubed. The data of Sunta et al.[46] agree with this formulation with glow peaks 7 (250°C) and 10 (395°C) included, although the maximum supralinearity of peak 10 is somewhat high. The fact that the maximum supralinearity is, under certain conditions, directly proportional to the cube of the peak temperature is certainly suggestive of some common mechanism, and Attix[27] related the T^3 dependence to carrier mobility as a function of T leading to enhanced track interaction effects and greater supralinearity as a function of T^3. The mobility of free carriers, however, is usually given by a T^{-n} dependence.[51,52] Another possibility is that vacancy mobilities and concentrations are playing a role at the higher temperatures since the ionic conductivity of TLD-100 increases exponentially as the temperature increases.[53,54] Jain and Ganguly[31] also report increasing supralinearity as a function of temperature, however, $f(D)_{max}$ = 20 for peak 7 compared to $f(D)_{max}$ = 3.5, 4, and 5 for peaks 3, 5, and 8, respectively. Linsley and Mason[47] report $f(D)_{max}$ = 3 (peak 5) and $f(D)_{max}$ = 60 for the higher temperature peaks at ~300°C. However, a double readout cycle was used to separate peak 5 from the 300°C composite peak, which implies cooling to lower temperatures between readout cycles and consequent possible trap rearrangement. Antinucci et al.[48] found $f(D)_{max}$ proportional to $T^{4.89}$.

Waligorski and Katz[49,50] also observed enhanced supralinearity for the high temperature peaks in TLD-700 inconsistent with T^3 behavior. Crittenden et al.[9] studied the TL dose response of BDH-LiF with 80 ppm Mg, 3 ppm Ti and 80 ppm Mg, 0.5 ppm Ti. In the former case peak 5 was linear to 10^3 Gy and peak "6" was strongly supralinear, whereas in the latter case both peaks were strongly supralinear. Obviously the "T^3" behavior observed by Podgorsak et al.[45] and Sunta et al.[46] does not apply to all forms of LiF:Mg,Ti nor to all the glow peaks. Experimental parameters (e.g., thermal treatments) also probably affect the temperature dependence of the supralinearity.

Driscoll and McKinlay,[55] attempting to link peak "6" with OH or O surface contamination, stated that the threshold of supralinearity for peak "6" occurs at higher dose than for peak 5 and connected this to the well-documented experimental observation that incorporation of OH^- ions depresses the supralinearity. Unfortunately, the exact opposite is true. Threshold of supralinearity for peak "6" occurs at lower dose than for peak 5.[32,33,47,49,50] Linsley and Mason[47] report the 210°C peak in TLD-700 supralinear from ~10 Gy and the "300°C" peak supralinear from ~0.1 Gy using ^{60}Co gamma rays. Busuoli et al.[33] similarly report the 210°C peak in TLD-100 supralinear above 10 Gy and the high temperature peaks supralinear above ~0.1 Gy using 46- and 210-keV X-rays. With ^{60}Co gamma rays peak 5 was linear to 100 Gy and the high temperature glow peak component was supralinear above ~5 Gy. Waligorski et al.[49,50] report peak 5 in TLD-700 supralinear above ~2 Gy and peak "6" supralinear above ~0.5 Gy using ^{60}Co gamma rays, and finally Jain[31,32] reports peak 7 (~250°C) supralinear above 1 Gy with peak 5 supralinear above 10 Gy. To the best of our knowledge there is no mention in the LiF-TLD literature of a higher threshold for supralinearity for high temperature TL (~250 to 300°C) compared to peak 5 TL.

Oliveri et al.[56,57] have analyzed the supralinearity of the main glow peak of $CaSO_4$:Dy based on their model of an approximately multi-Gaussian distribution of trap depths. The areas of the Gaussians show different supralinear behavior. The higher the trap energy, the greater the supralinearity and the lower the dose at which supralinearity begins. This behavior related to the trapping structures within a single glow peak is thus similar to the temperature behavior observed in LiF. In $CaSO_4$:Sm Bjarngard[58,59] also reports $f(D)_{max}$ = 3 and 6 for 2 peaks at 95 and 400°C, respectively. The rapidly increasing supralinearity with glow peak temperature is, however, apparently limited to LiF and $CaSO_4$. In CaF_2:Dy, Burgkhardt et al.[60] report $f(D)_{max}$ = ~2.4, ~2.5, ~3.9, and ~2 for peaks 1 (150°C), 2 (195°C), 3 (230°C), and 4 (260°C), respectively, and in Al_2O_3, Mehta and Sengupta[44] observe $f(D)_{max}$ = 2.5 and 1.2 for glow peaks occurring at 250 and 475°C, respectively, with a further peak at 625° showing sublinear behavior. In BeO, Scarpa et al.[61] report $f(D)_{max}$ = 10 for the main glow peak at 220°C and a higher temperature peak at 350°C which grows linearly with dose and then saturates at 10 Gy.

D. Supralinearity of Various TL Materials

The whimsical behavior of f(D) in LiF as a function of impurity composition and other factors extends to most or all of the other TLD materials that have been extensively investigated.

The dose response of $Li_2B_4O_7$ in its various forms has been studied by many groups.[6,24,62-65] Onset of supralinearity can vary from approximately 0.5 Gy for $Li_2B_4O_7$:Mn BDH[64] to ~20 Gy for $Li_2B_4O_7$:Mn,Si ribbons produced by Harshaw.[6] Kirk et al.[62] showed that the supralinearity of $Li_2B_4O_7$:Mn is a function of the compositional variation, n, used in the starting materials 1 Li_2O:nB_2O_3 centered around n = 2. Takenaga et al.,[65] on the other hand, report $Li_2B_4O_7$:Cu linear to 10^3 Gy. BeO is a very strongly supralinear material with $f(D)_{max} \simeq 14$ at ~10^2 Gy and onset of supralinearity at approximately 1 Gy[6] for ^{60}Co irradiation. Calcium fluoride in its natural form has been studied by Schayes et al.[66] who observed that peaks II, III, IV, and V were linear to 10^3 Gy, however, Podgorsak et al.[67] using high purity CaF_2 and 30 keV eff. X-rays observed supralinearity above 1 Gy (irradiation was carried out at liquid nitrogen temperatures). CaF_2:Mn has been studied by Marrone and Attix[68] and Ehrlich and Placious[69] who reported linear response to 3×10^3 Gy, whereas Gorbics et al.[12] and Burghardt et al.[60] observed supralinearity, with the latter reporting $f(D)_{max}$ = 1.6 at 5×10^2 Gy. Gorbics et al. observed a progressive decrease in the supralinearity with increasing sample thickness which was approximately accounted for by phosphor discoloration at high dose and resulting optical absorption of TL light within the phosphor itself. CaF_2:Mn powders of varying thickness also yielded similar results. CaF_2:Dy has been studied by Binder and Cameron[26] and Burgkhardt et al.[60] The former have found onset of supralinearity at approximately 6 Gy with $f(D)_{max} \simeq 2$ at 10^2 Gy, while the latter have reported onset of supralinearity at 1 Gy with $f(D)_{max}$ = 2.6 at 500 Gy. Lucas and Kapsar[70] have studied the behavior of CaF_2:Tm (TLD-300, Harshaw) and report a single high temperature peak at 250°C with maximum supralinearity a strong function of Tm concentration [$f(D)_{max}$ = 3 and 7 for 0.35 and 0.05% Tm concentration, respectively]. The TL dose response of $CaSO_4$:Dy has been studied by many groups.[56,60,71-74] Onset of supralinearity varies from 1 to 50 Gy with $f(D)_{max}$ varying from 1.6 to 3.6 in the 10^2- to 10^3-

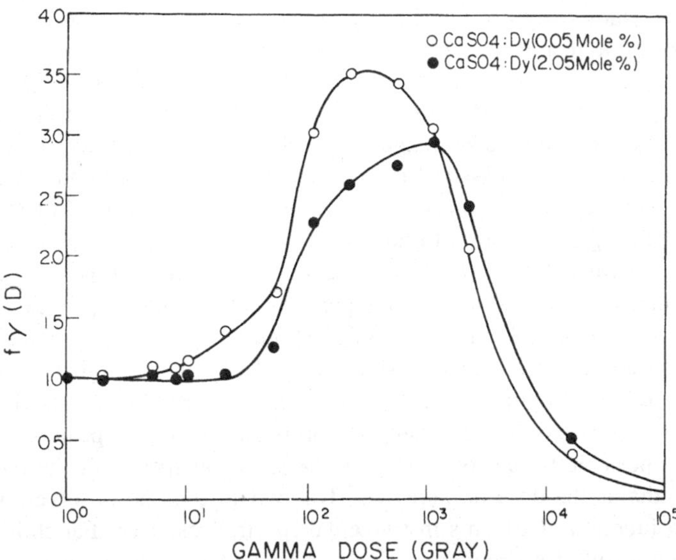

FIGURE 7. TL dose response as a function of Dy concentration in $CaSO_4$:Dy. (Adapted from Chandra, B. and Bhatt, R. C., *Nucl. Instrum. Methods*, 164, 571, 1979.)

Gy dose range. The supralinearity is strongly dependent on the Dy concentration (Figure 7). $CaSO_4$:Tm shows equal diversity[60,71] with reports of onset of supralinearity at 6 Gy and $f(D)_{max}$ = 2 at 10^2 Gy[71] and onset of supralinearity at 1 Gy and $f(D)_{max}$ = 3.3 at 5×10^2 Gy.[60] On the other hand, Bjarngard[59] reports $CaSO_4$:Mn linear to 10^2 Gy followed by saturation. Other supralinear materials which have been studied less extensively are Mg_2SiO_4:Tb,[75] $SrSO_4$:Dy[76] $BaSO_4$:Dy,[76] BaF_2:Dy,[77] and Al_2O_3.[78,79] Thus, supralinearity in thermoluminescence is certainly the rule rather than the exception. Unfortunately, it is obvious that the TL dose response is a nonuniversal characteristic even for different batches of the same host TL material with the same nominal doping. This observation has several important implications.

1. Every batch of dosimetric TL material must be individually calibrated under the identical experimental conditions for which its dosimetric use is intended.
2. The exact form of f(D) is certainly dependent on the impurity and defect mapping of the host lattice, as well as on a great number of other experimental factors (see following section).
3. Because of this dependence on a large number of factors, theoretical model building becomes an especially difficult and precarious game.
4. Since TST predicts relative HCP TL efficiencies via a convolution of f(D) with the ionization density around the HCP track, it follows that relative HCP and neutron TL yields can be expected to show the same variegated behavior as f(D) itself. This latter point will be developed in detail in Chapter 3.

E. Influence of Various Other Factors on Supralinearity

1. Emission Spectrum

Many groups have studied the TL emission spectrum of TLD-100.[9,45,80-83] Until the work of Fairchild et al.[83] it was universally concluded that no variation of the emission spectrum occurred with dose over the range of 10 to 10^4 Gy for any of the glow peaks. DeWerd et al.[81] extended the studies to even lower dose and reported that the ratio of the TL for high dose (3.4×10^2 Gy) to that of low dose (1 Gy) was the same for all wavelengths. In general, peaks 4 and 5 are connected to a single emission band centered at 4000 Å with a full width at half maximum (FWHM) of ~1200 to 1600 Å. The constancy of the emission spectrum over the linear and supralinear dose range is usually advanced as convincing evidence vs. models of supralinearity in LiF based on the creation of luminescent recombination centers.[84]

Bloch,[85] however, reported an emission spectrum which tended to shift to longer wavelengths with increasing dose from 4 to 60 Gy. One possible explanation of Bloch's results proposed by Zimmerman,[86] is that Bloch included a fraction of peaks 2 and 3 in his TL measurements since he used a 20-hr room temperature anneal following irradiation. Fairchild et al.[83] have, indeed, observed that the emission spectra of peaks 1 to 3 contain very strong 2.71 eV emission and that the emission shifts to 3.01 eV for the higher temperature glow peaks. Since the relative population of the higher temperature glow peaks also increases as a function of dose because of the increasing supralinearity with glow peak temperature (peak 3 is linear), it is possible to observe a shift in the emission spectra of the integral glow to longer wavelengths as the dose is increased. This would then be observed as varying supralinearity as a function of Bloch's instrumental spectral response. Fairchild et al.[83] using sophisticated equipment[87] for simultaneously determining the emission intensity and emission spectrum as a function of sample temperature studied ^{60}Co-irradiated TLD-100 from 5 to 3×10^5 Gy. The actual sample temperature differed from the oven temperature by at most 1°C. This was determined by inserting one junction of a fine wire differential chromel-alumel thermocouple into a hole drilled in a typical LiF crystal and fastening the other junction to the oven. Below 10^5 R (encompassing the supralinear region) the TL emission could be described by a single Gaussian-shaped band whose peak wavelength and full width varied irregularly with temperature and not in accord with well-known theoretical expressions relating emission spectra peak wavelength, FWHM, to sample temperature. To obtain agreement with the behavior with temperature three Gaussian-shaped bands are required whose approximate peak energies and full widths (in eV) are 3.01 (0.90), 2.90 (0.72), and 2.71 (0.96). At 6.7×10^6 R, bands at 4.0 and 2.3 eV are observed and above approximately 3×10^7 R, a fifth additional band at approximately 1.5 eV is also discernible. Resolution of the measured spectra for glow peaks 1 to 7 and 9 into individual Gaussian-shaped bands (10^5 R ^{60}Co irradiation) is shown in Figure 1 of Chapter 4 of Volume I and Figure 8 shows the glow peaks for the three prominent emission bands. Fairchild et al. suggest that the appearance of only one emission band at low temperature, three at intermediate temperature, and one at high temperature could arise from varying stability of electron or hole traps with temperature. For example, at low temperature only electron or hole traps are unstable, at intermediate temperature both electron and hole traps are unstable, and at high temperature only hole (or electron) traps are unstable. Obviously many other temperature-dependent mechanisms may also be operative. Although the evidence for multiple bands is not entirely conclusive (the bands are totally unresolved by eye) the evidence is certainly strongly suggestive of the presence of more than one kind of luminescent recombination center with even greater complexity above 10^5 R. From the point of view of supralinearity, however, Fairchild et al. are in agreement with previous work that no changes in the emission spectra of individual glow peaks occur over the linear-supralinear dose region. Other authors have also reported multiple bands in the emission spectrum of LiF at higher wavelengths and at higher dose in agreement with the above observations.[9,82,88,89]

Several authors have studied the TL emission spectrum of LiF following irradiation with densely ionizing HCP radiations.[88,90,91] The HCP TL emission spectrum is, in TST, a convolution of the emission spectrum as a function of dose with the dose distribution around the HCP track.[14] Both Strash et al.[88] and Horowitz et al.[91] report slightly greater emission in the long wavelength region (5500 Å) for protons and alphas, respectively, when compared

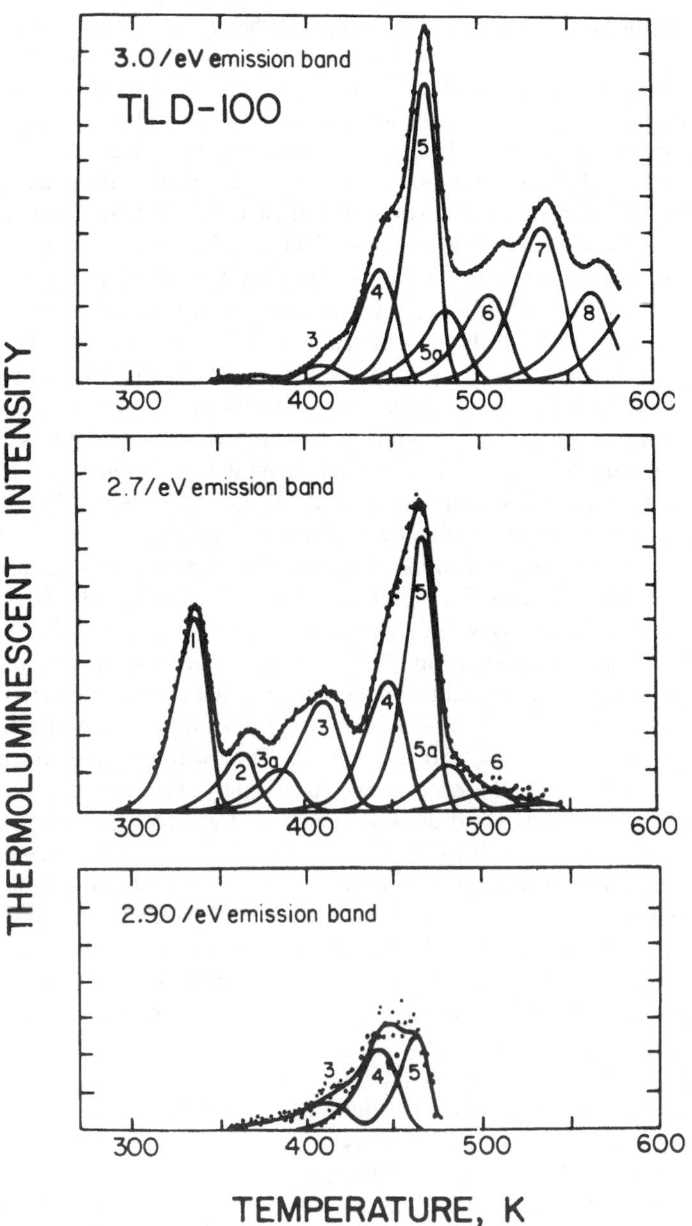

FIGURE 8. Emission spectra of the various glow peaks of TLD-100 measured at 10^5 R ^{60}Co. Also shown in the resolution of the measured spectra into individual Gaussian-shaped bands. (Adapted from Fairchild, R. G., Mattern, P. L., Langweiler, K., and Levy, P. W., *J. Appl. Phys.*, 49, 4512, 1978.)

to ^{60}Co gamma rays in the linear dose region. This observation is consistent, therefore, with the observations of Fairchild et al. who report new emission bands at lower energies for D $> 10^3$ Gy since the HCP dose distribution extends into the very high dose region. Oltman et al.[90] observed very intense additional peaks following alpha irradiation at 5200 and 6000 Å, however, his glow curves were not typical and his annealing procedures were carried out in air so that we agree with his speculation that the very large differences between the alpha and gamma spectral emission were probably caused by surface contamination.

2. Spectral Emission Influence on Supralinearity in Other TL Materials

Horowitz et al.[91] have studied the TL emission spectrum of $Li_2B_4O_7$:Mn,Si (TLD-800, Harshaw) following gamma and meV neutron irradiation. No differences were observed in the emission spectrum which is consistent with the observation (in TST) that the $Li_2B_4O_7$ emission spectrum is not significantly dependent on dose. Konschak et al.[89] have observed a shift in the wavelength of the emission maximum as a function of the absorbed dose for CaF_2:Mn. At 40 kGy the emission maximum shifts about 150 Å to lower wavelengths in comparison to 100 Gy (emission maximum at 5200 Å). The shift is slight, however, since it occurs in the supralinear dose region the TL dose response may be slightly dependent on reader spectral response for this material. Sunta[92] has observed dramatic changes in the TL emission spectrum of natural CaF_2 as a function of dose. At low dose there is a broad maximum peaking at 3700 Å with a small shoulder at 3250 Å. The emission spectrum remains stable over the linear dose response region to 10^2 Gy. In the sublinear dose region at 3.5×10^2 Gy the emission spectrum changes to roughly three equal peaks at 3250, 3450, and 3700 Å. It is slightly ironic (re: the creation model of luminescent centers for the explanation of supralinearity) that in both natural calcium fluoride and LiF the changes in the emission spectrum, associated with new luminescent centers, are closely correlated with the onset of the sublinear (rather than the supralinear) region of dose response. Binder and Cameron[26] have studied the emission spectrum of CaF_2:Dy (TLD-200, Harshaw) and observed identical emission following ^{137}Cs gamma rays, ^{90}Sr betas, and ^{244}Cm alpha particles. Again this observation is consistent (in TST) with no significant changes of the emission spectra with dose. Finally, Plichta and Neruda[93] showed that the supralinearity of aluminum phosphate glass activated with Mn was a very strong function of three different photomultipliers with peak response at 380, 420 and 550 nm. The maximum supralinearity observed at 5000 R was 1.9, 1.6, and 1.25, respectively. In summary, the emission spectra of several TL materials have been shown to be dose dependent (LiF, CaF_2:Mn, CaF_2:nat., and aluminum phosphate glass doped with Mn). $Li_2B_4O_7$:Mn and CaF_2:Dy show identical emission spectra following low dose gamma irradiation and HCP irradiation so that their emission spectra are essentially independent of dose; however, this requires differential studies as a function of dose to enable more accurate assessment. In LiF the emission spectrum of individual glow peaks is definitely stable over the linear and supralinear dose regions, and changes in the emission spectrum do not, therefore, play a role in the supralinear response.

3. Grain Size

Shiragai[94] using LiF single crystals (Institute of Applied Optics, Tokyo) observed f(D) increasing from 2.7 to 3.2 at 1000 R as the tyler mesh size was increased from 50 to 325. The IAO material showed far greater supralinearity than either Conrad-N (~1.8) or TLD-100 (~1.1) and onset of supralinearity was at 10 R in the IAO material. On the other hand, Zanelli[95] studied TLD-700 grains of size $\bar{d} < 1$ μm, $\bar{d} \simeq 4$ μm, and $\bar{d} \simeq 100$ μm, and observed no dependence of the supralinearity on grain size with $f(D)_{max} = 2.2$ at 500 Gy. The most probable conclusion is that the effects noted by Shiragai were peculiar to the very strongly supralinear IAO TL material rather than a general property of the more common commercially available dosimetric LiF.

4. Physical State

Gorbics et al.[12] and Piesch et al.[16] have observed significant changes in the supralinear behavior of TLD-700, $Li_2B_4O_7$:Mn, and CaF_2:Mn (Harshaw, Teledyne) as a function of the physical state of the TLDs (i.e., enclosed in thick or thin teflon®, chips, or powder). For $Li_2B_4O_7$:Mn the former report $f(D)_{max} = \sim 1.6$, ~2.5, and ~2.7 for chips, powder, and teflon®, respectively; for CaF_2:Mn, $f(D)_{max} = \sim 1.05$, ~1.3, and ~1.4 for chips, powder, and teflon®, respectively. Interestingly, of the three phosphors, LiF showed the greatest similarity between the various forms, all reaching $f(D)_{max}$ of 2.3 to 2.6 at approximately 10^5 R. Finally, Koczynski et al.[96] report that the supralinearity of $Li_2B_4O_7$:Mn,Si was a function of graphite loading.

5. Heating Rate and Annealing Conditions

In RPL-BGU using Harshaw LiF dosimeters we have observed large variations in f(D) as a function of T_2' and also as a function of gas environment during the 400°C anneal. Annealing in air with fairly rapid cooling tends to reduce the supralinearity, whereas annealing in inert N_2 with slower cooling (30 min to room temperature) significantly increases the supralinearity. Figure 3 shows the effect of the high temperature anneal (T_1 = 400 and 490°C) on the supralinearity and saturation characteristics of LiF:Mg,Na. Nakajima and Watanabe[97] and Nakajima[98] using TLD-100 reported strong influence of T_7' on supralinearity, $f(D)_{max}$ = 10 using 6°C/sec and $f(D)_{max}$ = 4 using 25°C/sec when the TL signal was integrated to 300°C. Both Nakajima[98] and Jain[99] tried to explain this dependence on the basis of peak shift to increasing temperatures as T_7' is increased. For example, at 25°C/sec peak 5 occurs at approximately 230°C and at 6°C/sec at approximately 215°C.[100] Thus the effect of T_8 = 300°C would be to integrate less of the TL signal arising from the high temperature peaks as T_7' were increased. Since at 10^5 R, however, the integral peak 5 intensity is still considerably greater than the combined intensity of all the high temperature peaks,[31,97] it is difficult to attribute the drastically increased $f(D)_{max}$ (~250%) to this effect. Using peak height measurements for peak 5 the influence of T_7' on the supralinearity was, however, greatly decreased and reversed in direction, $f(D)_{max} \simeq 2.5$ (5°C/sec) and ~2.8 (15°C/sec). A similar dependence for peak 5 (peak height) was observed by Jain.[99] Pradhan and Bhatt[101] also studied the effect of T_7' on peak 5 f(D) in TLD-100. For peak height measurements the supralinearity increased by ~10% from 4.5 to 33°C/sec, and for peak area measurements the effect was exactly reversed, i.e., a 10% decrease in supralinearity as the heating rate was increased. As previously discussed only a part of the discrepancy between these results and the results of Nakajima and Watanabe and Jain arises from the fixed T_8 used by the latter authors. It appears, also, that T_7' may influence very strongly the supralinear behavior of the higher temperature peaks and this possibility requires further experimental study. Such behavior would not be entirely unexpected since defect clustering may be occurring during recording of the glow curve.[102] No dependence on T_7' was found in the supralinear behavior of TLD-200, $f(D)_{max}$ = 2 at 10^3 Gy from 5 to 25°C/sec. It is worthwhile mentioning that a NTL-50p LiF phosphor (Aloka Co., Tokyo) was also studied by Nakajima and Watanabe[97] which showed an approximately linear response to 5000 R followed by saturation. No high temperature glow peaks were observed in the NTL-50p material even at 10^6 R. Increasing T_7' from 5 to 30°C/sec depressed the TL signal of peak 5 equally over the dose range from 500 to 10^4 R by ~40% in the opposite direction to the observed effect on LiF (Harshaw) where increase of the heating rate enhanced the TL signal in the supralinear region.

6. Effect of Deformation

Petralia and Gnani[103] have shown that the supralinear response occurs at higher exposure in deformed crystals than in undeformed crystals, and Bradbury et al.[53] interpreted this result as an indication for the formation of deep lying traps during deformation. Kos and Mieke,[104] however, have reported no dependence of f(D) on deformation in LiF:Ti. These results again emphasize the great variability of the TL properties of LiF as a function of impurity composition and defect structure.

In summary, the particular form of f(D) is a function of many complex interrelated factors including (1) the mapping of the impurity composition of the material, the mode of incorporation of the impurity into the crystal lattice, and the mapping of the intrinsic crystal defects; (2) the density and possibly the spatial correlation of the released charge carriers following irradiation; (3) thermal and irradiation pretreatments (see following section); (4) glow peak temperature; (5) grain size; (6) physical form; (7) glow curve heating rate; (8) annealing conditions; (9) deformation; and (10) possible surface effects (this effect may be necessary to explain the presence of thermal neutron-induced supralinearity in TLD-100 coupled with the lack of low energy alpha-induced supralinearity). Kalef-Ezra[105] has observed significant changes in f(D) for tritium beta rays and 20 kV_p X-rays (effective energy approximately 7 and 8 keV, respectively) for surface and volume irradiation, respectively. One can conclude without further comment that very precise TL dosimetry (precision of measurement, intercalibration between various radiation fields, intercomparison between various laboratories) requires very careful control and standardization of these various factors.

II. SENSITIZATION

Sensitization refers to an enhanced efficiency, $S/S_0(D')$, which is obtained in many TL materials via a high preirradiation dose, D', followed by an annealing procedure unique to each material. S is the TL efficiency following the sensitization procedure and S_0 is the TL efficiency in the untreated material. An additional characteristic is the increased range of linear response observed in all sensitized materials. Unfortunately low dose measurements in the mR range are hampered by a large background signal in the sensitized sample arising from phosphorescence from the filled deep traps. The deep centers interfere with low dose measurements in two ways.[106,107]

1. The deep centers cause high temperature TL peaks of their own whose low temperature tails can interfere with the lower temperature dosimetry peaks.
2. Over long periods of time charge carriers are transferred from the deep centers to the dosimetry centers and result in an increased background signal.

Typical data[108] on the sensitization in some common TLD phosphors are given in Table 1. As in the case of supralinearity there exists, however, great variability in the results for S/S_0 as reported by various groups. All the sensitized materials exhibit strong residual thermoluminescence (RTL) at high temperature and all the sensitized materials exhibit strong changes in their glow peak structure relative to the untreated material. Mayhugh and Fullerton[109] have described a technique in which the interfering deep centers are removed by exposure to 254 nm UV light during the 300°C sensitizing anneal. The UV treatment does not alter the increased sensitivity and hence extends the use of sensitized LiF-TLDs into the mR range. If the sensitized TLDs are exposed to dose greater than 0.1 Gy reapplication of the UV anneal is necessary before reuse. The best approach is to apply the UV anneal for sensitization and use the sensitized TLDs exclusively in the mR dose range so that no subsequent UV anneal is ever required. Also of importance, the UV anneal is applicable to large batches without damaging reproducibility. The standard deviation for 20 ribbons read before sensitization was ±2.8% and after sensitization ±2.5%.[109]

A. Sensitization in LiF

Jones[110] has compared many of the important dosimetric characteristics of untreated and sensitized TLD-100 and concluded that the sensitized TLDs show more uniform characteristics and retain their calibration factors more successfully than the unsensitized TLDs. These comparative characteristics are reproduced in Table 2. Bartlett and Sandford[111] and Charles et al.[112] have shown, however, that sensitization using either the thermal or the thermal/UV anneal does not allow absorbed dose reestimation at doses below approximately 0.2 Gy (2 S.D. above background — compare to approximately 0.01 Gy for unsensitized LiF phos-

Table 1
SENSITIZATION IN SOME COMMON TL MATERIALS

Material	Preannealing	Postannealing	T_s	S/S_0
TLD-100 (Harshaw)	400°C/1 hr 80°C/24 hr	280°C/1 hr	240°C	5
$CaSO_4$:Tm (DRP)	450°C/1 hr	280°C/1 hr	275°C	4
Mg_2SiO_4:Tb (DNT)	450°C/1 hr	280°C/1 hr	275°C	3.3
CaF_2:Dy (Harshaw)	500°C/1 hr	280°C/1 hr	270°C	1.85
$Li_2B_4O_7$:Mn (Harshaw)	300°C/0.5 hr	280°C/1 hr	260°C	0.85
CaF_2:Mn (Harshaw)	450°C/1 hr	400°C/1 hr	345°C	0.96

Note: Sensitizing exposure, 1.3×10^5 R; test exposure, 90 R.

Table 2
SOME COMPARATIVE CHARACTERISTICS OF UNTREATED AND SENSITIZED TLDS

	Untreated TLDs		Sensitized TLDs
	80°C/16 hr	No preannealing	(no preannealing)
Relative efficiency	0.56	1	3.8
Fading per month at 25°C (%)	0.69 ± 0.23	10 ± 2.6	0.9 ± 0.22
Maximum change in efficiency in the range of 0.04—1.25 MeV (%)	—	34	10
Relative efficiency after 100 reading cycles and 240 hr at 80°C (%)	—	0.93 ± 0.06	1.00 ± 0.05
Standard deviation after 100 cycles and 240 hr at 80°C (%)	—	4.3(9.1)[a]	8.7(6.6)[a]
Standard deviation after 100 cycles and 240 hr at 80°C, precalibrated (%)	—	5.6(4.7)[a]	3.7(3.0)[a]
Onset of 10% supralinearity (R)	—	250	3000
Persistent RTL after 100-R exposure (mR)	—	98	54

[a] Figures in brackets refer to thin (0.28-mm) TLDs.

phors). Thus, the reestimation capability of sensitized TLDs may be adequate for accident reassessments, but is at least an order of magnitude too large for routine use.

For TLD-100, the original extensive studies of sensitized TLDs were carried out by Cameron and co-workers[84] and further studies were carried out on the ionization density dependence by Suntharalingam and Cameron.[20] Several important features were described:

1. The optimal sensitization procedure was 10^3 Gy of ^{137}Cs gamma rays followed by a 280°C/30 min anneal. The usual 400°C/1 hr anneal removes the sensitization entirely and longer annealing at 400°C produces $S/S_0 = 0.8$ attributable to permanent damage.
2. Following the sensitization procedure the dose TL response exhibits a greatly extended range of linearity.

3. The enhanced TL efficiency shows (in gross form) a similar dependence on ionization density as supralinearity, i.e., S/S_0 increases with decreasing ionization density.
4. The emission spectra of sensitized and unsensitized material are identical.[113]
5. The dependence on dose of S/S_0 and f(D) (in the case of S/S_0 we refer to the preirradiation dose, D') is very similar up to approximately 20 Gy.

This latter resemblance was emphasized by Zimmerman.[86] Later work[114] reconfirmed the linearity of peak 5 in sensitized LiF [T:BARC (Bhabha Atomic Research Center)] and TLD-100 and showed that peak 5 saturates in the untreated and sensitized LiF at the same exposure of approximately 10^3 Gy. Although most authors have reported the range of linearity in sensitized LiF to extend to 10^5 R, Jones[110] using TLD-100 chips and a sensitizing routine of 8×10^4 R (^{60}Co), 290°C/50 min anneal, 254 nm UV light except for the last 5 min of the anneal, reported the sensitized product linear to only 3000 R. For the reasons outlined above (similar ionization density dependence, similar saturation dose, identical emission spectra) various authors have concluded that the phenomena of supralinearity and sensitization are slightly different manifestations of the same physical mechanisms underlying the TL process. Unfortunately, this hypothesis could be very wrong for the following reasons.

1. It is well known that the principal effect of thermal treatments in ionic solids (and especially in LiF with the great variability and unknown nature of its defect structures) is to alter the lattice defect equilibrium, including the concentration and arrangement of cation and anion vacancies, impurities, impurity vacancy associates, aggregates, and assorted electrons and holes[5] which may be associated with these various defects. Untreated LiF is irradiated after a 400°C anneal lasting 1 hr, however, sensitized LiF is irradiated after a 280°C anneal lasting 0.5 hr, following a high dose of ionizing radiation. These different procedures certainly bring about very different initial concentrations and configurations of the various traps in the host lattice. For example, Kos et al.[115] observed dramatic variations of the 3.3 eV, Z_2, Z_3, and F absorption bands with preirradiation temperature from 100 to 300°C. Obviously, any serious attempt to explain the sensitization phenomena should approach this problem. Direct support for this argument from TL-LiF data is presented by Niewiadomski[11] who, in a detailed study of sensitization involving many forms of LiF (including various commercial TLDs and home-grown doped LiF of varying Mg,Ti concentration), has shown that in sensitized LiF the contribution of peaks 2 and 3 is reduced relative to peak 5, and peak 4 is almost entirely absent — dramatic evidence for significant reorganization of the trapping structures. Jain et al.[114] and Lakshmanan et al.[108] have also observed significant changes in the glow curve structure of LiF-T(BARC) and other LiF-TLD materials as a function of the annealing temperature prior to the sensitization exposure.
2. Although f(D) and $S/S_0(D')$ are nearly equal to approximately 10^4 R, their behavior after this dose diverges dramatically. With TLD-100, Suntharalingam and Cameron[20] observed $(S/S_0)_{max}$ = 6.5 at 10^3 Gy, whereas $f(D)_{max}$ = 3 at 500 Gy. The details of the behavior with increasing ionization density are also significantly different. Using thermal neutrons as the test radiation they observed $(S/S_0)_{max}$ = 2.5 at 10^3 Gy, whereas the maximum supralinearity ever observed for thermal neutrons in LiF is approximately 1.2.
3. Niewiadomski[11] has shown that the maximum values, of the sensitization are dependent on Ti concentration, are always much higher than the $f(D)_{max}$ values, and always appear at much higher dose levels. For example, $f(D)_{max}$ = 2.5 is found to be independent of Ti concentration, whereas $(S/S_0)_{max}$ varies from 3.5 to 4.5 as the Ti concentration is increased from 2 to 20 ppm.

4. Suntharalingam and Cameron[20] report S/S_0 of peak 5 (peak height measurements) larger than when using the integral of peaks 4 to 7, S/S_0 = 6.5 and 4.5, respectively. This was interpreted as a preferential sensitization of the main dosimetry peaks over the high temperature peaks. On the other hand, Lakshmanan and Vohra[116] report $(S/S_0)_{max}$ = 28 for peak 6 (250°C) and S/S_0 = 7 for peak 5 (280°C/1 hr anneal following 10^3 Gy ^{60}Co). Further experiments are required to resolve this very dramatic discrepancy. Lakshmanan and Vohra interpret their data as confirmation of the proposal by Attix of increased range of charge carrier migration at higher glow peak temperatures. This may be true, but the ratios of sensitization for peaks 6 and 5 certainly do not agree with the T^3 supralinearity behavior as observed by Sunta et al.[46] and Podgorsak et al.[45] In any event, the glow peak temperature-dependent sensitization indicates that the mechanism of sensitization is no less complex than the mechanism of supralinearity.

Various authors[31,46,114,117] have correlated the sensitization effect with the presence of glow peak 10 at approximately 400°C. This correlation is based on the rapidly decreasing value of S/S_0 for peak 5 with annealing temperatures which approach 400°C (Figure 9). This behavior was further examined by Jain et al.[114] who showed that the variations of S/S_0 and peak 10 (height) with residual TL (RTL) as a function of annealing temperature (t = 1 hr) were very similar (Figure 10). Indeed, the two curves run almost parallel to each other over the whole range of RTL except when the latter is due to peak 10 itself. At this stage both S/S_0 and peak 10 are decreasing very rapidly with annealing temperature. Jain et al. have concluded that the phenomenon of sensitization is closely linked with the existence and degree of population of peak 10 and indeed, their correlated trapping center-luminescent center (TC-LC) model (Section III.G) identifies peak 10 as the radiation product of the combined TC-LC.

B. Sensitization in Other TL Materials

Further evidence illustrating the differences in the mechanisms leading to sensitization and supralinearity can be seen in the presence of sensitization in materials that do not show supralinearity and the converse, i.e., supralinear materials which cannot be sensitized can also be seen. For example, CaF_2:natural, which has a linear TL dose response, shows S/S_0 = 1.7 following ^{60}Co exposure of 10^5 R and a 450°C/30 min anneal;[66] and Lakshmanan and Bhatt[118] did not observe sensitization in supralinear $Li_2B_4O_7$:Mn (Harshaw) using a 280°C/1 hr anneal. In fact, these authors report (S/S_0) = 0.85 at D′ = 3 × 10^5 R indicating reduced sensitivity due to damage for $Li_2B_4O_7$:Mn (Harshaw) and (S/S_0) = 0.96 for CaF_2:Mn (Harshaw). The data on sensitization reported by various groups differ substantially. In $Li_2B_4O_7$, e.g., Kirk et al.[62] studied the sensitization of $Li_2B_4O_7$:Mn (NRL) and found $(S/S_0)_{max}$ = 4 for D′ = 10^5 R ^{60}Co followed by a 300°C/15 min anneal. Nakajima[119] and Lakshmanan and Vohra[116] studied the sensitization of Mg_2SiO_4:Tb [Dai Nippon Tokyo (DNT)] with significantly different results. Nakijima reported $(S/S_0)_{max}$ = 2.3 at D′ = 10^4 R, whereas Lakshmanan and Vohra reported $(S/S_0)_{max}$ = 2.8 at D′ = 2 × 10^5 R. It is likely, however, that Nakajima did not reach the correct optimal sensitization dose for his material, because, contrary to the usual behavior, his sensitized material was still supralinear. Nakajima also reported dependence on ionization density opposite to that observed in LiF. The results illustrated in Figure 3 of their work imply that 38 kV_p X-rays result in greater S/S_0 than that induced by ^{137}Cs gamma rays for the dose range of 1 to 50 Gy. It is possible to deduce that the sensitization anneal temperature for this particular result was approximately 270°C. However, at 270°C the high temperature components of the major dosimetric peaks, centered at approximately 220°C, have not been completely emptied. It is possible, therefore, that these results are a mixture of residual TL remaining from the sensitization exposure and true sensitized TL. Burgkhardt et al.[60] using a 400°C/1 hr sensitization anneal and D′

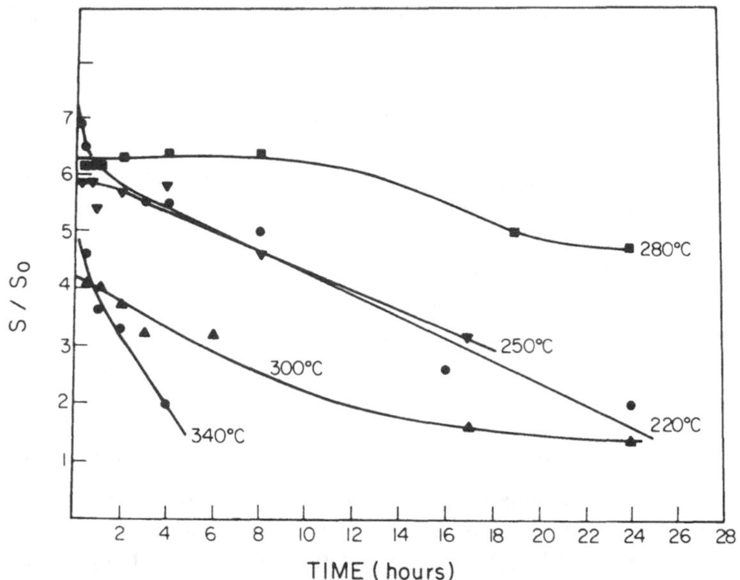

FIGURE 9. Change in the sensitization factor (S/S_0) for peak 5 with annealing time at different temperatures. (Adapted from Jain, V. K. and Ganguly, A. K., Some Aspects of Thermal, Radiation, and LET Effects in the TL of LiF, B.A.R.C./I-466, Bhabha Atomic Research Center, Bombay, 1977.)

up to 10^4 Gy report no sensitization in CaF_2:Dy (Harshaw), but Lakshmanan and Bhatt[118] report $(S/S_0)_{max}$ = 1.85 for a 280°C/1 hr anneal and $D' = 1.3 \times 10^5$ R. The sensitizing anneal temperature of 400°C employed by Burgkhardt et al. was, in all probability, too high. One can fairly confidently correlate the relatively low value of $(S/S_0)_{max}$ observed in CaF_2:Dy with the fact that the lower temperature peaks are preferentially populated with increasing exposure, opposite to the behavior with temperature observed in LiF and most other TL materials. Burgkhardt et al.[60] also report $(S/S_0)_{max} = 2.7$ at $D' = 4 \times 10^5$ R for $CaSO_4$:Tm (Matsushita). This sensitization is unique because no change in the response was found after additional annealing and reuse, which means that the sensitization is permanent and can be more readily used in the low dose region. Similar results have been reported by Blum and Bewley[120] and Lakshmanan and Bhatt.[118] Lakshmanan et al.[108,121] have studied rather extensively the gamma radiation-induced sensitization in $CaSO_4$:Dy and have reviewed the work of other groups. The maximum sensitization factor was 2.2 for $D' = 2.2 \times 10^5$ ^{60}Co gamma-ray exposure followed by a 300°C/1 hr sensitizing anneal. Unlike many other materials optimal sensitization does not yield a greatly enhanced range of linear response (linearity in sensitized $CaSO_4$:Dy was found to extend only to approximately 3×10^3 R); $f(D)_{max}$ in the untreated and sensitized material (T_8 = 300°C) was 2.7 and 1.3, respectively, occurring for both materials at $D' = 5 \times 10^4$ R. The glow peak structure shows a very complicated behavior for both untreated and sensitized material as a function of annealing parameters and dose. The degree of sensitization was found to critically depend on the intensity of a residual TL glow peak at 400°C which was exhibited by the sensitized samples. Unlike the situation in LiF, however, this deep trap is not supralinear but rather grows sublinearly with dose which is therefore suggestive of a competing trap mechanism in the sensitization of $CaSO_4$:Dy. Finally, Al_2O_3 [Indian Aluminium Co. (IAC)] has been studied by Mehta and Sengupta[44,79] with special attention to the supralinear and sensitization properties. The supralinearity decreases with increasing glow peak temperatures, whereas sensitization (10^5 R followed by a 525°C anneal) results in $(S/S_0)_{max}$ of approximately 2 and 5,

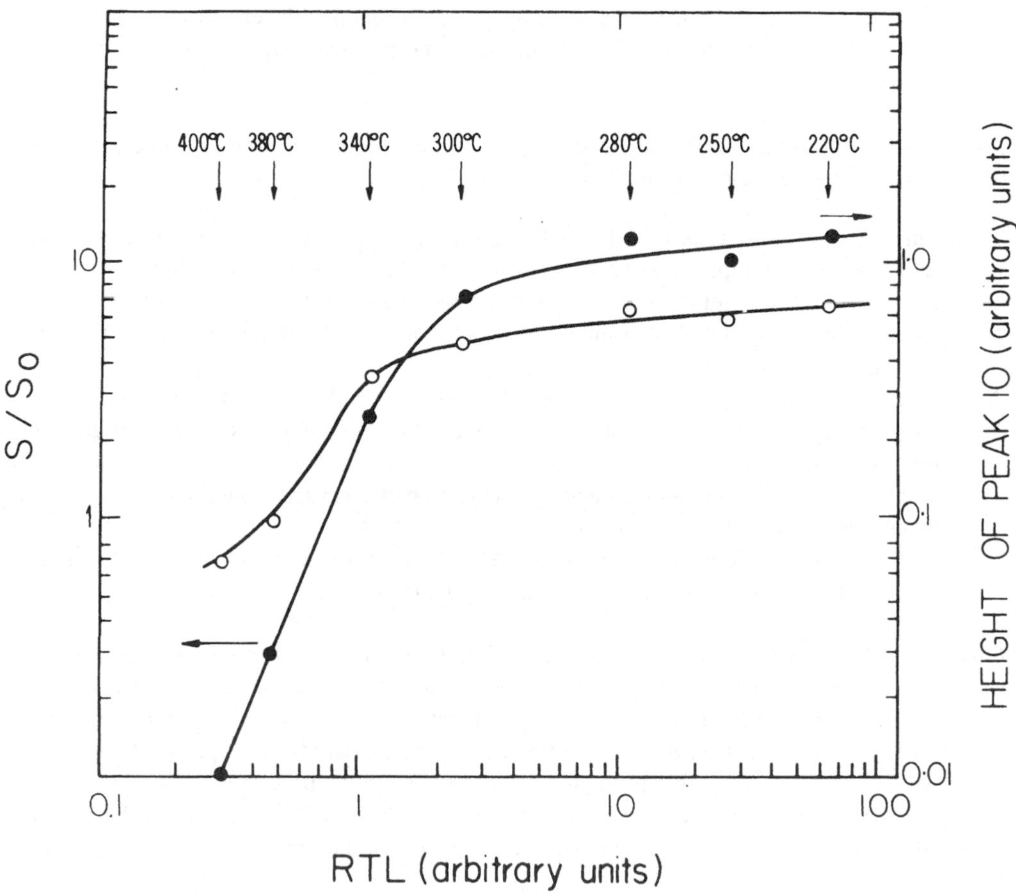

FIGURE 10. Variation of S/S_0 and height of peak 10 with change in residual thermoluminescence as a function of annealing temperature for t_1 = 1 hr. (Adapted from Jain, V. K. and Ganguly, A. K., Some Aspects of Thermal, Radiation, and LET Effects in the TL of LiF, B.A.R.C./I-466, Bhabha Atomic Research Center, Bombay, 1977.)

respectively, for the 250 and 475°C peaks. The TL dose-response curves of the sensitized peaks are nearly linear (the 250°C peak shows very slight supralinearity), and the entry into saturation closely resembles the saturation of the deep trap at 625°C. Thus, two of the main objections to the deep trap competition model of Suntharalingam and Cameron[20] present in LiF (see following section) are not present in Al_2O_3.

III. TL MODELS FOR SUPRALINEARITY AND SENSITIZATION

The complexity of the TL mechanism arises from the multistage nature of the conversion from the initial energy imparted by the radiation field to the final energy liberated as TL photons. This conversion occurs in three stages:

1. Absorption of energy from the radiation field via capture of charge carriers at defect trapping centers accompanied by possible creation or alteration of radiation-induced defects which participate or compete with the TL process
2. Absorption of thermal energy in nonisothermal annealing (glow curve heating) which liberates the charge carriers and probably brings about changes in the trapping structures and/or their spatial correlations

3. Dissipation of the thermal and radiant energy via diffusion of the charge carriers through the crystal lattice followed by recombination, a certain fraction of which results in TL photons

A detailed model of the thermoluminescence mechanism should therefore include a description of the following various factors:

1. The trapping structures present in the crystal lattice at the irradiation temperature (electron or hole traps, cross sections for charge carrier capture as a function of energy and temperature, energy level structure, spatial distribution within the crystal lattice — random or correlated with intrinsic defects, or other trapping structures or recombination luminescence centers)
2. Changes in the trapping structures as a function of irradiation and temperature
3. Kinetics of the charge carrier release upon heating, mechanisms of nonradiative recombination
4. Luminescent recombination structures present in the crystal lattice as a function of irradiation dose and temperature
5. The microscopic density of charge carriers in the absorption and heating stages due to the parameters of the radiation field and crystal lattice

Obviously our knowledge of the details and relative importance of these various factors is extremely limited. An additional difficulty is the very multidisciplinary nature of the problem, encompassing broad areas in radiation and solid-state physics, reaction kinetics, microdosimetry, and others. Thus, at the present time, there are no satisfactory quantitative models which deal with this formidable problem. At best we have qualitative models dealing with average quantities and limited aspects of the problem. The two following brief discussions will further illustrate the present level of our knowledge concerning the TL mechanism in LiF-TLD.

A. Are There Spatially Correlated Trapping Centers and Luminescent Centers?

Probably the most challenging aspect of the TL mechanism is the pheonomenon of supralinearity and its dependence on ionization density. In the framework of current microdosimetric theories this dependence is a clear indication that size is somehow playing a role in the efficiency of the TL process. Cameron et al.[84] may have been the first to make this point: "...increased sensitivity is somehow dependent on the activation of widely distributed crystal impurity sites, the low ion pair density is probably more effective in the activation of these sites..." Today, there exists a fairly large body of mainly circumstantial evidence suggesting spatial correlations between the trapping centers and the luminescent recombination centers. Moreover, if size plays no role in TL supralinearity it could be said with considerable confidence that the very dramatic dependence of supralinearity on ionization density cannot be understood in the context of current theories of electron slowing-down spectra.[122] These latter studies indicate only very slight differences in the electron slowing-down spectra originating from 1 MeV thru 10 keV electrons. Some of the evidence suggesting spatially correlated TC-LC is outlined below.

1. There is a dependence of supralinearity and sensitization on ionization density.
2. The radioluminescence data of Zimmerman[86] indicate that TL traps are localized near the luminescent centers, especially those centers which are responsible for increased photon emission after sensitization.
3. The TL-TAC (thermally activated conductivity) experiments of Fields and Moran[123] (these correlated TL-TAC experiments were carried out for glow peaks at approxi-

mately 160 and 270 K in TLD-100, but their results may have a general significance). The very large effective charge of the recombination centers suggests that local concentration of carriers near optically active TL recombination centers may be much higher than the average bulk carrier concentration measured in the TAC experiment. This leads to the speculation that in highly doped TL material trapped carriers and associated TL recombination centers are spatially correlated to a substantial degree.

4. The paramagnetic resonance studies of Bloch[124] observed satellite lines in the Ti resonance spectrum after exposure of LiF (Semimetals Ltd.) to 6000 R of ^{60}Co radiation. The satellite lines indicate that the Ti ion is strongly coupled to a trapped electron.
5. In the TL emission spectra measurements of Fairchild et al.[83] and Sunta,[125] Fairchild et al. observed three main emission centers at 2.71, 2.9, and 3.01 eV in TLD-100 with varying population of the different emission centers as a function of glow peak temperature. It was suggested that this may be due to the possibility that the stability of different traps (electron or hole) is temperature dependent because nonradiative transitions predominate at high temperatures. An equally possible alternate explanation could involve spatially correlated trapping and recombination centers. Sunta arrived at the same conclusion following his studies of CaF_2:Dy (TLD-200, Harshaw) in which he observed that the 380-nm emission associated with Tb^{3+} centers (present at very low concentration) is strongly correlated with the trapping centers of some of the high temperature peaks. ESR studies in lanthanide-doped CaF_2[126] indicate the presence of aggregate centers where electrons and holes are trapped adjacent to each other.
6. Theoretical studies of the pair correlation function for F centers in NaCl[127] indicate that at room temperature the probability for having an F center as a second neighbor is approximately 20 times higher than that corresponding to a random distribution.
7. The maximum of the X-ray-induced luminescence as a function of temperature occurs at approximately the same temperature as the major peak in LiF.[128] This suggests some kind of correlation between particular hole traps and corresponding luminescent centers.
8. The observation that Ti concentration has a dramatic influence on the glow curve structure of LiF suggests that either Ti is a component of the trapping structures or that Ti concentration somehow alters the spatial correlations between the trapping and luminescent centers.
9. Rossiter et al.[7] have pointed out that the 310-nm optical absorption band (often associated with peak 5) is only very slightly affected by preirradiation annealing time at 400°C contrary to the behavior of the intensity of peak 5. They suggested, therefore, that the main function of the 400°C anneal may be to bring back peak-5 electron traps into the vicinity of emission centers by a diffusion process. They further suggested that the greater availability of defects at grain boundaries may aid this process accounting for the more ready attainment of maximum efficiency in powdered specimens rather than in crystal slices.

B. Supralinearity — Absorption Stage or Recombination Stage?

The TL trapping centers in LiF:Mg,Ti have often been associated with optical absorption bands in the region of 200 to 400 nm. This association has come from fairly convincing circumstantial evidence based on thermal annealing, lifetime, and other studies. For example, glow peak 5 has been associated with the 310-nm optical absorption band in irradiated LiF[129] via the similarity in the rate of decay of this absorption band on thermal annealing with the rate of decay of peak 5. Grant and Jones[18] studied the optical absorption bands at 250 (F band) and 310 nm for ^{60}Co irradiation and 40 keV_{eff} X-rays. The F band was observed to grow at the same rate for both types of radiation, whereas the 310-nm band grew faster with ^{60}Co. This behavior parallels the variation of supralinearity with ionization density and can also be considered as corroborative evidence for the correlation of the 310-nm band with the TL of peaks 4 and 5. Other evidence has been reviewed by Stoebe and Watanabe.[5] There do exist problems, however, with the association of the 310-nm band with peaks 4 and 5. For example, all the optical absorption bands[7,18,130-133] identified in the 200- to 400-nm region grow linearly and then sublinearly with dose at approximately 10^3 Gy. Even for the 310-nm band itself, Claffy et al.[131] report a linear increase up to approximately 10^2 Gy followed by sublinearity. In no case do the optical absorption bands exhibit supralinear behavior. This single observation is dramatically inconsistent with the supralinearity of TL response observed over the dose region from approximately 10 to 10^3 Gy. Another study[134] of the 310-nm band shows that this band magnitude is unaffected by treatments which increase (sensitize) the TL efficiency. These observations have led various investigators to suggest that supralinearity and sensitization must, therefore, arise from dose-dependent competition in the recombination stage leading to greater luminescence efficiency at certain dose levels. Circumstantial evidence for the likelihood of increased competition in the emission stage is suggested by the work of Lucke[135] who showed that the TL efficiency for gamma-ray irradiation in the main glow peak and in the linear part of the TL dose-response curve is only 0.039% in TLD-100. Obviously very small radiation-induced changes in the energy dissipation mechanisms involving the remaining 99.96% could very easily benefit the TL yield by default. Many authors have quoted the work of Zimmerman[86] as presenting direct evidence for the increase of probability for photon emission to explain increased TL efficiency at high dose (supralinearity). Zimmerman, however, assumes identical TL sensitivity "S" for irradiated TL samples before and after glow curve "drainage" which is equivalent to the assumption that no defect or luminescent center rearrangement or other changes take place during the glow curve nonisothermal annealing. Furthermore, and of even greater importance, Zimmerman bases her arguments on dose dependence of the TL signal given by $L = S_0D + aD^2$ (increased probability of photon emission) and $L = S_0D + a/2\ D^2$ (increased probability of trapping), and then compares these two functions with her experimental data over a limited dose region of up to 20 Gy. This analysis ignores the abrupt transition from linear to supralinear behavior observed in most TL dose-response curves. Moreover, it is well known that the quadratic form is a good fit to TL dose-response curves over very limited nonlinear dose regions. Attempts to fit the quadratic form to the TL dose-response curve of up to even 100 Gy (well below the knee where damage effects may begin to be important) have mainly been unsuccessful. Finally, as previously stated, Niewiadomski[11] has shown that above 100 Gy the TL dose-response curves for sensitized and untreated LiF-TLDs are very different. Zimmerman's arguments are therefore of limited merit as supportive evidence for increased luminescence efficiency leading to supralinearity and sensitization. As for the possibility of the creation of new recombination centers via irradiation, the experimental data are unequivocal and unanimous that no changes in the emission spectra of LiF are taking place in the supralinear dose region. Only above 10^5 R, Fairchild et al.[83] observe new, comparatively low intensity bands at approximately 4.0, 2.98, 2.5, 2.3, and possibly approximately 1.5 eV. Furthermore, no changes in the emission spectra between sensitized and unsensitized samples are observed. These data are clearly inconsistent with the creation of new luminescent recombination centers that would be expected to have energy states different from the original centers (unless, of course, nature is playing one of its cruel practical jokes). Finally, it deserves mention that many published attempts to establish correlations between individual glow peaks and changes in the radiation-induced optical absorption spectra have been made with samples exposed to large doses for absorption and low doses for TL measurements. Inasmuch as the glow curves are both dose dependent and include contributions from probably three or more emission centers this procedure cannot be considered very reliable. On the other hand, it should also be remembered that the optical absorption studies (which show linear rather than supralinear growth with dose) do not reproduce the exact thermal treatment that occurs when a glow peak is generated. By invoking temperature-dependent effects, the linear rise of the optical absorption data could be neutralized and the major objection to absorption stage models of supralinearity and sensitization thereby removed. For example, Takeuchi et al.[166] used 150°C postirradiation pulse annealing to simulate glow curve heating and claimed to observe Z_3 optical band supralinearity. However, this work can be criticized on several grounds.

1. The experiments were carried out on LiF:Mg not LiF:Mg,Ti.
2. The Z_3 band appears as a very low intensity band on the high energy side of the F band and is totally unresolved visually. Parfianovitch et al.,[150] for example, have interpreted the F band in LiF to consist of at least ten overlapping peaks so that unambiguous interpretation of the optical absorption band structure in heavily doped LiF is extremely difficult.
3. The optical absorption data shown in Figure 2 of Reference 166 for 80°C/1 hr preirradiation and 150°C postirradiation annealing is linear with doses up to 200 Gy, well into the TL supralinear dose region.

In conclusion, there appears to be considerable circumstantial evidence suggesting the presence of spatially correlated trapping and recombination centers in LiF-TLD. Although no conclusive evidence has been presented to bear on the question of whether supralinearity arises from absorption stage or recombination stage mechanisms, again, considerable circumstantial evidence favors the latter.

In the following section we will briefly describe several of the models dealing with the TL mechanism in LiF-TLD. Two models (the V_k-center model and the Z-center model) deal mainly with the defect structures responsible for first stage trapping; a third is constructed around the possibility of complex spatially correlated trapping centers and luminescent centers; and two others (the deep trap competition model and the track interaction model) were developed specifically to explain supralinearity and sensitization and their dependence on ionization density.

C. The V_k-Center Model

This model developed by Mayhugh and collaborators[136-138] was based on optical absorption measurements and attempts to reconcile the evidence that suggests that the dosimetric glow peaks originate from the untrapping of electrons, but that the light emission comes from the recombination of mobile holes. The model postulates that electrons thermally released from the trap associated with peak 5 migrate until they recombine with a hole at a multiple hole-trapped center, V_3, converting it to a V_k center (self-trapped hole). The V_k center, even influenced by defects, is expected to be thermally unstable at room temperature,[139] and thus produces a mobile hole in the valence band. This hole then retraps in an activator (say, Ti)-related center and recombines with an electron tunneling from an F center. This is the recombination which produces the 3.01-eV luminescence. Although peak −4, which has been directly identified with the V_k center,[45] has its emission maximum at 250 nm rather than 400 nm, Christy and Mayhugh[138] argue that their emission mechanism does not require that the recombination emission of V_k-center holes near 150 K (peak −4) should have the same spectrum as that of subsequently freed holes above 400 K (peaks 4 and 5). In the first case it is possible that the recombination centers giving the V_k hole emission at 250 nm could be depleted before the indirect holes are subsequently released so that only 400-nm recombination centers are left for the high temperature TL emission. Second, it is, of course, possible that the recombination trapping or emission mechanisms could themselves be temperature dependent. It was also proposed that sensitization is correlated with the formation by radiation of Z_3 centers by either reducing the number of radiationless transition sites or more indirectly if the indirect holes recombine at irradiation formed or activated Z_3 centers to give the luminescence. Stoebe and Watanabe[5] have pointed out, however, that the Z_3 center can be optically destroyed although sensitization is not affected by light.

D. The Z-Center Model

The main idea of this model[115,132,133,140-144] is that during irradiation Z_2 and Z_3 centers are formed by capture of electrons at Z_0 centers (tentatively correlated with a 9.05-eV optical absorption band) via the reaction $Z_0 + e^- -- Z_3$, $Z_3 + e^- -- Z_2$. During heating for TL readout, the Z_2 centers ($Mg^{2+} - F'$ pairs) related to the 310-nm optical absorption band are converted to Z_3 centers ($Mg^{2+} - F$ pairs) related to the 225-nm optical absorption band, and the electrons liberated by this conversion cause TL peak 5. Further, high temperature TL (peak ''6'') is caused by thermal annealing of the Z_3 centers. The Z-center model has been critized by Horowitz[145] via the following arguments. In the terminology of microdosimetric track structure theory the 310-nm band is formed by ''two-hit'' processes (capture of two electrons) and the 225-nm band is formed by a ''one-hit'' process.[146] The application of cumulative Poisson statistics[147] can then make some interesting predictions about the general behavior of the intensity as a function of dose of the 310 nm-two hit and 225 nm-one hit bands. The one-hit process should result in a linear dose-response curve and the two-hit process should result in a supralinear dose-response curve following gamma-ray irradiation. Unfortunately, all the optical absorption bands[7,18,130,131,140,141] identified in the 200- to 400-nm region grow linearly and then sublinearly in the dose region up to 10^3 Gy. Even for the 310-nm band itself, Claffy et al.[131] report a linear increase up to approximately 100 Gy followed by saturation. This single observation is, therefore, dramatically inconsistent with the expected behavior of the 310 nm-two hit band. Since the two-hit process is predicted by Poisson statistics to be much more likely at high ionization density, one would expect that the 310-nm band would grow relative to all the other bands as the dose increases. This, of course, is not observed since all the optical absorption bands grow linearly with dose.

If we turn our attention to the TL dose response of peaks 5 and ''6'' an additional inconsistency becomes apparent. The TL dose response of peak ''6'' (associated with single-electron capture) shows greater supralinearity and has a lower threshold for supralinearity than peak 5 (associated with two-electron capture). The linear growth with dose of the optical absorption bands generally associated with the TL process has suggested to many investigators (Section III.B) that the supralinearity of the TL dose response is associated with dose-dependent luminescent efficiency in the recombination stage. Nonetheless, the lesser supralinearity observed for peak 5 in comparison with peak ''6'' is again in contradiction to the Z-center model which relates the supralinear TL dose response of the glow peaks to absorption stage mechanisms. Based on this latter assumption, Waligorski and Katz[49,50] carried out a multihit analysis of LiF (TLD-700) supralinearity and arrived at sensitive trap radii (in the statistical ''multihit'' model of Katz and collaborators, radiation detectors are characterized by a sensitive target radius and a macroscopic dose, D_0, at which there is an average of one hit per target) of approximately 100 Å for peak 5 and 400 Å for peak ''6''. These enormous sensitive sites cannot possibly be reconciled with the usual models[148] for Z_2 and Z_3 structures. Very large radius electron states in alkali halide phosphors are known to exist; however, even in the largest of these (e.g., the F-center excited electron state in KI at room temperature) the effective radius is equal to only approximately 15 lattice constants. The effective trap radii of 100 and 400 Å in TLD-700 for peaks 5 and ''6'' would correspond to 50 and 200 lattice spacings, respectively. Since the intensity of the optical absorption bands at 225 and 310 nm as observed by Nink and Kos is approximately equal and since readout is assumed to further convert Z_2 centers to Z_3 centers it would appear that peak ''6'' should be considerably more intense than peak 5 rather than the reversed lower intensity always observed at low and intermediate dose. It is possible that the reversed intensity arises from a diminished availability of recombination centers following peak-5 emission and prior to peak-"6" emission. However, a limited supply of recombination centers at low dose is difficult to reconcile with the linear dose response of peak 5 over many orders of magnitude of dose. The counter argument that the luminescent centers may be created via irradiation is doubtful since this process would hardly be expected to yield a linear dose response at low dose. Yet another possibility is that the recombination centers are unstable at elevated temperatures, but the spectral emission characteristics of peaks 5 and "6" are essentially identical indicating that the same luminescent centers are involved. Cooke[149] using experimental data for the ratio of peak-"6" to peak-5 intensities for high ionization density radiation (negative pions) and low ionization density radiation (positive pions) arrived at approximately 60% conversion of electrons (liberated in Z_2 200°C $Z_3 + e^-$) to TL photons and only 1.6% conversion to TL photons for electrons liberated in the reaction Z_3 285°C $Z_0 + e^-$. Thus the reversed intensity might arise from the relative increase in nonradiative transitions with increasing temperature. This very rapid dependence of the luminescent efficiency on temperature is hardly consistent with the increasing supralinearity observed with increasing glow peak temperature nor with the dominance of peak 5 (200°C) when compared to the lower temperature peaks. Nonetheless, it deserves mention that thermal quenching studies[100] indicate for TLD-100 a *maximum* in the TL efficiency of peak 5 at approximately 100°C min^{-1} so that LiF-TLD may not exhibit the generally assumed behavior that the probability of nonradiative transitions rises rapidly with temperature. In any event the maximum thermal quenching effect observed by Gorbics et al.[100] was a factor of two. Obviously, it would be of considerable interest to study the thermal quenching characteristics of peak "6" in LiF-TLD.

There are additional difficulties with the Z-center model. The absorption spectra on which this model is primarily based show very little resolved structure and consequently must be deconvoluted into the hypothesized respective absorption bands using a model-dependent approach. Parfianovich et al.,[150] e.g., have interpreted the F band in LiF to consist of at least 10 overlapping bands. In most other LiF-TLD materials (e.g., TLD-100) peak "6" shows strong supralinearity in the dose range of up to 10^3 Gy followed by saturation. No data are reported for peak "6" by Kos and Nink[142] below 10^3 Gy, a serious deficiency for a model which attempts to correlate TL peaks with radiation or thermally induced defects. The unique behavior of peak "6" (increasing linear response between 10^3 and 10^4 Gy) in their material (home-grown LiF — 300 ppm Mg, 3 ppm Ti) thus raises serious doubts as to the possibility of achieving a single universal model for TL mechanisms in LiF:Mg,Ti. Finally, Kos and Nink[142] proposed in the context of their model that the sensitization mechanism is mainly due to the conversion of Z_3 centers to Z_2 centers by trapping of electrons thus leading to a higher concentration of Z_2 centers (peak 5) in sensitized than in unsensitized material. This hypothesis, however, contradicts the results of Lakshmanan et al.[151,152] which show peak "6" with $(S/S_0)_{max} = 28$ (Z_3 centers) compared to $(S/S_0)_{max} = 7$ for peak 5.

Recently, Lakshmanan[167] has suggested that peaks 7 and 10 in LiF:Mg,Ti are the Z_2 center and Z_3 center, respectively. The TL peak at 400°C (peak 10) has been correlated with the Z_3 band at 225 nm (5.5 eV) by many investigators and, as previously pointed out, the Z_2 band has apparently not been observed so far (probably because of its greatly reduced intensity relative to the F band) so that OA arguments cannot exclude the identification of peak 7 with the Z_2 band. Lakshmanan argues that the TL properties of peaks 7 and 10 correlate well with their identification with Z_2 and Z_3 centers and in the following we critically examine his arguments.

Both peaks 7 and 10 are supralinear with ^{60}Co-gamma dose from threshold, peak 10 behaves more supralinearly than peak 7 (in accordance with the approximate T^3 behavior of $f(D)_{max}$ observed for the glow peaks of LiF:Mg,Ti), but peak 7 is approximately two orders of magnitude more intense than peak 10. It is reasonable to attribute the difference in TL efficiency of the two peaks to thermal quenching since they are separated by 140°C, however, on microdosimetric statistical arguments we would expect peak 7 (associated with the "2-hit" Z_2 center) to behave more supralinearly than peak 10, which is not observed. Of course, the multistage nature of the TL process allows the argument that different, yet unidentified, dose dependent competition mechanisms in the luminescence recombination stage somehow depress the supralinearity of peak 7 or enhance the supralinearity of peak 10, however, there is, as yet, no direct evidence to support this hypothesis. The reversed supralinearity of peaks 7 and 10 is, therefore, *not* corroborative evidence for the association of peak 10 with a "1-hit" OA band and peak 7 with a "2-hit" OA band.

Another interesting observation is that the supralinearity of both peaks 7 and 10 is ionization density dependent, i.e., they are linear in their dose response to alpha particles. Lakshmanan attempts to attribute this somewhat vaguely to the hypothesis that at high ionization density the competitors are either damaged less efficiently, or the damage is less dependent on dose. Horowitz,[163,168] however, has shown quantitatively that decreasing supralinearity observed with increasing ionization density is explained by track interaction effects and has little to do with the nature of the trapping centers. The TST-track intersection model developed by Horowitz shows that heavy charged particle (HCP) supralinearity does not exist because the highly localized nature of the HCP track radial dose distribution suppresses the probability of significant track intersection at dose levels where gamma ray or electron supralinearity is normally encountered. The model calculates the TL yield due to intersecting alpha particle tracks by folding in the summed dose from two alpha particle tracks with the TL dose response generated by low energy electrons matched as closely as possible in energy to the primary electron spectrum liberated by the alpha particles. The calculations show that there is no intertrack distance for which the total TL yield for two intersecting alpha particle tracks is greater than twice the TL yield of a single alpha particle track. The immediate conclusion is that alpha particle supralinearity cannot exist for glow peaks with electron TL dose response curves normally encountered in most TL materials. Decreasing supralinearity with ionization density can, therefore, not be considered as corroborative evidence for LET-dependent or dose-dependent competition mechanisms in the recombination stage and certainly has no direct bearing on the nature of the trapping centers associated with peaks 7 and 10.

The relative TL response, $\eta\alpha\gamma$, (the TL yield per Gy for alpha particles compared to gamma rays), is less than unity for peak 7. This again is contrary to statistical microdosimetric arguments which predict $\eta\alpha\gamma$ greater than unity for a "two-hit" mechanism. Lakshmanan argues that this may be attributed to the dominant effect of energy "wastage" due to electron-hole recombination losses for high ionization density alpha particles. This is a reasonable statement but leaves open the question of how peak 6 in LiF:Mg,Ti achieves values of $\eta\alpha\gamma$ greater than unity. The fact that $\eta\alpha\gamma$ is less than unity for peak 7, therefore, contradicts the association of peak 7 with a "two-hit" trapping center.

In conclusion, it can be seen that neither peaks 5 and 6 nor peaks 7 and 10 meet the criteria imposed by microdosimetric track structure theory concepts for "two-hit" and "one-hit" mechanisms, respectively. If anything, their behavior is contrary to expected on the basis of these identifications. Since the unique feature of thermoluminescence in LiF and other TL materials is the ionization density dependence of S/S_0 and f(D) it is unfortunate that more attention has not been paid to the dependence of the optical absorption band intensities on ionization density. It would also be of considerable interest to study the dose response of the OA bands in the alkali halides where their identification is considered more definite (NaCl, KCl, and KBr fall in this category). Supralinear bahavior would confirm their identification as two electron capture centers, strengthen our convictions as to the applicability of microdosimetric statistical theory to the formation of these centers, and furnish us with a reliable and easily applied method to identify multiple charge carrier trapping centers. Nonsupralinear behavior would raise doubts as to the correct identification of the OA bands as due to Z_2 centers and/or to the applicability of statistical microdosimetric concepts to the probability of formation of these centers. Conclusive evidence could arise from identification and study of the Z_2 OA band in LiF:Mg,Ti presumably in the 265 to 280-nm region and the determination of its dose response characteristics. This would be especially true if observation of supralinear OA bands associated with Z_2 centers in NaCl, KCl, and KBr would conclusively establish supralinear OA behavior as a dose-behavior-signature for multiple charge carrier trapping centers.

E. The Deep Trap Competition Model

This model[20,84] postulates the presence of deep traps (released at a temperature higher than that used in normal readout) which compete strongly with the dosimetric traps at low doses in electron or hole capture during irradiation. The reason for assuming the competing trap to be deep is the fact that this trap remains filled after an annealing of 1 hr at 280°C (sensitization anneal), but is emptied after the standard high temperature anneal of 1 hr at 400°C. For doses above approximately 10 Gy the number of empty deep traps becomes significantly reduced and the probability of charge carrier capture by the dosimetric traps (which have a smaller cross section for electron capture) rises, resulting in an increase in the TL efficiency. Since the cross sections for charge carrier capture vary from trap to trap, their supralinearity factors are different. It is also necessary to assume that the total number of deep traps is much smaller than the number of TL traps. As the ionization density of the radiation-TLD interaction increases more electron-hole pairs are created in a given volume which results in a high probability of filling the normal TL traps while also filling the deep traps. When very high ionization density radiation is used the filling of the competing traps uses relatively few charge carriers and a linear or near linear response is observed. Since the model does not assume creation of any new traps or light centers the emission spectra of both sensitized and untreated material are identical. Kristianpoller et al.[153] and Chen and Bowman[1] have investigated the kinetics of this model both analytically and numerically to determine the conditions that would lead to supralinearity from zero dose and supralinearity preceded by a linear region. A simple energy level diagram is assumed as shown in Figure 11, in which N_T, N_C denotes the concentrations of the dosimetric trap and the competitor, respectively; n_T, n_C represents the concentration of charge carriers in these traps; and N_L, n_L represents the equivalent variables for the luminescent centers. It is further assumed that at the end of the irradiation the concentration of trapped charge carriers is much less than the number of charge carriers trapped at luminescence centers so that the glow peak intensity or area is proportional to n_T. Neglecting band-to-band or band-to-center recombination, the equations governing the TL process (i.e., charge carriers are raised by the irradiation from the valence band into the conduction band and fall into either N_T or N_C) are

$$dn_T/dt = A_T(N_T - n_T)n_C' \quad (1)$$

$$dn_C/dt = A_C(N_C - n_C)n_C' \quad (2)$$

$$dn_C'/dt = X - dn_T/dt - dn_C/dt \quad (3)$$

where n_C' is the concentration of electrons in the conduction band, A_T and A_C are the transition probabilities into N_T and N_C, respectively, and X is the rate of creation of electron-hole pairs. The condition defining supralinearity ($d^2n_T/dD^2 > 0$) leads immediately to the condition that $A_C > A_T$, i.e., that the retrapping probability of the competitor must be larger than that of the trapping center. Similarly, for $A_C > A_T$, n_C must grow sublinearly with dose. The introduction of band-to-band recombination at high dose (which is assumed to

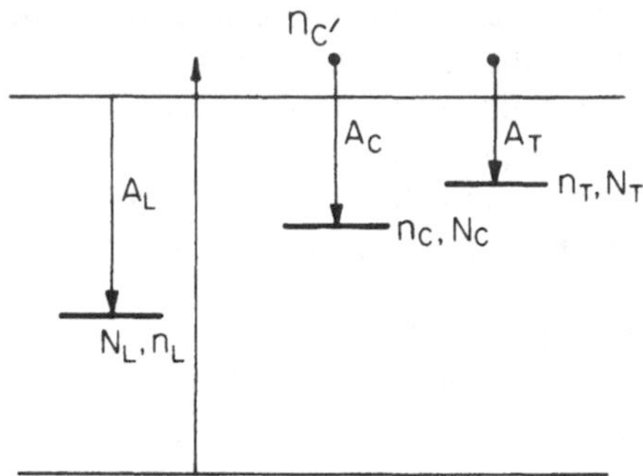

FIGURE 11. Energy level scheme for the deep trap competition model.

lead to saturation) is incorporated by the addition of a band-to-band transition probability, A_K, into Equation 3

$$dn_C'/dt = X - dn_T/dt - dn_C/dt - A_K n_C' p \tag{4}$$

where p is the concentration of holes in the valence band. This leads to the following necessary conditions for supralinearity

$$A_C > A_K$$

$$\frac{N_C A_C^2}{N_T A_T} > A_K$$

$$A_C > A_T$$

In general, if recombination into the centers during irradiation is negligible ($A_K \simeq 0$) and $A_C > A_T$, supralinearity exists for the entire growth curve. The supralinearity is such, however, that at lower and higher dose the dependence is very nearly linear with a supralinear region in between. At high doses introduction of A_K causes a rather early saturation, however, the dose ranges can decrease, increase, or disappear by the appropriate choice of N_T,N_C, etc. Thus it is possible to observe linear-supralinear-saturation behavior or even supralinear-linear-saturation behavior[154] depending on the choice of parameters. It is interesting to note that the supralinear behavior can start at the lowest doses if $N_C - n_{CO}$ is sufficiently small. This means that supralinearity at low dose followed by saturation is not necessarily indicative of trap creation via irradiation (cf. peak 10 in the TC-LC model, following section).

The main failures of the deep trap competition model in LiF-TLD are as follows.

1. There is linear growth with dose of all the optical absorption bands in the energy region usually associated with peak 5.
2. The behavior of the high temperature glow peaks in LiF (TLD-100) is supralinear, just the opposite of what is required for the deep traps. The temperature of glow peak 10 (~400°C) is in fairly good agreement with the temperature prediction of approximately 500°C[155] determined via thermal stability measurements of the increased sen-

sitivity of TLD-100. Unfortunately, contrary to the requirements of the competing trap hypothesis, the 400°C peak does not saturate early (when peak 5 begins to show supralinearity), but continues to exhibit supralinearity even after peak 5 has entered saturation. Higher temperature glow peaks above 400°C have yet to be observed. The speculation that the deep trap is not observed in the glow curve is definitely possible, however, the objection has been raised that the deep trap must be in the vicinity of the dosimetric traps in order to compete so that it would be expected to have equal access to the luminescent recombination centers.

3. The model provides only a very vague framework in which to explain the glow peak temperature dependence of the supralinearity. It deserves mention that some of these objections are absent in Al_2O_3 TLD studied by Mehta and Sengupta.[44,79] The deep trap observed at 625°C is sublinear throughout, and the other two peaks at 250 and 475°C do not exhibit any well-defined temperature dependence in their supralinear behavior. Unfortunately, the optical absorption properties of this material have not yet been studied to determine whether the absorption bands grow in a supralinear manner, nor has the ionization density dependence of the supralinearity been determined.

F. The Track Interaction Model

This model postulates that electrons and holes produced during irradiation are trapped near the track of the ionizing particle and that some of the trapped charge carriers produce F centers which act as the luminescent recombination center. Claffy et al.[131] specified the light-emitting centers as being F centers on the basis of earlier considerations by Klick et al.,[156] but the identification of the trapping and light centers is not required by the track interaction model. A major premise of the track interaction model is that TL occurs at luminescent recombination centers rather than transition emission luminescent centers. The distances calculated by Attix[27] between the tracks at low doses of ^{60}Co irradiation are separated far enough for the recombination during phosphor heating to occur between the charge carriers and the luminescent recombination centers in the same track. The TL dose-response curve is linear in this dose region. At higher dose (approximately 10 Gy) when the average distances between the tracks (approximately 660 Å) become comparable with the average separation of the centers along each track, the probability that a charge carrier generated in one track will recombine with a center produced along another track increases. TL efficiency thereby increases and the TL dose response rises more rapidly than linearly. As the incident energy of the gamma or electron radiation field is decreased, the tracks become more and more localized requiring even greater dose levels to initiate the track interaction. Since the supralinear dose region is immediately followed by a dose region in which saturation effects occur the result is a decreasing supralinearity with increasing ionization density (decreasing incident electron energy). Further calculations indicate that most of the luminescent centers are not reached by the released charge carriers and survive the readout cycle (and even long-term annealing at temperatures below 300°C) so that the response is enhanced either at high doses or at later exposure and readout because of the added population of luminescent centers remaining from the sensitization exposure. The sensitized TLD is linear because the large number of additional luminescent centers depresses the importance of the track interaction. Any serious attempt to test the track interaction model quantitatively would require a "full-blown" Monte Carlo calculation of great complexity. For simplicity of computation, Attix assumed that the Compton electron tracks are linear, parallel, and that all of the liberated electrons are of equal energy. Sunta et al.[46] and Podgorsak et al.[45] have suggested that the experimentally observed temperature dependence of TL supralinearity in LiF-TLD from glow peak to glow peak tends to support the track interaction model. The maximum supralinearity factor observed by these investigators varies as the cube of the absolute temperature. This implies that the range of charge carrier migration may be proportional to the absolute temperature since the number of luminescent centers within a sphere having the migrating range for a radius would then vary as T^3. Although no theoretical or experimental justification for this suggestion has been advanced, and in fact the mobility of free carriers is usually given by a T^{-n} dependence[51] a linear dependence on T is possible in heavily doped LiF:MgTi. Dobson and Midkiff[157] have formulated a simple phenomenological approach to the track interaction model. The integrated TL light is given by the expression

$$I(D) = I_0 \{1 - \exp - (\lambda D/D_0)\}(1 - \gamma^k) \quad (5)$$

where I_0 is the integrated light output at saturation and D_0 is the dose value at which a second track is close enough to the trapped charge carrier to compete on equal terms with the original track. The constant λ is proportional to the charge carrier trapping probability and $\gamma = 1 - p$ where p is the average probability of an interaction between a trapped charge carrier and a nearby track. The effective number of interacting tracks is given by k, where $k = 1 + D/D_0$. Although Equation 5 gives a reasonable fit to the supralinearity data of Cameron et al.,[84] the value of $D_0 = 30$ Å is more than an order of magnitude smaller than the 660 Å calculated by Attix for the average separation of tracks for a gamma-ray exposure of 10^3 R (onset of supralinearity). In general, the experimentally observed transition from linear to supralinear behavior in other materials as well as LiF-TLD is too rapid to be accurately fitted by Equation 5. This is a general failing of all the semiphenomenological analytic approaches.[158]

Niewiadomski[11] has pointed out that the experimental result that supralinearity and sensitization appear at lower exposures and are more pronounced in phosphors containing low Ti concentration is not entirely consistent with the calculation by Attix which predicts that for 15 ppm Ti (12.7-nm average separation of Ti centers) and 3 ppm Ti (21.8-nm average separation of Ti centers) the supralinearity should not be strongly dependent on Ti concentration because in both cases the average separation is considerably lower than the 35-nm average separation of electron-hole pairs created during irradiation. This objection, however, takes these average simplistic calculations far too seriously. Furthermore, they are based on the assumption that each Ti ion is associated with a luminescent center whereas, in fact, there may be many Ti ions which do not take part in the luminescent recombination so that it does not necessarily follow that there is a direct linear relationship between Ti concentration and the number of active luminescent centers. The suggestion by Attix that the luminescent centers survive readout and even the sensitization anneal contradicts the findings of Davies[159] who showed that the $(Ti^{3+}F_3^-O_3^{2-})$ centers associated with luminescence are no longer observed following readout. Sensitization is therefore not adequately explained by the track interaction model as formulated by Attix,[27] however, this does not detract from its attractive aspects with respect to supralinearity.

Both Suntharalingam and Cameron[20] and Attix[27] predicted on the basis of their models that the increased TL efficiency obtained due to increase in ionization density as the dose is increased (supralinearity) should be observed to a similar extent as the gamma ray or electron energy is decreased from 1 MeV to approximately 50 keV. The increase in TL efficiency at low energy, however, when observed is approximately 15%[160] rather than the several hundred percent observed due to supralinearity. The correlation between these two aspects of TL efficiency as a function of ionization density is indeed hinted at in the sensitization studies of Mayhugh and Fullerton[161] who observed that sensitization not only results in a linear TL dose response (i.e., reduced TL efficiency in a certain range of ionization density), but also reduces the response relative to ^{60}Co in the 30- to 135-keV region by approximately 40%. Sensitization thus alters the energy dependence of LiF-TLD without any change in atomic number! Another example is the 50% enhancement observed by Mieke and Nink[162] in LiF:Ti at low photon energies in a material which was considerably more supralinear than the usual commerical products. Microdosimetric arguments, however, suggest why the greatly increased TL efficiency observed in supralinearity may not be observed at low electron energy relative to ^{60}Co. The average distance between electron-hole pairs following a single ^{60}Co interaction in LiF is approximately 350 Å,[27] however, calculations which attempt to take into account the three-dimensional aspects of the electron track indicate that this figure may be considerably greater, perhaps equal to approximately 2000 Å which corresponds macroscopically to only 0.15 Gy. The figure of 2000 Å is a biased average which neglects electron-hole pairs created at very high ionization density where the TL efficiency is very small. Decreasing the X-ray or electron energy from 1 MeV to 50 keV increases the microscopic average ionization density by approximately 1 order of magnitude which corresponds to the still linear dose region of approximately 1.5 Gy. The approximate 15% increase in η_X for low energy X-rays arises because a slightly greater fraction of the X-ray microscopic dose distribution lies in the supralinear region of ionization density. Thus in LiF-TLD where the onset of supralinearity is approximately 5 Gy we do not expect to observe η_X significantly greater than unity at low X-ray energies.

Additional substantiation for track interaction effects as well as spatially correlated TC-LC is obtained from the recent work by Horowitz et al.[163] which successfully relates the saturation in low energy alpha TL dose-response curves to track intersection. In this model, which is based on calculations using the Monte Carlo method as well as analytical methods, the alpha particle track is described as a cylindrical volume whose height corresponds to the range of the alpha in the crystal and whose effective radius (r_{eff}) is a single free parameter, determined by fitting the theoretical dose response to experimental results. The calculations are based on the assumption that the sublinear behavior is a result of overlapping tracks in which the overlapping areas are completely in saturation. The values of r_{eff} determined from the data of Jain and Ganguly[31,32] (5-MeV alphas stopping in LiF:Mg,Ti) for the peaks appearing at 130, 190, 255, and 390°C are 100, 150, 85, and 75 Å, respectively, and from the experimental results of Aitken et al.[41] (4-MeV alpha particles stopping in LiF:Mg,Ti) 200, 75, and 70 Å for the glow peaks appearing at 230 (cf. 190°C), 290 (cf. 255°C), and 460°C (cf. 390°C), respectively. The value of 150 to 200 Å for the effective radius associated with peak 5 is in excellent agreement with calculations of the radial dose distribution for 3-MeV alpha particles in LiF.[14,28]

This indicates that charge carrier diffusion during glow peak heating for peak 5 is negligible or not more than the order of tens of Angstroms, i.e., the thermally liberated charge carriers recombine to form TL photons at sites close to the initial trapping centers. Possible reasons for the lower effective radius obtained for higher temperature peaks are discussed in Chapter 3. The assumption that every overlapping volume is totally saturated is a simplification that does not allow the model to account for the supralinear part of the alpha TL dose-response curve observed for peak 8. To explain low energy heavy charged particle supralinearity Horowitz and collaborators are developing a more accurate model that takes into account the radial dose distribution around the axis of the track of the HCP,[168] (see Chapter 3, Section V.B for preliminary results using a variable ionization density model).

Various other experimental observations can also be easily accomodated in the track interaction model:

1. The sensitized TLD is linear because the large number of additional luminescence recombination centers activated by the sensitization irradiation depresses the importance of the track interaction.
2. The lack of supralinearity observed in LiF:Mg,Ti heavily doped with OH could arise because the increase in the number of competitors makes it unlikely that charge carrier

migration will be effective at low dose levels. At higher dose levels when the tracks are even closer, saturation effects begin to dominate the TL dose response.

3. The lack of supralinearity in radioluminescence can also be interpreted as substantiation for track interaction effects.

Since the radioluminescent process occurs at room temperature the abundance of competing Mg associated traps (yet unfilled by the irradiation dose) severly limits the migrating range of the charge carriers. It is worthwhile noting that the average distance between Mg centers for 100 ppm/w Mg doping is 55 Å, whereas for Ti centers for 5 ppm/w doping it is 200 Å. Thus one expects to see little dose dependence of radioluminescence consistent with the observations of Attix[27] who reported RL supralinearity only 10% as great as in TL and occurring over the same dose range. This observation ties in with the increased supralinearity as a function of glow peak temperature. At the higher temperatures many of the traps competing with the luminescence centers are inactive thereby increasing the migrating range.

A similar mechanism, co-existing with and possibly determining the importance of track interaction, may arise from localization between trapped electrons and holes due to spatially correlated TC and LC. A high degree of localization would imply that each released charge carrier had a high probability of recombination with its "own locally trapped hole" and a much lower probability of recombination with the remaining population of LCs. In this case the TL dose response is proportional to the number of locally paired electrons and holes, i.e., to the dose. At higher dose, the "sphere of capture" includes other LCs in nearby tracks and the TL dose response begins to behave supralinearly. The linearity of TL dose response at low dose is thus strong indication of localization between trapped electrons and holes and the degree of spatial correlation may determine the dose threshold at which track interactions begin to be significant.

In conclusion, the track interaction model specifically incorporating dose dependent efficiency of recombination in the luminescence recombination stage is our preferred candidate to explain supralinearity in LiF:Mg,Ti because of its ability to quantitatively predict the dependence of supralinearity on ionization density and its natural consistency with a large number of experimental observations.

G. The Correlated Trapping Center-Luminescent Center (TC-LC) Model

Proposed by Jain et al.,[31,114,164] this model is actually a converted hybrid of earlier models proposing trap and luminescent center creation via irradiation. Nonetheless, it is the first model to postulate complex-correlated TC and LC, the evidence for which was previously discussed in Section III.A. The TC-LC are tentatively associated with an optical absorption band at 137 nm which is known to be significantly reduced by 10^5 R ^{60}Co irradiation.[136] It is further postulated that irradiation in addition to filling lower temperature traps, splits these complex centers into peak 10 trapping centers (possibly Z_3 centers associated with the 225-nm optical absorption band), peak 5 trapping centers (partially filled), and additional filled luminescent recombination centers. The 400°C anneal is assumed to regenerate these complex centers and the 280°C sensitization anneal is assumed to empty the LC which were formed from breakup of the complex center. The model thus explains both supralinearity and sensitization by an increase in active luminescent centers. The sensitized phosphor is linear since the entire population of complex centers has been depleted by the 10^4 R sensitizing exposure. The schematics of the model may be described as follows

$$\xrightarrow[\text{anneal}]{400°\text{C}} (\text{TC,LC}) \xrightarrow{10^4\text{R}} (\text{TC}^* + \text{LC}^*) + \text{TC} + \text{LC}^* \xrightarrow[\text{anneal}]{280°\text{C}}$$

$$(\text{TC}^* + \text{LC}^*) + \text{TC} + \text{LC} \xrightarrow[\text{anneal}]{400°\text{C}} (\text{TC,LC})$$

where TC* and LC* indicate filled tenth peak traps and luminescent centers. Supportive evidence for the model comes from the close correlation between residual TL (arising from peak 10) and sensitization as a function of annealing temperature (Figure 10) and also from the fact that supralinearity and the presence of peak 10 were not observed in unannealed LiF. The supralinear behavior of peak 10 throughout the dose range studied (5×10^3 to 5×10^5 R) is also suggestive of trap creation via irradiation,[117] however, this point would be considerably strengthened by extending the study of peak 10 to lower doses in the linear range of peak 5. Moreover, as previously pointed out, the supralinear dose response can arise from kinetics[1] or track interaction effects. The objection by Lakshmanan et al.[151] to the TC-LC model based on their own experimental observations that the sensitization was found to decrease systematically below 300°C annealing temperature is untenable since the TC-LC model can easily accommodate temperature dependent sensitization. In any event, in their later work, Lakshmanan and Bhatt[152] report that glow curve area remains constant with preannealing temperature from 220 to 300°C. In its present form, however, the TC-LC model does not incorporate any mechanism to explain the ionization density dependence of supralinearity and sensitization. Somehow, high energy electrons are more capable by far (over a certain dose range!) of breaking up this complex TC-LC than are lower energy electrons. Although this is conceivable it would be comforting to have some direct understanding and evidence for such a mechanism. A combined track interaction and extended TC-LC model might be capable of explaining most of the phenomena associated with supralinearity and sensitization in LiF-TLD.

In conclusion, on the basis of existing data, there is an inclination among many researchers to explain supralinearity via a dose-dependent reduction of competitive processes in the recombination stage leading to higher TL efficiency over certain dose regions. The most convincing evidence for this mechanism is the linear rise with dose of all the optical absorption bands associated with the TL peaks. The track interaction model, in our opinion, is the most attractive candidate in this category. Further effort must be expended in the elucidation of trapping center-luminescent center spatial correlations which appear very probably to exist because of the ionization density dependence of both supralinearity and sensitization. On the basis of various considerations we have shown that it is possible to demote to a certain extent some of the models of LiF-TL which explain supralinearity as arising exclusively from mechanisms in the radiation absorption stage. These include the deep trap competition model of Suntharalingam and Cameron,[20] the trap creation model of Cameron et al.[84] and the multihit-multitrap model of Katz et al.[146] It should be emphasized, however, that no model can be entirely eliminated on the basis of existing data. The multihit model of Katz and collaborators has attractive aspects and, as has been previously discussed, the idea of multiple capture of charge carriers is a well-known phenomenon (e.g., the Z-center model). As a further example of the many possibilities, Townsend et al.[165] explained the nonclassical kinetics of the LiF-TLD glow peaks via loosely bound $(Mg + Li_{vac})_3$ dipoles + V_3 aggregates which can capture two electrons leading to competitive, nonradiative decay paths. Single-electron capture leads to the same radiative decay path as postulated by Mayhugh.[136] Unfortunately, the multihit model postulated by Katz et al. to explain supralinearity cannot in its present form accommodate the ionization density dependence of supralinearity. Moreover, the enormous sensitive sites required by this model to fit supralinearity require extremely long-range forces or correlations not yet alluded to in models of trapping centers in alkali halides.

REFERENCES

1. **Chen, R. and Bowman, S. G. E.,** Superlinear growth of TL due to competition during irradiation, *J. Eur. Study Group Phys. Chem. Math. Tech. Appl. Archaeol.*, 2, 216, 1978.
2. **Jasinska, M., Niewiadomski, T., and Ryba, E.,** Micro and single crystalline LiF:Mg,Ti phosphor and its application to radiation dosimetry, *Nukleonika*, 9, 995, 1969.
3. **DeWerd, L. A. and Stoebe, T. G.,** The influence of hydroxide impurities on TL in LiF, in Proc. 3rd Int. Conf. Luminescence Dosimetry, IAEA/AEC, Risö, Denmark, 1971, 78.
4. **Vora, H. and Stoebe, T. G.,** Role of OH^- ions in thermoluminescence of LiF, in Proc. 5th Int. Conf. Luminescence Dosimetry, Sao Paulo, Physikalisches Institut, Giessen, 1977, 3.
5. **Stoebe, T. G. and Watanabe, S.,** Thermoluminescence and lattice defects in LiF, *Phys. Status Solidi A*, 29, 11, 1975.
6. **Burgkhardt, B., Piesch, E., and Singh, D.,** High dose characteristics of LiF and $Li_2B_4O_7$ thermoluminescent dosimeters, *Nucl. Instrum. Methods*, 148, 613, 1978.
7. **Rossiter, M. J., Rees-Evans, D. B., and Ellis, S. C.,** Preparation of thermoluminescent LiF, *J. Phys. D*, 3, 1816, 1970.
8. **Rossiter, M. J., Rees-Evans, D. B., Ellis, S. C., and Griffiths, J. M.,** Titanium as a luminescence center in thermoluminescent lithium fluoride, *J. Phys. D*, 4, 1245, 1971.
9. **Crittenden, G. C., Townsend, P. D., Gilkes, J., and Wintersgill, M. C.,** LiF dosimetry. II. The effects of Mg and Ti on the thermoluminescent emission spectrum of LiF, *J. Phys. D*, 7, 2410, 1974.
10. **Vana, N., Aigenger, H., and Hager, A.,** Supralinearity in LiF, in Proc. 4th Int. Conf. Luminescence Dosimetry, Institute of Nuclear Physics, Krakow, 1974, 123.
11. **Niewiadomski, T.,** Confrontation of TL Models with Experimental Data, Rep. No. 936/D, Institute of Nuclear Physics, Krakow, 1976.
12. **Gorbics, S. G., Attix, F. H., and Kerris, K.,** TL dosimeters for high dose applications, *Health Phys.*, 25, 499, 1973.
13. **Rossiter, M. J.,** The use of precision TL dosimetry for intercomparison of absorbed dose, *Phys. Med. Biol.*, 20, 735, 1975.
14. **Kalef-Ezra, J. and Horowitz, Y. S.,** Heavy charged particle dosimetry: track structure theory and experiments, *Int. J. Appl. Radiat. Isot.*, 11, 1085, 1982.
15. **Eggermont, G., Jacobs, R., Jansens, A., Segart, O., and Thielens, G.,** Dose relationship, energy response and rate dependence of LiF-100, LiF-7 and $CaSO_4$:Mn from 8 keV to 30 MeV, in Proc. 3rd Int. Conf. Luminescence Dosimetry, IAEA/AEC, Risö, Denmark, 1971, 444.
16. **Piesch, E., Burgkhardt, B., and Kabadjova, S.,** Supralinearity and re-evaluation of different LiF dosimeter types, *Nucl. Instrum. Methods*, 126, 563, 1975.
17. **Naylor, G. P.,** Thermoluminescent phosphors: variation of quality response with dose, *Phys. Med. Biol.*, 10, 564, 1965.
18. **Grant, R. M. and Jones, D. E.,** TLD supralinearity and optical absorption of LiF:Mg, in Proc. 2nd Int. Conf. Luminescence Dosimetry, U.S. A.E.C. CONF-680920, NTIS, Springfield, Va., 1968, 310.
19. **Hendee, W. R., Ibbott, G. S., and Gilbert, D. B.,** Effects of total dose on energy dependence of TLD-100 LiF dosimeters, *Int. J. Appl. Radiat. Isot.*, 19, 431, 1968.
20. **Suntharalingam, N. and Cameron, J. R.,** Thermoluminescent response of LiF to radiations with different LET, *Phys. Med. Biol.*, 14, 397, 1969.
21. **Barber, D. E., Moore, R., and Hutchinson, T.,** Response of LiF to 1.0 to 4.0 keV electrons, *Health Phys.*, 28, 13, 1975.
22. **Lasky, J. B. and Moran, P. R.,** Thermoluminescent response of LiF (TLD-100) to 5 to 30 keV electrons and the effect of annealing in various atmospheres, *Phys. Med. Biol.*, 22, 852, 1977.
23. **Lasky, J. B. and Moran, P. R.,** Thermoluminescent response of LiF (TLD-100) to 0.1 to 5.0 keV electrons: an energy range relationship and comparison of the TL glow with TSEE glow curves, *J. Appl. Phys.*, 50, 4951, 1979.
24. **Tochilin, E., Goldstein, N., and Lyman, J. T.,** The quality and LET dependence of three thermoluminescent dosimeters and their potential use as secondary standards, in Proc. 2nd Int. Conf. Luminescence Dosimetry, U.S. A.E.C. CONF-680920, NTIS, Springfield, Va., 1968, 424.
25. **Tochilin, E., Goldstein, N., and Miller, W. G.,** BeO as a thermoluminescent dosimeter, *Health Phys.*, 16, 1, 1969.
26. **Binder, W. and Cameron, J. R.,** Dosimetric properties of CaF_2:Dy, *Health Phys.*, 17, 613, 1969.
27. **Attix, F. H.,** Further considerations of the track interaction model for thermoluminescence in LiF (TLD-100), *J. Appl. Phys.*, 46, 81, 1975.
28. **Horowitz, Y. S. and Kalef-Ezra, J.,** Relative TL yields of heavy charged particles: theory and experiment, *Nucl. Instrum. Methods*, 175, 29, 1980.

29. **Piesch, E., Burgkhardt, B., and Sayed, A. M.,** Supralinearity and re-evaluation of TLD-600 and TLD-700 in mixed neutron-gamma ray fields, in Proc. 4th Int. Conf. Luminescence Dosimetry, Institute of Nuclear Physics, Krakow, 1974, 1201.
30. **Nash, A. E. and Johnson, T. L.,** LiF (TLD-600) thermoluminescence detectors for mixed thermal neutron and gamma ray dosimetry, in Proc. 5th Int. Conf. Luminescence Dosimetry, Sao Paulo, Physikalisches Institute, Giessen, 1977, 393.
31. **Jain, V. K. and Ganguly, A. K.,** Some Aspects of Thermal, Radiation and LET Effects in the TL of LiF, B.A.R.C./I-466, Bhabha Atomic Research Center, Bombay, 1977.
32. **Jain, V. K.,** High temperature peaks in LiF-TLD: dependence on LET, *Nucl. Instrum. Methods*, 180, 195, 1981.
33. **Busuoli, G., Cavallini, A., Fasso, A., and Rimondi, O.,** Mixed radiation dosimetry with LiF (TLD-100), *Phys. Med. Biol.*, 15, 673, 1970.
34. **Dua, S. K., Boulenger, R., Ghoos, L., and Martens, E.,** Mixed neutron-gamma ray dosimetry, in Proc. 3rd Int. Conf. Luminescence Dosimetry, IAEA/AEC, Risö, Denmark, 1971, 1074.
35. **Mason, E. W.,** The effect of thermal neutron irradiation on the TL response of CONRAD type 7 LiF, *Phys. Med. Biol.*, 15, 79, 1970.
36. **Majborn, B., Botter-Jensen, L., and Christensen, P.,** TL dosimetry applied to areas with mixed neutron-gamma radiation fields, in Proc. Dosimetry in Agriculture, Industry, Biology and Medicine, SM-160/25, IAEA, Vienna, 1973, 169.
37. **Wingate, C. L., Tochilin, E., and Goldstein, N.,** Response of LiF to neutrons and charged particles, in Proc. 1st Int. Conf. Luminescence Dosimetry, U.S. A.E.C. CONF-650637, NTIS, Springfield, Va., 1965, 421.
38. **Ayyangar, K., Lakshmanan, A. R., Chandra, B., and Ramadas, K.,** A comparison of thermal neutron and gamma ray sensitivities of common TLD materials, *Phys. Med. Biol.*, 19, 665, 1974.
39. **Horowitz, Y. S., Kalef-Ezra, J., Moscovitch, M., and Pinto, H.,** Further studies on the non-universality of the TL-LET response in thermoluminescent LiF and $Li_2B_4O_7$: the effect of high temperature TL, *Nucl. Instrum. Methods*, 172, 479, 1980.
40. **Jahnert, B.,** TL research of protons and alpha particles with LiF (TLD-700), in Proc. 3rd Int. Conf. Luminescence Dosimetry, IAEA/AEC, Risö, Denmark, 1971, 1031.
41. **Aitken, M. J., Tite, M. S., and Fleming, S. J.,** TL response to heavily ionizing radiations, in Proc. 1st Int. Conf. Luminescence Dosimetry, U.S. A.E.C. CONF-650637, NTIS, Springfield, Va., 1965, 490.
42. **Lakshmanan, A. R., Rajendran, K. V., and Ayyangar, K.,** Thermal neutron and gamma ray mixed field dosimetry with $Li_2B_4O_7$:Mn, *Health Phys.*, 30, 489, 1976.
43. **Kitahara, A., Saitoh, M., and Harasawa, H.,** Analysis of the TL response of LiF powder to thermal neutrons and gamma ray exposures, *Health Phys.*, 31, 41, 1976.
44. **Mehta, S. K. and Sengupta, S.,** Al_2O_3 phosphor for thermoluminescence dosimetry, *Health Phys.*, 31, 176, 1976.
45. **Podgorsak, E. B., Moran, P. R., and Cameron, J. R.,** Thermoluminescent behaviour of LiF (TLD-100) from 77°K to 500°K, *J. Appl. Phys.*, 42, 2761, 1971.
46. **Sunta, C. M., Bapat, V. N., and Kathuria, S. P.,** Effects of deep trap on supralinearity, sensitization and optical thermoluminescence, in Proc. 3rd Int. Conf. Luminescence Dosimetry, IAEA/AEC, Risö, Denmark, 1971, 146.
47. **Linsley, G. S. and Mason, E. W.,** Re-estimation of dose in LiF, in Proc. 3rd Int. Conf. Luminescence Dosimetry, IAEA/AEC, Risö, Denmark, 1971, 157.
48. **Antinucci, M., Cevolani, M., Degli Esposti, G. C., and Petralia, S.,** An analysis of the thermoluminescent peaks of LiF (TLD-100), *Lett. Nuovo Cimento*, 18, 393, 1977.
49. **Waligorski, M. P. R. and Katz, R.,** Supralinearity of peak 5 and 6 in TLD-700, *Nucl. Instrum. Methods*, 172, 463, 1980.
50. **Waligorski, M. P. R. and Katz, R.,** Supralinearity of peaks 5 and 6 in TLD-700, *Nucl. Instrum. Methods*, 175, 48, 1980.
51. **Lax, M.,** Cascade capture of electrons in solids, *Phys. Rev.*, 119, 1502, 1960.
52. **Asche, M. and Von Borzeskowski, J.,** On the temperature dependence of the hole mobility in Si, *Phys. Status Solidi*, 37, 433, 1970.
53. **Bradbury, M. H., Nwosu, B. C. E., and Lilley, E.,** The effect of cooling rate on the performance of thermoluminescent dosimeter crystals (TLD-100) and LiF crystals, *J. Phys. D*, 9, 1009, 1976.
54. **Bradbury, M. H. and Lilley, E.,** Effect of solution treatment, temperature and ageing on TLD-100 crystals, *J. Phys. D*, 10, 1267, 1977.
55. **Driscoll, C. M. H. and McKinlay, A. F.,** Particle size effects in thermoluminescent lithium fluoride, *Phys. Med. Biol.*, 26, 321, 1981.
56. **Oliveri, E., Fiorella, O., and Mangia, M.,** High-dose behaviour of $CaSO_4$:Dy thermoluminescent phosphors as deduced by a continuous model for trap depths, *Nucl. Instrum. Methods*, 154, 203, 1978.

57. **Oliveri, E., Fiorella, O., and Mangia, M.,** A multi-reading method for dosimetric TL measurements with $CaSO_4$:Dy phosphors, *Nucl. Instrum. Methods,* 163, 569, 1979.
58. **Bjarngard, B. E.,** The Radiothermoluminescence of $CaSO_4$:Sm and its use in Dosimetry, Rep. No. AE-167, Aktiebolaget, Atomenergi, Stockholm, 1963.
59. **Bjarngard, B. E.,** The Properties of $CaSO_4$:Mn Thermoluminescence Dosimeters, Aktiebolaget, Atomenergi, Stockholm, 1963.
60. **Burgkhardt, B., Singh, D., and Piesch, E.,** High dose characteristics of CaF_2 and $CaSO_4$ thermoluminescent dosimeters, *Nucl. Instrum. Methods,* 141, 363, 1977.
61. **Scarpa, G., Benincasa, G., and Ceravolo, L.,** Further studies on the use of BeO as a TL material, in Proc. 3rd Int. Conf. Luminescence Dosimetry, IAEA/AEC, Risö, Denmark, 1971, 427.
62. **Kirk, R. D., Schulman, J. H., West, E. J., and Nash, A. E.,** Studies on thermoluminescent lithium borate for dosimetry, in Proc. Symp. Solid State Chem. Radiat. Dosimeters Med. Biol., SM-78/23, IAEA, Vienna, 1967, 91.
63. **Thompson, J. J. and Ziemer, P. L.,** The thermoluminescent properties of lithium borate activated by silver, *Health Phys.,* 25, 435, 1973.
64. **Langmead, W. A. and Wall, B. F.,** A TLD system based on lithium borate for the measurement of doses to patients undergoing medical irradiation, *Phys. Med. Biol.,* 21, 39, 1976.
65. **Takenaga, M., Yamamoto, O., and Yamashita, T.,** Preparation and characteristics of $Li_2B_4O_7$:Cu phosphor, *Nucl. Instrum. Methods,* 175, 77, 1980.
66. **Schayes, R., Brooke, C., Kozlowitz, I., and L'heureux, M.,** TL properties of natural calcium fluoride, in Proc. 1st Int. Conf. Luminescence Dosimetry, U.S. A.E.C. CONF-650637, NTIS, Springfield, Va., 1965, 138.
67. **Podgorsak, E. B., Fuller, G. E., and Moran, P. R.,** TL supralinear response as an optical emission effect for CaF_2 V_k-centers, *Radiat. Res.,* 59, 446, 1974.
68. **Marrone, M. J. and Attix, F. H.,** Damage effects in CaF_2:Mn and LiF-TLD, *Health Phys.,* 10, 431, 1964.
69. **Ehrlich, M. and Placious, M.,** Termoluminescent response of CaF_2:Mn in polytetrafluoroethylene to electrons, *Health Phys.,* 15, 341, 1968.
70. **Lucas, A. C. and Kapsar, B. M.,** The thermoluminescence of thulium doped calcium fluoride, in Proc. 5th Int. Conf. Luminescence Dosimetry, Sao Paulo, Physikalisches Institut, Giessen, 1977, 131.
71. **Yamashita, T., Nada, N., Onishi, H., and Kitamura, S.,** Calcium sulfate activated by thulium or dysprosium for thermoluminescence dosimetry, *Health Phys.,* 21, 195, 1971.
72. **Schmidt, K., Linemann, H., and Giessing, R.,** Influences of preparation and annealing on the properties of $CaSO_4$:Dy thermoluminescent phosphor, in Proc. 4th Int. Conf. Luminescence Dosimetry, Institute of Nuclear Physics, Krakow, 1974, 237.
73. **Aypar, A.,** Studies on thermoluminescent $CaSO_4$:Dy for dosimetry, *Int. J. Appl. Radiat. Isot.,* 29, 369, 1978.
74. **Chandra, B. and Bhatt, R. C.,** Effect of dysprosium concentration on the TL characteristics of $CaSO_4$:Dy phosphors, *Nucl. Instrum. Methods,* 164, 571, 1979.
75. **Mikado, T., Tomimasu, T., Yamazaki, T., and Chiwaki, M.,** TL response of Mg_2SiO_4:Tb in electron fields, *Nucl. Instrum. Methods,* 157, 109, 1978.
76. **Dixon, R. L. and Ekstrand, K. E.,** The systematics of TL in rare earth activated sulfate lattices, in Proc. 4th Int. Conf. Luminescence Dosimetry, Institute of Nuclear Physics, Krakow, 1974, 481.
77. **Lucas, A. C. and Kapsar, B. M.,** The thermoluminescence of BaF_2:Dy, in Proc. 4th Int. Conf. Luminescence Dosimetry, Institute of Nuclear Physics, Krakow, 1974, 507.
78. **McDougall, R. S. and Rudin, S.,** Thermoluminescent dosimetry of aluminum oxide, *Health Phys.,* 19, 281, 1970.
79. **Mehta, S. K. and Sengupta, S.,** Gamma dosimetry with Al_2O_3 thermoluminescent phosphor, *Phys. Med. Biol.,* 21, 955, 1976.
80. **Harris, A. M. and Jackson, J. H.,** The emission spectrum of thermoluminescent dosimetry grade LiF, *J. Phys. D,* 3, 624, 1970.
81. **DeWerd, L. A. and Stoebe, T. G.,** The emission spectrum of LiF-TLD at low and high exposures, *Phys. Med. Biol.,* 17, 187, 1972.
82. **Jain, V. K., Bapat, V. N., and Kathuria, S. P.,** TL spectrum of heavily irradiated LiF (TLD-100), *J.Phys. C,* L343, 1973.
83. **Fairchild, R. G., Mattern, P. L., Langweiler, K., and Levy, P. W.,** Thermoluminescence of LiF (TLD-100): emission spectra measurements, *J. Appl. Phys.,* 49, 4512, 1978.
84. **Cameron, J. R., Suntharalingam, N. and Kenney, G. N.,** *Thermoluminescent Dosimetry,* University of Wisconsin Press, Madison, 1968.
85. **Bloch, P.,** Supralinearity in LiF, in Proc. 2nd Int. Conf. Luminescence Dosimetry, U.S. A.E.C. CONF-680920, NTIS, Springfield, Va., 1968, 341.

86. **Zimmerman, J.**, The radiation induced increase of thermoluminescence sensitivity of the dosimetry phosphor LiF (TLD-100), *J. Phys. C*, 4, 3277, 1971.
87. **Mattern, P. L., Lengweiler, K., and Levy, P. W.**, Apparatus for the simultaneous determination of TL intensity and spectral distribution, *Mod. Geol.*, 2, 293, 1971.
88. **Strash, A. M. and Madey, R.**, Thermoluminescence spectrum and energy conversion efficiency of LiF, in Proc. 2nd Int. Conf. Luminescence Dosimetry, U.S. A.E.C. CONF-680920, NTIS, Springfield, Va., 1968, 607.
89. **Konschak, K., Pulzer, R., and Hubner, K.**, The emission spectra of various TL phosphors, in Proc. 3rd Int. Conf. Luminescence Dosimetry, IAEA/AEC, Risö, Denmark, 1971. 249.
90. **Oltman, B. G., Kastner, J., and Paden, C.**, Spectral analysis of TL glow curves, in Proc. 2nd Int. Conf. Luminescence Dosimetry, U.S. A.E.C. CONF-680920, NTIS, Springfield, Va., 1968, 623.
91. **Horowitz, Y. S., Fraier, I., Kalef-Ezra, J., Pinto, H., and Goldbart, Z.**, Nonuniversality of the TL-LET response in thermoluminescent $Li_2B_4O_7$: the effect of batch composition, *Nucl. Instrum. Methods*, 165, 27, 1979.
92. **Sunta, C. M.**, Thermoluminescence spectrum of gamma irradiated natural CaF_2, *J. Phys. C*, 3, 1978, 1970.
93. **Plichta, J. and Neruda, J.**, To the problem of the non-linear response-exposure dependence for the TL glass detectors, in Proc. 4th Int. Conf. Luminescence Dosimetry, Institute of Nuclear Physics, Krakow, 1974, 459.
94. **Shiragai, A.**, Effects of grain size and initial trap density on supralinearity of LiF-TLD, *Health Phys.*, 18, 728, 1970.
95. **Zanelli, G. D.**, Particle size and supralinearity in LiF, *Phys. Med. Biol.*, 17, 99, 1972.
96. **Koczynski, A., Wolska-Witer, M., Botter-Jensen, L., and Christensen, P.**, Graphite mixed non-transparent LiF and $Li_2B_4O_7$:Mn dosimeters combined with a two-side reading system for beta-gamma dosimetry, in Proc. 4th Int. Conf. Luminescence Dosimetry, Institute of Nuclear Physics, Krakow, 1974, 641.
97. **Nakajima, T. and Watanabe, S.**, Influence of heating conditions on the TL sensitivity of TLD-LiF phosphors, *Int. J. Appl. Radiat. Isot.*, 27, 113, 1976.
98. **Nakajima, T.**, On the competing trap model for the non-linear TL response, *Jpn. J. Appl. Phys.*, 15, 1179, 1976.
99. **Jain, V. K.**, The dependence of TL intensity on heating rate and the deep trap in LiF, *Jpn. J. Appl. Phys.*, 17, 949, 1978.
100. **Gorbics, S. G., Nash, A. E., and Attix, F. H.**, Thermal quenching of luminescence in six thermoluminescent dosimetry phosphors, *Int. J. Appl. Radiat. Isot.*, 20, 843, 1969.
101. **Pradhan, A. S. and Bhatt, R. C.**, Influence of heating rates on the response of TLD phosphors, *Int. J. Appl. Radiat. Isot.*, 30, 508, 1979.
102. **Booth, L. F., Johnson, T. L., and Attix, F. H.**, LiF glow peak growth due to annealing, *Health Phys.*, 23, 137, 1972.
103. **Petralia, S. and Gnani, G.**, Thermoluminescence of plastically deformed LiF (TLD-100), *Lett. Nuovo Cimento*, 4, 483, 1972.
104. **Kos, H. J. and Mieke, S.**, Effect of deformation on the thermoluminescence of Ti doped lithium fluoride, *Phys. Status Solidi A*, 50, K165, 1978.
105. **Kalef-Ezra, J.**, Study of the Applicability of Track Structure Theory to Thermoluminescence, Ph.D. thesis, Ben Gurion University of the Negev, Beersheva, Israel, 1980.
106. **Pearson, D. W. and Cameron, J. R.**, Supralinearity of UV repopulated TL from TLD-100, U.S. A.E.C. Rep. No. COO-1105-156, NTIS, Springfield, Va., 1966.
107. **Portal, G., Berman, F., Blanchard, P., and Prigent, R.**, Improvement of sensitivity and linearity of radiothermoluminescent LiF, in Proc. 3rd Int. Conf. Luminescence Dosimetry, IAEA/AEC, Risö, Denmark, 1971, 410.
108. **Lakshmanan, A. R., Chandra, B., and Bhatt, R. C.**, Gamma radiation induced sensitization in $CaSO_4$:Dy TLD phosphor, *Nucl. Instrum. Methods*, 153, 581, 1978.
109. **Mayhugh, M. R. and Fullerton, G. D.**, Thermoluminescence in LiF, sensitization useful at low exposures, *Health Phys.*, 28, 279, 1975.
110. **Jones, A. R.**, The application of sensitized LiF-TLDs to personnel and environmental dosimetry, *Nucl. Instrum. Methods*, 175, 145, 1980.
111. **Bartlett, D. T. and Sandford, D. J.**, Incompatibility of sensitization and reestimation of lithium fluoride thermoluminescent phosphor, *Phys. Med. Biol.*, 23, 332, 1978.
112. **Charles, M. W., Khan, Z. U., and Mistry, H. D.**, The theory and practice of simultaneous sensitization and reestimation in LiF, *Nucl. Instrum. Methods*, 175, 51, 1980.
113. **Pearson, D. W. and Cameron, J. R.**, Emission Spectra of Sensitized and Unsensitized LiF (TLD-100), U.S. A.E.C. Rep. No. COO-1105-123, NTIS, Springfield, Va., 1966.

114. **Jain, V. K., Kathuria, S. P., and Ganguly, A. K.,** Supralinearity and sensitization in LiF-TLD phosphor, *J. Phys. C*, 7, 3810, 1974.
115. **Kos, H. J., Mieke, S., and Nink, R.,** Dependence of LiF:Mg optical absorption on temperature during X-ray irradiation, *Health Phys.*, 38, 228, 1980.
116. **Lakshmanan, A. R. and Vohra, K. G.,** Gamma radiation induced sensitization and phototransfer in Mg_2SiO_4:Tb TLD phosphor, *Nucl. Instrum. Methods*, 159, 585, 1979.
117. **Mason, E. W. and Linsley, G. S.,** Properties of some deep traps in LiF, in Proc. 3rd Int. Conf. Luminescence Dosimetry, IAEA/AEC, Riso, Denmark, 1971, 164.
118. **Lakshmanan, A. R. and Bhatt, R. C.,** Gamma ray induced sensitization and residual TL in common TLD phosphors, *Int. J. Appl. Radiat. Isot.*, 29, 353, 1978.
119. **Nakajima,T.,**On the sensitivity factor mechanism of some TL phosphors, in Proc. 3rd Int. Conf. Luminescence Dosimetry, IAEA/AEC, Risö, Denmark, 1971, 466.
120. **Blum, E. and Bewley, D. K.,** Calcium sulfate phosphor for clinical neutron dosimetry, *Health Phys.*, 30, 257, 1976.
121. **Lakshmanan, A. R. and Bhatt, R. C.,** Further consideration of the competing deep trap model for thermoluminescence in $CaSO_4$:Dy, *Nucl. Instrum. Methods*, 164, 215, 1979.
122. **Hamm, R. N., Wright, H. A., Katz, R., Turner, J. E., and Ritchie, R. H.,** Calculated yields and slowing down spectra for electrons in liquid water: implications for electron and photon RBE, *Phys. Med. Biol.*, 23, 1149, 1978.
123. **Fields, D. E. and Moran, P. R.,** Analytical and experimental check of a model for correlated TL and thermally stimulated conductivity, *Phys. Rev.*, 9, 1836, 1974.
124. **Bloch, P.,** Paramagnetic resonance in LiF, in Proc. 2nd Int. Conf. Luminescence Dosimetry, U.S. A.E.C. CONF-680920, NTIS, Springfield, Va., 1968, 317.
125. **Sunta, C. M.,** Associated luminescence centers and traps in the thermoluminescence of CaF_2:Dy (TLD-200), *J. Phys. C*, L47, 1977.
126. **Hall, T. P. P., Laggeat, A., and Twidell, J. W.,** The structure of some trapped hole centers in CaF_2, *J. Phys. C*, 3, 2352, 1970.
127. **Agullo-Lopez, F. and Aguilar, M.,** On the space distribution of F centers — the pair correlation function, *Phys. Status Solidi B*, 92, K43, 1979.
128. **Miller, L. D. and Bube, R. H.,** Luminescence, trapping and F centers in LiF crystals, *J. Appl. Phys.*, 41, 3687, 1970.
129. **Jackson, J. H. and Harris, A. M.,** Annealing effects and optical absorption of thermoluminescent lithium fluoride, *J. Phys. C*, 3, 1967, 1970.
130. **Zimmerman, D. W. and Cameron, J. R.,** Supralinearity of LiF versus dose, in *Thermoluminescence of Geological Materials*, McDougall, D. J., Ed., Academic Press, London, 1968, 485.
131. **Claffy, E. W., Klick, C. C., and Attix, F. H.,** TL processes and colour centers in LiF:Mg, in Proc. 2nd Int. Conf. Luminescence Dosimetry, U.S. A.E.C. CONF-680920, NTIS, Springfield, Va., 1968, 302.
132. **Kos, H. J. and Nink, R.,** Correlation of TL and optical absorption of Z_3 centers in LiF:Mg,Ti crystals, *Phys. Status Solidi A*, 56, 593, 1979.
133. **Kos, H. J. and Nink, R.,** Dipole → Z center conversion in Mg doped LiF, *Phys. Status Solidi A*, 57, 203, 1980.
134. **Caldas, L. V. E.,** Ph.D. thesis, University of Sao Paulo, 1973.
135. **Lucke, W. H.,** Intrinsic efficiencies of some thermoluminescent dosimetry phosphors, *J. Appl. Phys.*, 42, 3004, 1971.
136. **Mayhugh, M. R.,** Colour centers and the TL mechanism in LiF, *J. Appl. Phys.*, 41, 4776, 1970.
137. **Mayhugh, M. R., Christy, R. W., and Johnson, N. M.,** TL and colour center correlations in dosimetry LiF, *J. Appl. Phys.*, 41, 2968, 1970.
138. **Christy, R. W. and Mayhugh, M. R.,** Thermoluminescence mechanism in dosimetry LiF, *J. Appl. Phys.*, 43, 3216, 1972.
139. **Delbecq, C. J., Hayes, W., and Yuster, P. H.,** Absorption spectra of F_2^-, Cl_2^-, Br_2^- and I_2^- in the alkali halides, *Phys. Rev.*, 121, 1043, 1961.
140. **Nink, R. and Kos, H. J.,** On the role of Z centers in the trapping mechanism of thermoluminescent LiF, *Phys. Status Solidi A*, 35, 121, 1976.
141. **Nink, R. and Kos, H. J.,** Lithium fluoride dosimetry. I. The Z center model — a new concept for the description of trapping processes in dosimetric LiF:Mg,Ti, *Nucl. Instrum. Methods*, 175, 16, 1980.
142. **Kos, H. J. and Nink, R.,** Z_1 centers in Mg doped LiF, *Phys. Status Solidi A*, 41, K157, 1977.
143. **Kos, H. J. and Nink, R.,** LiF dosimetry. II. The Z center model — origin and relevance to some dosimetry phenomenae, *Nucl. Instrum. Methods*, 175, 24, 1980.
144. **Gartia, R. K.,** Thermoluminescence of Z_3 centers in X-irradiated Mg doped LiF crystals, *Phys. Status Solidi A*, 44, K21, 1977.
145. **Horowitz, Y. S.,** Criticism of the Z center model in LiF-TLD, *Phys. Status Solidi A*, 69, K29, 1982.

146. **Katz, R., Sharma, S. C., and Homayoonfar, M.,** The structure of particle tracks, in *Topics in Radiation Dosimetry,* Suppl. 1, Attix, F. H., Ed., Academic Press, New York, 1972, 317.
147. **Larsson, L. and Katz, R.,** Supralinearity of thermoluminescent detectors, *Nucl. Instrum. Methods,* 138, 631, 1976.
148. **Ohkura, H.,** Z_2 and Z_3 colour centers in KCl and KBr, *Phys. Rev.,* 136, A446, 1964.
149. **Cooke, D. W.,** Trapping mechanism of thermoluminescent lithium fluoride based on Z centers, *Phys. Status Solidi A,* 58, K167, 1980.
150. **Parfianovitch, I. A., Alekseeva, E. P., Soterdotova, G. V., and Petrov, A. L.,** *Opt. Spektrosk.,* 41, 73, 1976.
151. **Lakshmanan, A. R., Bhatt, R. C., and Vohra, K. G.,** Gamma radiation induced sensitization, supralinearity and photostimulated thermoluminescence in LiF (TLD-100), *Phys. Status Solidi A,* 53, 617, 1979.
152. **Lakshmanan, A. R. and Bhatt, R. C.,** Re-investigation of the sensitization mechanism in LiF (TLD-100), *Phys. Status Solidi A,* 57, 323, 1980.
153. **Kristianpoller, N., Chen, R., and Israeli, M.,** Dose dependence of thermoluminescence peaks, *J. Phys. D,* 7, 1063, 1974.
154. **Aitken, M. J., Huxtable, J., Wintle, A. G., and Bowman, S. G. E.,** Age determination of TL: review of progress at Oxford, in Proc. 4th Int. Conf. Luminescence Dosimetry, Institute of Nuclear Physics, Krakow, 1974, 1005.
155. **Wilson, C. R., DeWerd, L. A., and Cameron, J. R.,** Stability of the increased sensitivity of LiF(TLD-100) as a function of temperature, U.S. A.E.C. Rep No. COO-1105-116, NTIS, Springfield, Va., 1966.
156. **Klick, C. C., Claffy, E. W., Gorbics, S. G., Attix, F. H., Schulman, J. H., and Allard, J. G.,** Thermoluminescence and colour centers in LiF:Mg, *J. Appl. Phys.,* 38, 3867, 1967.
157. **Dobson, P. N. and Midkiff, A. A.,** Explanation of supralinearity in the thermoluminescence of LiF in terms of the interacting track model, *Health Phys.,* 18, 571, 1970.
158. **Yamazaki, T., Tomimasu, T., Mikado, T., and Chiwaki, M.,** On the supralinearity of Mg_2SiO_4:Tb TLDs, *J. Appl. Phys.,* 49, 4929, 1978.
159. **Davies, J. J.,** EPR and ENDOR of titanium doped lithium fluoride, *J. Phys. C,* 7, 599, 1974.
160. **Budd, T., Marshall, M., People, L. H. J., and Douglas, J. A.,** The low and high temperature response of LiF dosimeters to X-rays, *Phys. Med. Biol.,* 24, 71, 1979.
161. **Mayhugh, M. R. and Fullerton, G. D.,** Altering the energy dependence of LiF-TLDs by pre-irradiation, *Med. Phys.,* 1, 275, 1974.
162. **Mieke, S. and Nink, R.,** LiF:Ti as a material for thermoluminescence dosimetry, *J. Luminescence,* 18/19, 411, 1979.
163. **Horowitz, Y. S., Moscovitch, M., and Dubi, A.,** Heavy charged particle induced thermoluminescence response curves: interpretation via track structure theory, *Phys. Med. Biol.,* 27, 1982.
164. **Jain, V. K. and Kathuria, S. P.,** Z_3 center thermoluminescence in LiF-TLD phosphor, *Phys. Status Solidi A,* 50, 329, 1978.
165. **Townsend, P. D., Taylor, G. C., and Wintersgill, M. C.,** An explanation of the anomalously high activation energies of TL in LiF (TLD-100), *Radiat. Eff.,* 41, 11, 1979.
166. **Takeuchi, N., Inabe, K., and Nakamura, S.,** Post irradiation annealing effect on the growth of the Z_3 band in LiF:Mg crystal, *Phys. Status Solidi A* , 65, K33, 1981.
167. **Lakshmanan, A. R.,** Modified Z center model for TL in LiF TLD phosphor, in *Proc. 7th Int. Conf. Solid State Dosimetry,* Nuclear Technology Publishing, England, in press.
168. **Horowitz, Y. S. and Moscovitch, M.,** Track Structure Theory- track intersection model for heavy charged particle TL dose response curves, in *Proc. 7th Int. Conf. Solid State Dosimetry,* Nuclear Technology Publishing, England, in press.

Chapter 2

THERMOLUMINESCENT RADIATION DOSIMETRY

Yigal S. Horowitz

TABLE OF CONTENTS

I. X-RAY DOSIMETRY

Radiobiology and radiotherapy studies have demonstrated that differences of 10% in absorbed dose will produce clearly observable variations in biological response. In general, cell survival studies do not allow the prediction of variations in absorbed dose determinations smaller than 5%. It has been suggested, therefore, that an accuracy of 5 to 6% and a precision of 2 to 3% are required for the determination of absorbed dose in biological applications.[1]

It should be mentioned that these requirements are far more severe than those generally encountered in radiation protection or environmental radiation dosimetry. There are claims in the literature to accuracy surpassing 5 to 6%,[2] however, the serious discrepancies in the TLD literature concerning the extent of the overresponse of LiF and other TL materials to X-rays in the energy range of 5 to 150 keV suggest that this accuracy is, in fact, very difficult to achieve. A figure greater than 10% is more than likely to be the best attainable until a more precise understanding of the response of TLDs to both high and low energy gamma rays is achieved. The central question in TLD application to X-ray dosimetry is thus whether there is indeed an overresponse for LiF and other TL materials arising from ionization density effects or whether the observed overresponse arises from experimental dosimetric inaccuracies. A corollary question of equal importance is to what extent the ionization density effects (if they exist) are batch and material dependent.

A. LiF-TLD Low Energy Response (4 to 1250 keV)

Greening[3] discussed the application of LiF-TLD in low energy X-ray dosimetry. The variation with X-ray energy of the TLD response per unit exposure depends on many factors: (1) the mass energy absorption coefficient of the TL material; (2) the proportion and type of intentional dopants and nonintentional impurities; (3) the grain size of the TL material; (4) the material surrounding the grains; (5) the dose level (if the linear dose-response region is exceeded); (6) the possibility of variations in TL efficiency due to ionization density effects; and (7) X-ray attenuation in the TLD or its container. Items (3) and (4) refer, of course, to the incorporation of the TL material into different matrices in various geometries, which changes Z_{eff} and changes the response.[4-7] Unfortunately, the change in response will be particle size and energy dependent in a manner not easily amenable to accurate calculation. A further difficulty is that the particle size distribution is difficult to measure accurately and is a function of the annealing history. Pradhan and Bhatt[8-10] compared their experimental data for $CaSO_4$:Dy powders embedded in teflon® or surrounded by various metal filters with the theoretical calculations of Bassi et al.[7] shown in Figure 1. The large discrepancies between the calculated and the experimental energy dependence show that the calculated data (which did not take into account grain size) have little relevance for TL materials imbedded in teflon®. Figure 1 also illustrates the necessity of precise reproduction of the TL phosphor grains, failing which severe batch-to-batch or even dosimeter-to-dosimeter energy response variations may occur. On the other hand, the variation of response of the approximately tissue-equivalent phosphors with particle size at low photon energies is not expected to play a very significant role for large grain sizes commonly encountered. Burlin et al.[4] and Chan and Burlin[5] e.g., have calculated for LiF a decrease in response of less than 1% for grain sizes greater than 100 μm at 50 keV and an increase in response of approximately 1% for $Li_2B_4O_7$ at 60 keV (100 μm) and 4% (30 μm). For the high Z_{eff} phosphors the underresponse is considerably greater, approximately 10% for 100-μm grains of CaF_2 at 100 keV reaching 30% for 30 μm-grain diameters. It should be emphasized, however, that these cavity theory calculations are based on the assumption that the thickness of the air around the grain is sufficient to establish electronic equilibrium, i.e., they are approximately valid for monolayer TLD samples. The corrections are greatly overestimated for the multilayer TLD samples usually employed. Item (7) is a general dosimetric problem not specific to TLD, and indeed

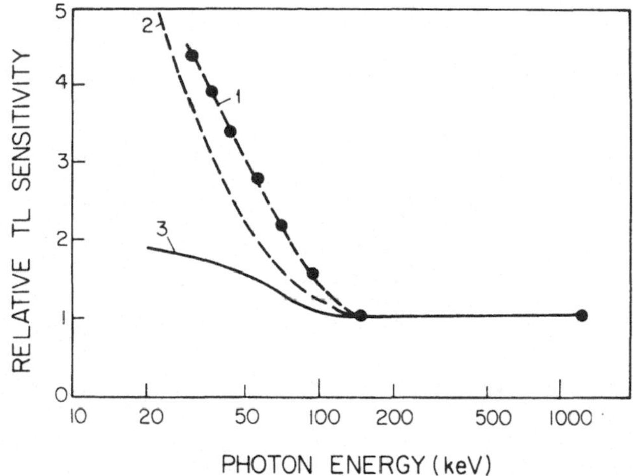

FIGURE 1. Photon energy dependence of $CaSO_4$ in teflon®. (1) Experimental curve of teflon® discs $CaSO_4$:Dy (5 *M* %) with grain size <1 μm. (2) Calculated curve following Chan and Burlin.[5] (3) Calculated curve following Bassi et al.[7] (Adapted from Pradhan, A. S., Kher, R. K., Dere, A., and Bhatt, R. C., *Int. J. Appl. Radiat. Isot.*, 29, 243, 1978.)

many authors report a sharp drop in response below approximately 25 keV attributable to attenuation of the low energy photons in the TLD. Endres et al.,[6] e.g., have pointed out that for 0.9-mm thick LiF-TLDs the attenuation of 10-keV photons is approximately 70% decreasing to 30% at 0.25 mm. On the other hand, the efficiency of the approximately tissue-equivalent materials (LiF, $Li_2B_4O_7$, BeO, MgB_4O_7) is sufficient to allow doses of fractions of a mGy to be measured from masses small enough not to cause significant attenuation[11] so that this aspect with proper dosimeter design can be essentially eliminated. With respect to item (5), Jayachandran[12] has noted that supralinearity is not a problem since calibrations are always carried out. In fact, the dependence of the supralinearity on X-ray energy brings about the requirement that the calibration and measurement be carried out with identical energy spectra *at the TLD*, a requirement that is very difficult to fulfill to high accuracy.

Following this brief review we see that the principal problems arise from our lack of knowledge concerning items (1), (2), and (6), and we now address ourselves to a detailed consideration of these subjects. More specifically, we require to know to what extent the observed overresponse of many TLD materials reported in the literature is truly due to an ionization density dependence. The only other reasonable alternative is hidden dosimetric inaccuracies which obviously compromise the validity of the experimental measurements if left unaccounted for.

The photon energy dependence of TLD materials is given by

$$S(E) = \frac{(\mu_{en}/\rho)_{TLD}}{(\mu_{en}/\rho)_{air}} \eta(E) = \mu'(E)\eta(E) \tag{1}$$

where S(E) is the energy-dependent relative response of a bare TLD assuming negligible self-absorption and the existence of electronic equilibrium, μ_{en}/ρ are the mass energy absorption coefficients, which are a measure of the average fractional amount of incident photon energy transferred to kinetic energy of charged particles. Thus, this imparted charged

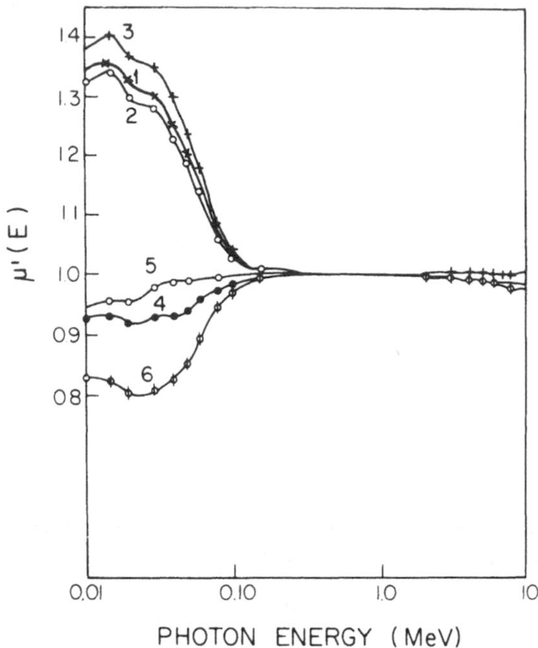

FIGURE 2. Energy dependence of some "tissue-equivalent" materials relative to the absorbed dose in air and normalized to 1-MeV photons. (1) TLD-100 (0.09% Si, 0.3% Mg, 0.03% P). (2) TLD-100 (0.04% Mg). 3. LiF (0.04% Mg, 1% NaF). (4) $Li_2B_4O_7$ (Mn 0.1%, Si 0.5%). (5) $Li_2B_4O_7$ (Mn 0.34%). (6) BeO Thermalox 995.

particle kinetic energy is a measure of the amount of energy made available for the production of TL. Figure 2 shows $\mu'(E)$ for various approximately "tissue equivalent" materials following cross-section data from Storm and Israel.[13] Since the photoelectric cross section is both energy and Z dependent[14,15] the ratio $\mu'(E)$ significantly differs from unity below 100 keV. Millar and Greening[14,15], e.g., have shown that a cubic polynomial of the form $\ell n\tau = \sum_{i=1}^{3} C_i(\ell nE)^i$ accurately describes the relationship between the photoelectric cross section and the photon energy. $\eta(E)$ is the relative TL response (the energy-dependent relative TL efficiency) usually normalized to the TL efficiency at 1.25 MeV. It should be emphasized at the outset that $\eta(E)$ can be different from unity in LiF and other TL materials. Budd et al.[16] have, e.g., measured $\eta \simeq 2$ for peaks 6 and 7 in TLD-100 (Figure 3) over the X-ray energy range from 30 to 100 keV.* Thus, there is no *a priori* reason to assume $\eta = 1$ in TLD for any peak-material combination. Hubbell[17,18] has carefully described the interrelationship between μ/ρ (the mass attenuation coefficient), μ_a/ρ (the mass absorption coefficient), μ_{tr}/ρ (the mass energy transfer coefficient), and μ_{en}/ρ. For example, the difference between the mass energy absorption coefficient and the mass energy transfer coefficient is that the former does not count energy transferred to electrons if that energy is subsequently lost as bremsstrahlung X-rays. The mass attenuation coefficient is a measure of the average number of interactions between incident photons and matter that occur in a given mass per

* To enhance the intensity of peaks 6 and 7 relative to peaks 4 and 5 the TLDs were annealed by Budd et al. at 400°C for 1 hr and then cooled slowly ($T_2' = 3°C\ min^{-1}$) to 80°C (rather than normal rapid cooling) and held at 80°C for 12 hr.

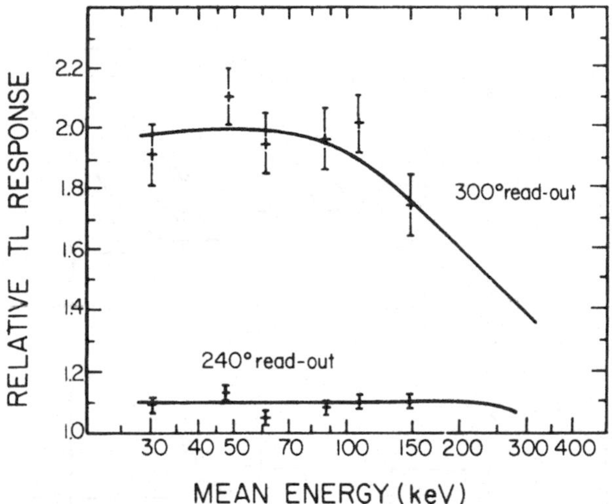

FIGURE 3. Relative TL sensitivity of peaks 6 to 8 (T_8 = 300°C) as a function of X-ray energy. Also shown is the relative TL sensitivity for T_8 = 240°C. (Adapted from Budd, T., Marshall, M., People, L. H. J., and Douglas, J. A., *Phys. Med. Biol.*, 24, 71, 1979.)

unit area thickness of material traversed. Carlsson[19] has discussed the effect of neglect of L shell fluorescence escape on mass energy transfer and mass energy absorption coefficients. For approximately tissue-equivalent materials the error is negligible, but increases to as much as 10% for the heaviest elements and photon energies just above the K edge. The relationship between these various coefficients is shown in Figure 4. The mass energy absorption coefficients have been calculated by Hubbell[17,18] according to the following expression

$$\mu_{en}/\rho = \sigma_{en}(N_A/A_r) \tag{2}$$

in which N_A is Avogadro's number (6.022045×10^{23} mol^{-1}) and A_r is the relative atomic mass (atomic weight). σ_{en} is the atomic energy absorption cross section taken as

$$\sigma_{en} = \sigma_{incoh} \cdot f_{incoh} + \tau \cdot f_\tau + \kappa \cdot f_\kappa \tag{3}$$

in which the fractions f_{incoh}, f_τ, and f_κ weight the incoherent scattering cross section, σ_{incoh}, the photoeffect cross section, τ, and the pair (including triplet) production cross section, κ, respectively, by the ratio of charged particle kinetic energy to incident photon energy resulting for each of these interactions as indicated schematically in Figure 4. The calculations by Hubbell[18] were carried out by numerical integration of the expression

$$\sigma_{incoh} \cdot f_{incoh} = (1 + \Delta_{KN}^{M}) \int_0^{\pi} (d\sigma_{KN}/d\Omega) \cdot S(q,z) \cdot (T/k) \cdot (1 - G_{br}(T,z)\, d\Omega \tag{4}$$

in which Δ_{KN}^{M} is a double Compton and radiative correction, $d\sigma_{KN}/d\Omega$ is the Klein-Nishina Compton scattering differential cross section, S(q,z) is the incoherent scattering function, T is the outgoing Compton electron initial kinetic energy, k is the incident photon energy, and $G_{br}(T,z)$ is the bremsstrahlung radiation yield.

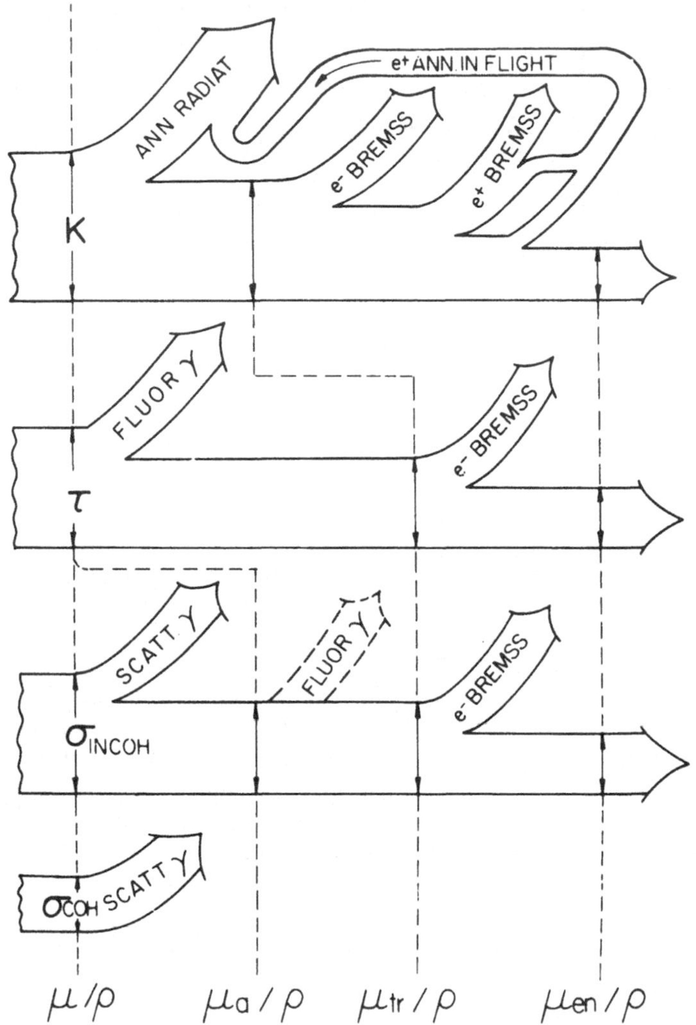

FIGURE 4. Schematic representation of the mass attenuation coefficient μ/ρ, the mass absorption coefficient μ_a/ρ, the mass energy transfer coefficient μ_{tr}/ρ, and the mass energy absorption coefficient μ_{en}/ρ in terms of the cross sections for coherent (σ_{coh}) and incoherent (σ_{incoh}) scattering, photoeffect (τ), and pair production (κ). The upward branching arrows represent fractions of the initial photon energy lost to the volume of interest in the form of secondary photons such as positron annihilation radiation (ANN RAD), bremsstrahlung (e^+, e^- BREMSS), fluorescence X-rays (FLUOR γ), and scattered photons (SCATT γ). The enhancement of annihilation photon energies due to positron annihilation in flight (e^+, ANN IN FLIGHT) at the expense of positron bremmstrahlung is also shown. (Adapted from Hubbell, J. H., *Int. J. Appl. Radiat. Isot.*, 11, 1269, 1982.)

Several compilations exist for μ_{en}/ρ for pure LiF and these are reproduced in Table 1. Jayachandran[12] based his calculations on pre-1965 data and Bassi et al.[7] used only data from Storm and Israel,[13] so that both these sets of data are inferior to the ICRU-Greening data, which identically compared and smoothed data from several compilations including that of Storm and Israel. It nonetheless deserves mention that Millar[15] has assessed recent total attenuation and photoelectron cross-section data and concluded that the photoelectric cross

Table 1
MASS ENERGY ABSORPTION COEFFICIENTS OF PURE LiF[a] RELATIVE TO AIR[b]

Energy (keV)	$(\mu_{en}/\rho)_{LiF}/(\mu_{en}/\rho)_{air}$					TLD-100 (0.04 wt % Mg)
4	1.244			1.37	1.277	
5	1.243			1.37	1.285	
6	1.262			1.36	1.291	
8	1.279			1.35	1.304	
10	1.276	1.349	1.31	1.34	1.304	1.31
15	1.286	—	1.34	1.30	1.308	1.34
20	1.281	1.263	1.29	1.27	1.304	1.30
30	1.273	1.209	1.28	1.23	1.285	1.28
40	1.235	1.196	1.22	1.19	1.248	1.23
50	1.166	1.162	1.18	1.15	1.197	1.19
60	1.127	1.148	1.14	—	1.144	1.14
80	1.056	1.060	1.06	—	1.069	1.06
100	1.019	1.012	1.03	1.02	1.034	1.03
150	1.000	—	1.00		1.008	1.00
200		1.022	1.01		1.003	1.01
400		1.005	1.00		1.000	
600		1.010	1.00		1.000	
800		1.014	1.00		1.000	
1000		1.002	1.00		1.000	
1250		1.000	1.00	1.00	1.000	
1500		0.998	1.00			
Ref.	3, 21	22	7	20	17,18	7

[a] The values for LiF are applicable to materials such as TLD-100 which contains the naturally occurring mixture of lithium isotopes, or to TLD-700 which contains virtually 100% ^{7}Li. For TLD-600 which is highly enriched in ^{6}Li, the values should be increased by approximately 3.5%.

[b] The composition of air is taken as N_2 (75.5), 0 (23.2), and Ar (1.3) parts by weight.

section is a less smooth function of atomic number than of energy. This conclusion could be of significance to the ICRU-Greening data, which smoothed atomic number data from several compilations. The data presented by Reddy and Mehta[20] are based on updated mass energy absorption coefficients for air as calculated by Hubbell.[18] These latter data are in very good agreement with the highly accurate mass attenuation coefficient measurements for air reported by Millar and Greening[14] and Millar:[15] approximately 1% from 5 to 20 keV increasing to approximately 2 to 3% deviation at 4 keV. At 30 to 40 keV the Reddy and Mehta data are approximately 4% lower than the ICRU-Greening data and indeed Hubbell[18] has commented that the μ_{en}/ρ values arrived at in ICRU-17 are at 30 keV, 5% lower than the average of all the other compilations. At 100 keV the maximum deviation between the various compilations of (μ_{en}/ρ) for air is less than 2%. At 20 keV the Reddy and ICRU-Greening data agree to within 1%, but the former reports a far more rapid increase with decreasing energy so that the discrepancy is approximately 10% at 4 keV. Since the Hubbell, ICRU, and Storm and Israel air data agree to within 1% at 3 and 10 keV, the source of this discrepancy must arise from the LiF data used by Reddy and Mehta. Unfortunately, these authors make no attempt to explain this discrepancy. Photoelectron bremsstrahlung losses (estimated to be less than 2%) were neglected by Storm and Israel so that the Hubbell compilation is probably the most reliable to date. Hubbell estimates the uncertainties in

μ/ρ to be of the order of ±5% below 5 keV and ±2% up to 10 MeV. For light elements in the photon energy range of 200 keV to 5 MeV, the uncertainties in μ/ρ are estimated to be 1% or better. Uncertainties in μ_{en}/ρ are estimated to be slightly greater, as the additional factors such as given in Equation 4 are theoretically less well known and less amenable to direct measurement. The not insignificant differences shown in Table 1 between the various compilations certainly underline the conceptual and experimental difficulties that exist at the 5% level in low energy X-ray dosimetry. As we shall see in Section I.A.2, the approximate 5% discrepancies in $\mu'(E)$ in the energy range of 20 to 60 keV do not critically affect arguments concerning the validity of $\eta > 1$ since the greatest percentage overresponse observed by most authors occurs for X-ray energies between approximately 80 and 150 keV where the maximum deviation between the various compilations is approximately 2% (80 keV) decreasing to less than 1% (100 keV).

1. Effect of Impurity Composition

Approximately 100 ppm/w of high Z material as a dopant is sufficient to significantly increase by ~1% the response at energies below 100 keV. Bassi et al.[7] have calculated $\mu'(E)$ for four LiF compositions: TLD-100 (Harshaw®, 0.9 wt % 0, 0.3 wt % Si, 0.03 wt % Mg, and 0.025 wt % P[23]); TLD-100 (Harshaw, 0.04 wt % Mg[24]); LiF-PTL (CEC, 0.04 wt % Mg, 0.55 wt % Na, 0.45 wt % P[25]); and pure LiF. In TLD-100 these impurities increase $\mu'(E)$ by roughly 1 to 2% in the 10- to 60-keV energy interval, whereas the LiF-PTL is approximately 6% greater than pure LiF in this energy interval. Uncertainty in the exact impurity composition at the parts per million level thus certainly can lead to significant underestimation (up to approximately 5%) of $\mu'(E)$ for LiF-TLD.

2. Experimental Determinations of $\eta(E)$

A detailed compilation of many of the previous experimental results (Figure 5) has been carried out by Budd et al.,[16] which clearly illustrates the overresponse of peaks 4 and 5 in LiF-TLD in the energy region of 30 to 150 keV leading to $\eta(E) \simeq 1.1$. Of particular interest are experiments which report S(E) for several different forms of LiF under identical X-ray beam conditions. For example, Endres et al.[6] obtain $S(E) \simeq 1.25$ from 20 to 50 keV ($\eta \simeq 1 \pm 0.05$) for TLD-100 chips, but $S(E) \simeq 1.4$ ($\eta \simeq 1.15$) for ^{7}LiF powder for the same X-ray energies. Grain sizes for the powder were not recorded, but in any event the cavity theory corrections would be too small to significantly affect the discrepancy between the two sets of data. We thus have clear indication of a batch-dependent ionization density effect. Most of the experiments reporting $\eta > 1$ have been carried out with continuous X-ray spectra, and therefore, Jayachandran[12] attributed the overresponse in the 30- to 150-keV region to be most likely due to the spectral spread and the effect of scattered X-rays due to the TLD surroundings. Rossiter,[26] however, reported identical S(E) values measured for TLD-700 under extreme heavy filtration and light filtration conditions; moreover, even monoenergetic gamma-ray experiments have yielded $\eta > 1$. For example, Liu et al.[27] report $\eta = 1.064 \pm 0.022$ for ^{99m}Tc gamma rays (140 keV) in TLD-100. At this high energy the 6.4% overresponse cannot be due to inaccuracies in μ_{en}/ρ or impurity composition and must arise from ionization density effects or hidden dosimetric inaccuracies. Many authors[26,28-31] reported overresponse leading to $\eta = 1.2$ to 1.3. In at least one case the TL was integrated to 310°C so that the high overresponse might be due to the inclusion of the high temperature peaks ($\eta^{6,7} \simeq 2$, $\epsilon_{T>} \simeq 0.05$ results in a 50% increase in the overresponse). Gorbics and Attix[29] however, used peak height measurements so that the influence of the high temperature peaks could not exceed a few percent at low dose.

The result $\eta \simeq 1.1$ or greater must be reconciled with a subset of other experiments carried out with high accuracy and in some cases monoenergetic X-rays between 5 and 20 keV[11,32-34] and with another subset of experiments[12,35] which obtained η less than or equal

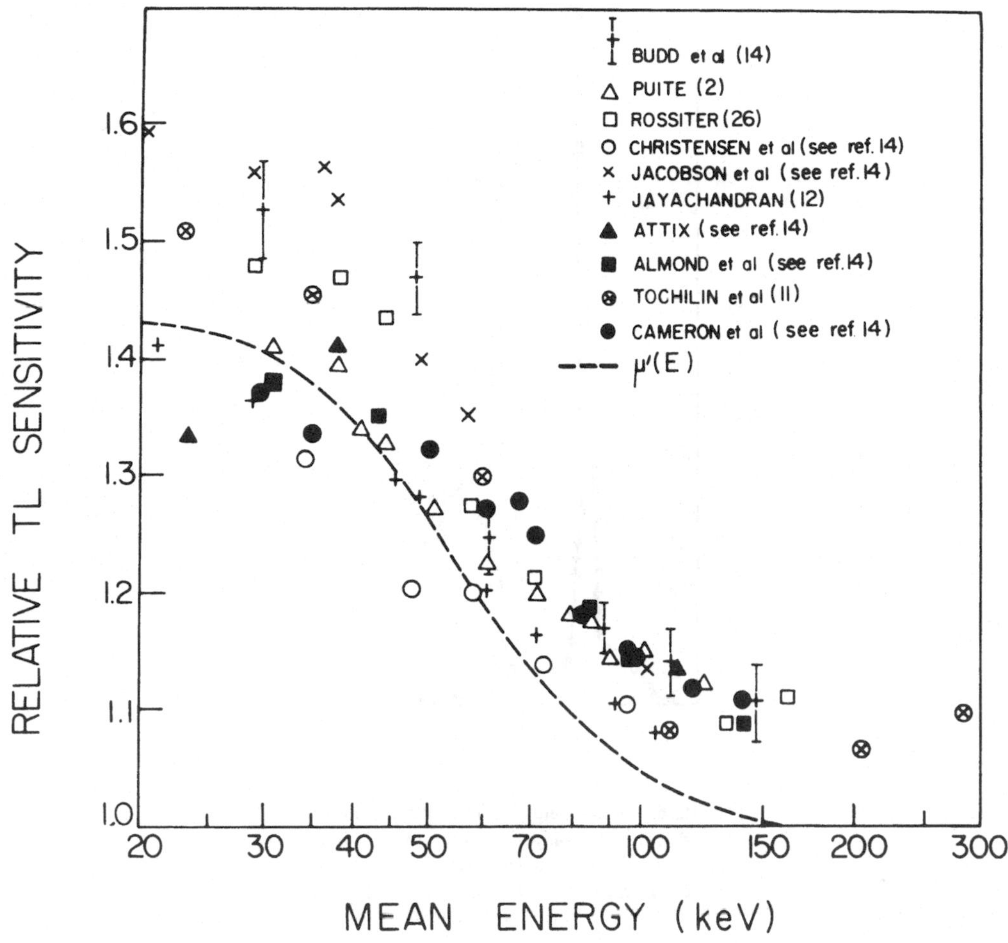

FIGURE 5. Relative TL sensitivity of peaks 4 and 5 (T_8 = 240°C) as a function of X-ray energy. (Adapted from Budd, T., Marshall, M., People, L. H. J., and Douglas, J. A., *Phys. Med. Biol.*, 24, 71, 1979.)

to unity over the entire range of X-ray energies from 20 to 150 keV. Storm et al.[35] obtained S(E) consistent with unity at approximately 20 to 100 keV ($\eta \simeq 0.8$) for EG & G mini-LiF-TLDs, whereas Jayachandran[12] using monoenergetic photons obtained values of η consistent with unity over the same energy interval with mass energy absorption coefficients including 350 ppm Ca and 400 ppm Mg as determined by spectrochemical analysis. At even lower X-ray energies, Law[32] obtained η (11 keV) = 0.97 ± 0.05 for TLD-700 with additional results from 13 to 26 keV_{eff} also consistent with unity. Horowitz and Kalef-Ezra[34] used monoenergetic 4-keV X-rays and calibrated the X-ray fluence with a Si(li) detector in identical geometrical configuration as in the TLD-600 and TLD-100 irradiations. A 500 mCi H-3 source was used as the approximately monoenergetic low energy X-ray generator. The tritium is absorbed in a layer of Ti on a Cu backing so that attenuation of the beta particles leads to X-rays characteristic to Ti and Cu as well as bremsstrahlung radiation with a combined fluence approximately three orders of magnitude less than the primary beta fluence. A 19-μm thick mylar foil or a 1.52 Mg cm^{-2}-thick Ni foil was inserted between the source and the TLD which stopped the beta particles while producing additional bremsstrahlung and characteristic X-rays. A typical X-ray energy spectrum used by Horowitz and Kalef-Ezra[34] is shown in Figure 6. The mass energy absorption coefficients of the LiF-TLD

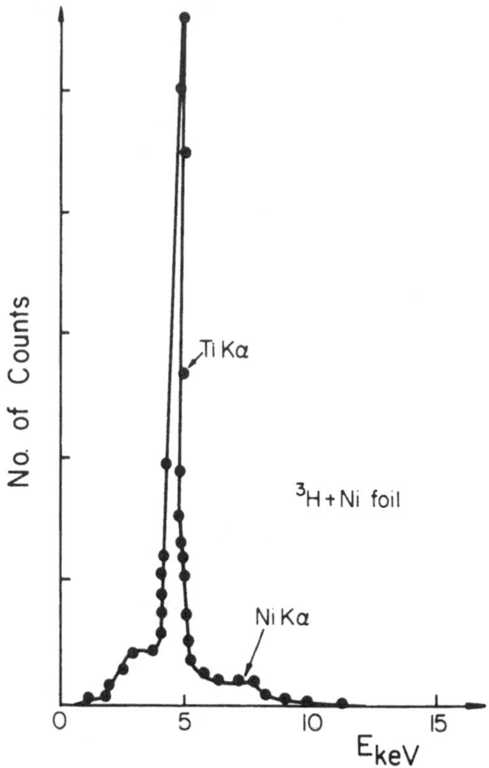

FIGURE 6. X-ray energy spectrum of the tritium source covered with a Ni foil as measured by a 5-mm Si(li) detector. (Adapted from Horowitz, Y. S. and Kalef-Ezra, J., *Nucl. Instrum. Methods*, 188, 603, 1981.)

were calculated using (in ppm/w) Mg(560), Al(15), Si(25), Ti(7), Cr(13) as measured using atomic spectrochemical analysis, and contributed approximately 2% to the coefficients averaged over the X-ray interaction spectrum. Horowitz and Kalef-Ezra[34] obtained $\eta_{x\gamma} = 1.03 \pm 0.05$ for the TL glow curve integrated to 300°C: $\epsilon_{T>} \simeq 0.1$ and $\eta^{6\text{-}7} \simeq 2$ yields an estimate of $\eta^{4\text{-}5} \simeq 0.93 \pm 0.07$ at 4 keV. Finally, Tochilin et al.[11] measured η between 0.8 and 1 for monoenergetic 8.0-, 11.9-, and 17.5-keV X-rays. Monolayered grains of unspecified size were used for the energies below 20 keV so that the results are free from self-absorption and the maximum cavity chamber correction for even 10-μm grains is approximately 1 to 2%.[5] At the higher energies their results are, however, consistent with $\eta \simeq 1.1$ and since the glow peak measurements were via peak height, the contribution to $\eta^{4\text{-}5}$ from the anomalous behavior of the higher temperature peaks was minimal.

a. Conclusions

The experimental data indicate an ionization density effect in LiF-TLD leading to $\eta \simeq 1.1$ in the energy interval from approximately 40 to 150 keV followed by a subsequent decrease in the relative TL response leading to $\eta \simeq 1$ in the energy interval of 4 to 20 keV with the possibility that $\eta \simeq 0.9$ in the very low energy region of 4 to 10 keV. This picture is consistent with the results of Lasky and Moran[36,37] who studied the relative TL response of TLD-100 to low energy electrons and obtained η(30 keV) = 0.95, η(10 keV) = 0.88, and η(5 keV) = 0.80. The question remains how to explain this ionization density behavior in the low dose linear region (increasing then decreasing relative TL response as the X-ray energy decreases from approximately 500 to 5 keV) with the opposite behavior observed in the supralinear dose region (decreasing supralinearity as the X-ray energy decreases from 500 to 5 keV). A possible qualitative explanation has been outlined by Horowitz.[38] In the supralinear dose region the decreasing TL efficiency with X-ray energy is due to the reduction in the interaction between tracks due to the more localized energy dissipation at the lower X-ray energies. In the linear dose region there are no track interactions and as the X-ray energy is decreased from 500 to approximately 50 keV, the increase in η must therefore be due to ionization density effects within a single track. The microscopic dose distribution along a single track has shifted to higher doses and a slightly greater part of that distribution now lies in the supralinear region of ionization densities leading to values of η slightly greater than unity. As the X-ray energy decreases even further to 4 to 10 keV a significant portion of the microscopic dose distribution *along a single track* becomes locally saturated and η begins to decrease reaching values perhaps 10% smaller than unity. These ideas are supported in the following section by evidence linking the onset and degree of supralinearity ($f(D)_{max}$) with the extent of the deviation of η from unity for low energy X-rays in the energy region of 30 to 150 keV. We have already encountered one example where $\eta_x \simeq 2$ for the very strongly supralinear high temperature peaks in LiF-TLD. In other words we suggest that the existence of enhanced ionization density effects via track interaction (supralinearity) implies a likelihood of observing enhanced ionization density effects along single tracks. The correlation $\eta_x > 1$ with the degree of supralinearity also accommodates the possibility of batch dependencies in η_x in order to explain the smaller subset of experiments that have yielded $\eta_x = 1$. Obviously this discussion is of a speculative nature. Further experimentation along the following lines may be of benefit.

1. Future measurements of η_x in LiF-TLD should be accompanied by measurements of f(D) at the corresponding X-ray energy in order to investigate the possible correlation between f(D) and η.
2. Additional careful experiments are required with monoenergetic X-rays to reduce the possibility of significant attenuation effects. Ideally the TL efficiency experiments should be accompanied by attentuation experiments on the identical sample material and compared with theoretical estimates of the attenuation based on existing compilations and experimental measurements of the TLD sample impurity composition.
3. More data are required on the unexpected behavior of the high temperature peaks (6 and 7). All future data analysis should attempt to measure the relative TL response as a function of X-ray energy for the dosimetric and high temperature peaks separately.

Obviously in designing a practical TL dosimetry system, published photon energy dependence data cannot be expected to be universally applicable at the 10 to 20% accuracy level, and the energy dependence of the dosimeter system must be determined experimentally. Care must be taken to exclude the high temperature peaks as completely as possible because of their undesirable nontissue-equivalent properties. In this respect peak-5 height measurements may be superior to integral measurements because of the low temperature tail of peaks 6 and 7. Obviously LiF-TLD batches with low values of $\epsilon_{T>}$ are decidedly more tissue equivalent. Since the grain size distribution of LiF powders may vary significantly with time (decreasing particle size with annealing), frequent recalibration may be necessary when optimal accuracy is required.

3. Correlation between η_x and f(D)

As previously mentioned, several experiments have indicated the possibility of a significant correlation between strongly supralinear behavior and X-ray overresponse.

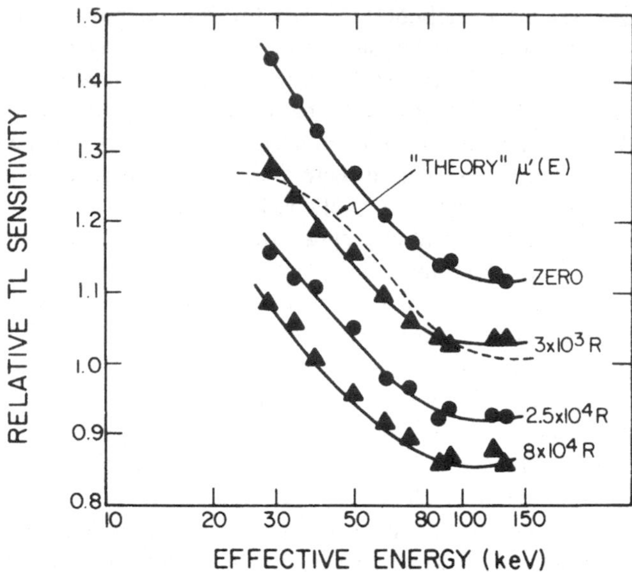

FIGURE 7. Relative TL sensitivity of TLD-700 following various sensitizing exposures; annealing at 290°C for 1 hr. (Adapted from Mayhugh, M. R. and Fullerton, G. D., *Med. Phys.*, 1, 275, 1974.)

1. Budd et al.[16] have measured $\eta_x \simeq 2$ for the strongly supralinear high temperature peaks (6 and 7) in TLD-100.
2. Mieke and Nink[39] have reported S(E) = 1.86 for low photon energies ($\eta_x \simeq 1.6$) in home-made LiF-Ti material. This same material shows enhanced supralinearity relative to TLD-100 with onset of supralinearity at approximately 0.5 Gy.
3. Mayhugh and Fullerton,[40] Shinde and Shastry,[41] and Lakshmanan and Bhatt[42] have shown that η_x is significantly reduced in sensitized TL materials [TLD-100, Al_2O_3:Si,Ti, Mg_2SiO_4:Tb, and $CaSO_4$:Tm (Dy)]. In TLD-100 the decrease in relative TL response is approximately 40% over the energy range of 30 to 130 keV, resulting in values of η significantly less than unity (Figure 7). These experiments dramatically and unambiguously prove that η need not equal unity in TL materials. Since the sensitized materials almost always show greatly enhanced linear response with no supralinear region, we expect a decrease in η corresponding to the reduced TL efficiency in the fraction of the microscopic single track dose distribution which falls in the supralinear dose region. Of course, it is possible, even likely, that the proposed correlation between η and f(D) represents only part of the underlying mechanisms controlling the ionization density dependence. In other words, the ionization density dependence in the interaction between tracks may be somehow different from the ionization density dependence within a single track. An interesting advantage of the behavior of the sensitized LiF materials is that the energy response can be altered to be in principle ideally tissue equivalent. In the nontissue-equivalent materials the reduction in relative TL response of approximately 30% results in only a marginal improvement in the energy response characteristics.
4. Tochilin et al.[11,43] have observed a very strong overresponse in BeO with η = 1.6 from 6 to 30 keV. BeO is very strongly supralinear with $f(D)_{max}$ greater than 10, and onset of supralinearity in the material studied by Tochilin et al. occurs at approximately 0.5 Gy.

Table 2
MASS ENERGY ABSORPTION COEFFICIENTS OF $Li_2B_4O_7$ OF VARIOUS COMPOSITIONS RELATIVE TO AIR

Dopant energy	Pure	0.3% Mn	Pure	0.34% Mn	0.1% Mn 0.5% Si	Pure	0.1% Mn	—[a]	Pure
4	0.883	0.885				0.897	0.899		0.883
5	0.875	0.879				0.882	0.884		0.876
6	0.869	0.870				0.868	0.871		0.870
8	0.865	0.932				0.849	0.873		0.866
10	0.853	0.933		0.94	0.92	0.829	0.857	0.914	0.857
15	0.838	0.940	0.846	0.96	0.93	0.798	0.831	0.907	0.846
20	0.833	0.948	0.838	0.95	0.91	0.779	0.816	0.913	0.837
30	0.834	0.970	0.820	0.98	0.92	0.767	0.808	0.928	0.833
40	0.847	0.978	0.836	0.98	0.93	0.782	0.822	0.939	0.845
50	0.869	0.977	0.874	0.98	0.94	0.820	0.854	0.954	0.871
60	0.907	0.990	0.906	0.99	0.96			0.964	0.903
80	0.958	0.998	0.958	0.99	0.97			0.981	0.951
100	0.979	0.998	0.972	0.99	0.98			0.992	0.975
150	1.000	1.000	—	1.00	0.99	0.962	0.970	0.999	0.993
200			1.006	1.00	1.00			0.998	0.997
400			1.005	1.00	1.00			1.002	0.999
600			1.005	0.99	0.99			1.003	1.000
800			1.005	0.99	0.99			1.000	1.000
1000			1.001	1.00	1.00			1.000	1.000
1250			1.000	1.00	1.00	1.000	1.000		1.000
1500			0.999	0.99	0.99			1.000	
Ref.	3,21		12		7		20	46	17,18

[a] Including Al (400), Mg (200), Mn (2000), Cu (10), Si (100), Fe (20), and Ag (10).

B. Relative TL Response of Lithium Borate

The considerations of the previous sections concerning LiF apply equally well to the other approximately tissue-equivalent materials including the question of the validity of $\eta = 1$. The various compilations of $\mu'(E)$ for $Li_2B_4O_7$ are shown in Table 2. Jayachandran[12] and Bassi et al.[7] noted that $\mu'(E)$ for $Li_2B_4O_7$ (0.34 wt % Mn) theoretically matches the response of air very closely below 100 keV. Of course, the agreement at this particular Mn dopant level is somewhat academic because of the presence of ionization density effects leading to η different from unity. Many authors have observed $S(E) > 1$ implying, as in the case of LiF, $\eta > 1$. Tochilin et al.[11] studied $Li_2B_4O_7$:0.1 wt % Mn (Harshaw and Riso) and observed $S(E) = 1.0 \pm 0.05$ from 6 to 35 keV resulting in $\eta \simeq 1.15$. Above 35 keV a slight increase in S(E) results in a gradually declining value of η reaching unity at approximately 600 keV. Brunskill[44] used 0.15 wt % Mn and reported $\eta \simeq 1.1$ between 20 and 40 keV. Binder and Cameron[45] also reported S(E) significantly greater than unity from 20 to 100 keV. The overresponse observed by the former two investigators is at a significantly lower average energy than the energy at which the overresponse is observed in LiF. Since both Tochilin et al. and Binder and Cameron studied both LiF and $Li_2B_4O_7$ under similar experimental conditions, the shift to lower X-ray energies is therefore indicative of a real ionization density-material dependent effect rather than an artifact arising from experimental inaccuracies as has been suggested by Jayachandran[12] and others. The overresponse is too great to be attributed to be mainly due to the presence of reasonable quantities of high Z impurities as has been suggested by Thompson and Ziemer.[46] Nonetheless, several other groups[6,12,46] report data consistent with $\eta \simeq 1$. Thompson and Ziemer[47] also studied the energy dependence of $Li_2B_4O_7$:Ag. Activation with Ag causes a sharp increase in the relative response at 25.5 keV, the K edge of Ag. The most favorable energy response occurred at 0.1 *M%* Ag yielding $S(E) \simeq 1$ down to approximately 30 keV dropping sharply at the silver K edge to a maximum underresponse of $S(E) = 0.84$. Jayachandran[12] studied the response of three different lithium borates containing 0% Mn ($C_5H_{40}O_{18}N$ tissue equivalent), 0.1% Mn (water equivalent), and 0.34% Mn (air equivalent), and in each case obtained results for S(E) consistent with $\eta = 1$. The phosphors prepared by Jayachandran[12] were strongly supralinear ($f(D) > 1$ above 0.3 Gy) so that the result of $\eta = 1$ in this case is inconsistent with the proposed correlation between η and f(D) since the phosphors used by Tochilin[11] ($\eta \simeq 1.15$) were supralinear above approximately 5 Gy.

1. Conclusion

The ability to adjust the energy response curve of lithium borate to any desired tissue equivalence by using various concentrations of Mn activator certainly makes $Li_2B_4O_7$:Mn potentially the best tissue-equivalent TLD material. As in the case of LiF:Mg,Ti further carefully controlled experiments are required to settle the issue of the ionization density dependence of η in $Li_2B_4O_7$.

C. Relative TL Response of BeO and MgB_4O_7

BeO has attracted less attention as a tissue-equivalent phosphor mainly because of its extreme light sensitivity, pyroelectric properties, and highly toxic properties in powder form. Bassi et al.[7] have tabulated values of $\mu'(E)$ for pure BeO and Thermalox® 995 (Fe—0.01%; Mg—0.0945%; Na—0.01%; Si—0.215%). For the latter material $\mu'(E) = 0.86$ at 20 keV, 0.90 at 50 keV, and 0.98 at 100 keV. Tochilin et al.[11,43] measured $S(E) = 1.3$ to 1.4 for BeO (Thermaload) equivalent to $\eta = 1.5$ to 1.6 from 10 to 100 keV. The almost flat response below 35 keV rules out any possibility of high Z impurities being responsible for the large values of S(E). Scarpa[48] studied three batches of BeO (slipcast, nuclear quality, and hot pressed from Consolidated Be Ltd.) and observed systematic differences leading to $\eta = 1.4$ to 1.5, 1.3, and 1.1 to 1.2 for these three materials, respectively (Figure 8), again clear evidence for material-dependent ionization density effects. Crase and Gammage[49] studied Thermalox® 995 and also obtained $\eta > 1.4$ at 100 keV and equal to approximately 1.3 from 20 to 50 keV. Thus, contrary to the situation in LiF and $Li_2B_4O_7$, there is unanimous agreement on the overresponse of BeO to low energy X-rays leading to values of η between 1.1 and 1.5. It is tempting to correlate these large values of η with the very strong supralinearity of BeO. The large value of the overresponse allows the clear observation of batch effects so that experimental measurements of f(D) and η might easily establish this correlation. Barbina et al.[50] have measured S(E) for MgB_4O_7:Dy(Tm), $S(E) = 2.2$ at 50 keV compared to $S(E) \simeq 1.3$ for LiF at the same energy so that the tissue equivalence is significantly worse than that of LiF.

D. Relative TL Response of CaF_2:Dy and $CaSO_4$:Dy

The relevant values of $\mu'(E)$ and S(E) are shown in Table 3 for CaF_2:Dy (TLD-200). Only the results of Endres et al.[6] are consistent with $\eta = 1$. At 100 keV (where there is only a very small probability of attenuation effects) the results of Binder and Cameron[45] and Nollman and Thomasz[31] imply $\eta = 1.3$ to 1.4 at 100 keV. The underresponse at 30 keV[31,51] is probably due to attenuation. On the other hand, Gorbics and Attix[29] obtained results consistent with $\eta = 1$ for CaF_2:Mn (hot pressed, EG & G). The situation for $CaSO_4$ powders doped with various impurities is shown in Table 4. Again at 100 keV there is considerable overresponse of the order of 20% for the data of Yamashita et al.[53] and Pradhan et al.[8] and approximately 100% overresponse for the data reported by Lakshmanan and Bhatt.[42] Pradhan

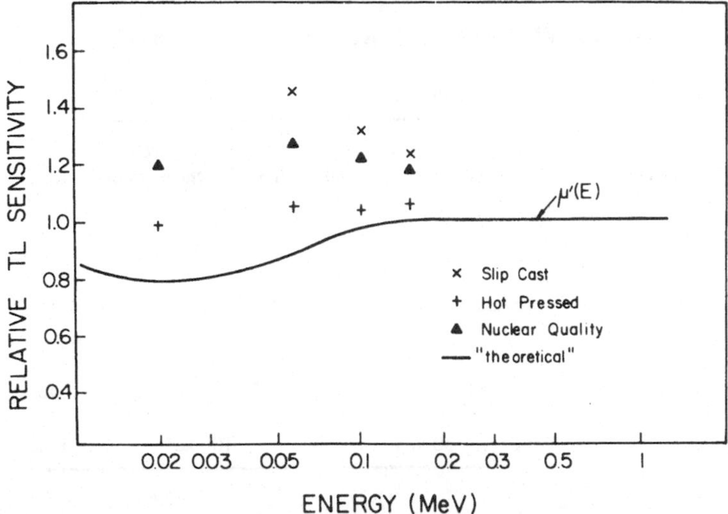

FIGURE 8. Relative TL sensitivity of various forms of BeO. A theoretical curve following Bassi et al.[7] is also included. (Adapted from Scarpa, G., *Phys. Med. Biol.*, 15, 667, 1970.)

Table 3
RELATIVE X-RAY TL RESPONSE OF CaF_2:Dy (TLD-200)

	30 keV		100 keV		
Material	**S(E)**	**μ'(E) theoretical**	**S(E)**	**μ'(E) theoretical**	**Ref.**
Pure CaF_2		14.48		2.99	7
CaF_2 (1% Dy)		15.39		3.85	7
TLD-200[a] powder	17		5		45
TLD-200 chip (0.125 × 0.125 × 0.125 in.)	15		4		6
TLD-200 chip	13		—		51
	12		5.2		31
TLD-200 chip (3.2 × 3.2 × 0.9 mm)	16		3		52

[a] Unspecified grain size.

and Bhatt[9] studied the effectiveness of various metal filters of aluminum, stainless steel, copper, cadmium, tin, and lead to compensate for the photon energy dependence of the response of $CaSO_4$:Dy teflon® TLD discs. A combined metal filter (0.55 mm Sn + 0.35 mm stainless steel) makes the energy response almost independent of energy for photon energies greater than 80 keV (Figure 9). Of course, the major disadvantage of this technique is the considerable angular dependence (Figure 10). Nonetheless TLD badges of this type meet the requirements of personnel monitoring of X- and gamma radiation fairly adequately.

Table 4
RELATIVE X-RAY TL RESPONSE OF $CaSO_4$

	50 keV		100 keV		
Material	**S(E)**	**μ'(E) theoretical**	**S(E)**	**μ'(E) theoretical**	**Ref.**
Pure $CaSO_4$		8.61		2.45	7
$CaSO_4$ (0.22% Tm)		8.83		2.64	7
$CaSO_4$:Tm (0.1%, 100—200 mesh)	9.5—10		3		53
$CaSO_4$:Tm powder	9.5—10		5.5		42
	30 keV		100 keV		Ref.
$CaSO_4$ (0.21% Dy)		11.27		2.63	7
$CaSO_4$:Dy (0.05% Dy, 0—74 μm)	10.52		3.2		8
$CaSO_4$:Dy (0.1 *M* % Dy)	~10		2.2		57
$CaSO_4$:Dy (Radiation Detection Co.)			3.0		58
$CaSO_4$:Dy (TLD-900)	~9		2.2		59

Significant overresponse from 60 to 200 keV has also been observed for BaF_2[54] in a 0.65-mm thick crystal (Harshaw, optical grade) and in $SrSO_4$:Dy (0.4 mol %).[55] In the latter experiment substantial effort was expended to reduce self-attenuation (the TLD sample was approximately 1 grain thick of 75- to 180-μm diameter) and backscattering (the sample was attached to a thin suspended sheet of plastic foam). The values of S(E) and μ'(E) overlapped adequately in the case of $BaSO_4$:Dy over the entire energy range, but under identical irradiation and experimental conditions the $SrSO_4$:Dy overresponse was approximately 35% between 30 and 60 keV, well outside the limits of both experimental error and uncertainties in the theoretical calculations. Application of cavity theory corrections (not applied by Dixon et al.[55]) to these monolayer samples would significantly increase the 35% overresponse in $SrSO_4$:Dy and would also result in values of $\eta > 1$ for $BaSO_4$:Dy. Finally, significant overresponse has been reported in Al_2O_3.[42,56] For example, at 100 keV, S(E) = 2 and 2.5 to 2.8 compared to μ'(100 keV) = 1.34.

E. Use of S(E) to Measure Radiation Quality

The dependence of S(E) on Z_{eff} can be used to measure the effective energy of a continuous X-ray spectrum by a double phosphor technique where the response ratio of the two materials (chosen to differ greatly in Z_{eff}) will vary fairly rapidly with energy over certain energy intervals. Obviously the sensitivity and useful energy range of the technique increases the greater the difference in the mass energy absorption coefficient of the two materials. The presence of ionization density effects leading to S(E) ≠ μ'(E) may seriously compromise the accuracy of this technique unless calibration measurements of S(E) are carried out for both materials using monoenergetic X-rays. Cameron et al.[60] discussed early applications of this technique using LiF and CaF_2. Because S(E) for both materials is approximately

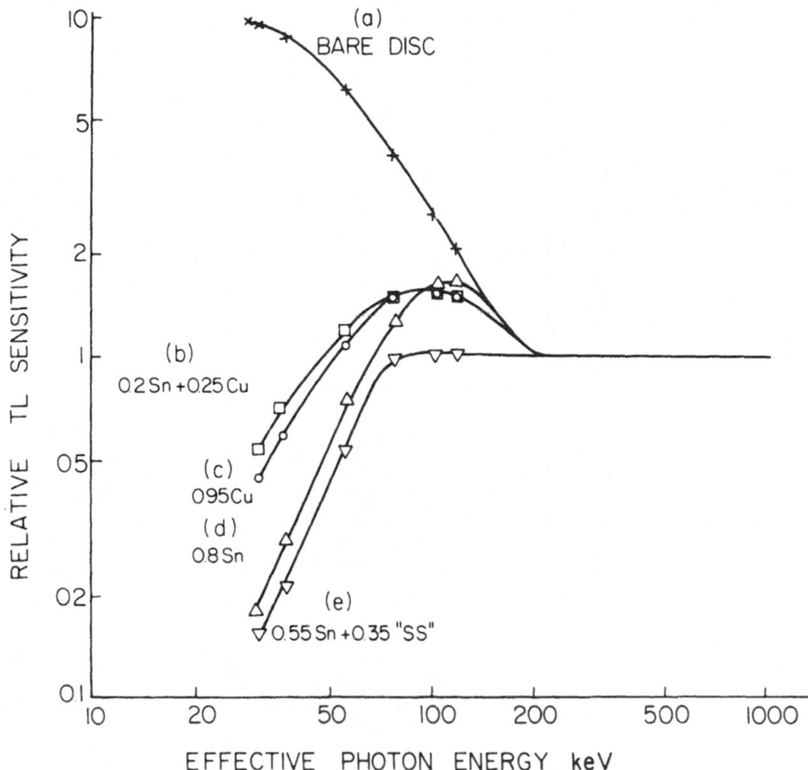

FIGURE 9. Photon energy dependence of $CaSO_4$:Dy teflon® discs under various filters; for b to e additional filtration of 0.8 mm Al was also used. Numbers are the thickness of filters in millimeters. (Adapted from Pradhan, A. S. and Bhatt, R. C., *Nucl. Instrum. Methods*, 166, 497, 1979.)

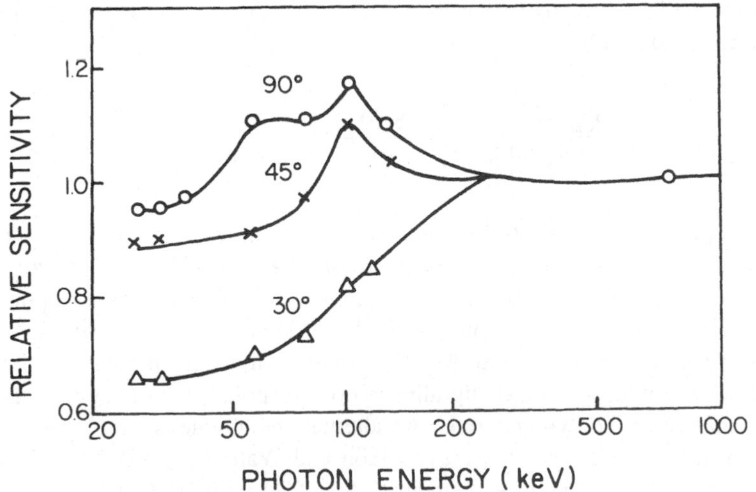

FIGURE 10. Angular dependence of the relative TL sensitivity of $CaSO_4$:Dy PTFE discs under a 0.55 mm Sn and 0.35-mm stainless-steel filter. (Adapted from Pradhan, A. S. and Bhatt, R. C., *Nucl. Instrum. Methods*, 166, 497, 1979.)

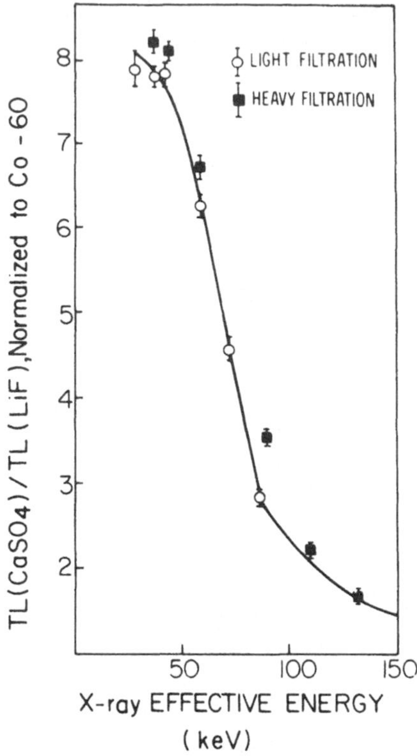

FIGURE 11. "Tandem" ratio curve (95% confidence limits are shown) under light and heavy filtration conditions. (Adapted from Rossiter, M. J., *Phys. Med. Biol.*, 20, 735, 1975.)

constant below 40 keV, this particular choice of materials is useful between approximately 50 and 200 keV and yields

$$\left.\frac{S(E)_{CaF_2}}{S(E)_{LiF}}\right|_{50\ \mathrm{keV}} \bigg/ \left.\frac{S(E)_{CaF_2}}{S(E)_{LiF}}\right|_{200\ \mathrm{keV}} \simeq 7{:}1 \qquad (5)$$

Accuracy of approximately 10% (95% confidence limit) in the measurement of the ratio of S(E) translates to a most probable error of approximately 10 keV in the effective energy.[52] Another problem is that even the use of 20-mg samples creates a nonunique response function between 10 and 100 keV due to attenuation in the TLD sample at the lower energies, with a maximum in the sensitivity ratio at 30 keV. Another difficulty is that the response ratio will be a function of dose in the supralinear dose-response region due to the different supralinear behavior of the two materials. An alternate technique is simply to measure S(E) in the beam of unknown effective energy. Dixon and Watts[54] used BaF_2, (S(E) = 100 at 60 to 70 keV) and obtained good agreement (2.5 to 5%) with HVL (mm Cu) as measured by "good geometry" transmission experiments. Similar experiments with CaF_2 yielded larger differences (2.5 to 15%) due to the reduced values of S(E). Rossiter[26] studied the "tandem" ratio TL($CaSO_4$:Dy)/TL(TLD-700) for heavy and light filtration conditions (Figure 11). The "light filtration" results are consistently 5 to 10% lower than the "heavy filtration" results at the same effective energy. Spurny et al.[61] also reported that the tandem ratio was strongly

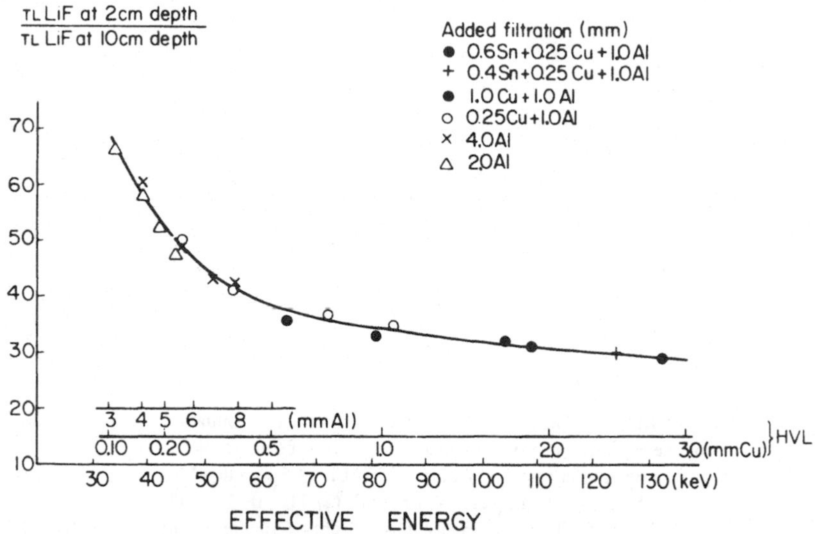

FIGURE 12. Ratio of the relative TL sensitivities of LiF at 2- and 10-cm depth in water for various effective X-ray energies. (Adapted from Puite, K. J. and Crebolder, D. L. J. M., *Phys. Med. Biol.*, 19, 341, 1974.)

dependent on beam spectral distribution and Puite and Crebolder[62] experienced difficulty in obtaining realistic estimates of E_{eff} within an irradiated phantom. These difficulties have not been resolved and are probably due to beam-TLD geometry and ionization density effects. In a later study, Puite[2] carried out a careful study of two TLD techniques distinguishing between beams intended for deep therapy (HVL > 1.1 mm Cu, E_{eff} > 85 keV) and surface or superficial therapy (HVL between 0.11 and 1.4 mm Cu, effective energy between 34 and 95 keV). For the former, because of the width of the X-ray spectrum, the response ratio, CaF_2/LiF, measured at a depth of 5 cm in water, is almost constant from 30 to 80 keV_{eff} so that the method is useful only above approximately 80 to 90 keV_{eff}. For the latter, Puite measured the response ratio (TL LiF at 2 cm depth/TL LiF at 10 cm depth) making use of the strong attenuation of the low energy X-rays in water (Figure 12). The percentage uncertainties in the determination of the effective energies are shown in Figure 13 based on ±1.8% error (1 S.D.) in both the LiF and CaF_2 readings. It can be seen that in the latter case the uncertainty in the determination of the effective energy of the incident beam is ±3 to ±7% depending on energy. The two techniques complement each other over the entire energy range of 30 to 120 keV. In conclusion, it is possible to measure with adequate accuracy the absorbed dose in a water phantom, together with the quality of the incident beam, using TLD. Due to variations in the energy dependence of various TLD batches great care, however, must be taken in the use of published energy dependence data.

Finally, Spurny[63] has studied the use of very high Z_{eff} phosphors [CdI_2 (52), $CeCl_3$ (55), and $BaPt(CN)_4$ (71)]. At Z_{eff} = 55 the response ratio at 100 keV is roughly 50:1 and the useful range of energies could be increased to 800 keV. Obviously the lower useful energy limit would also be increased due to attenuation problems in even very thin samples but the new TLDs could be of advantage in environmental dosimetry where the energy of terrestrial radiation is between 250 and 800 keV. Further optimization studies of these materials are required, however, because of fading, low sensitivity, and light sensitivity problems.

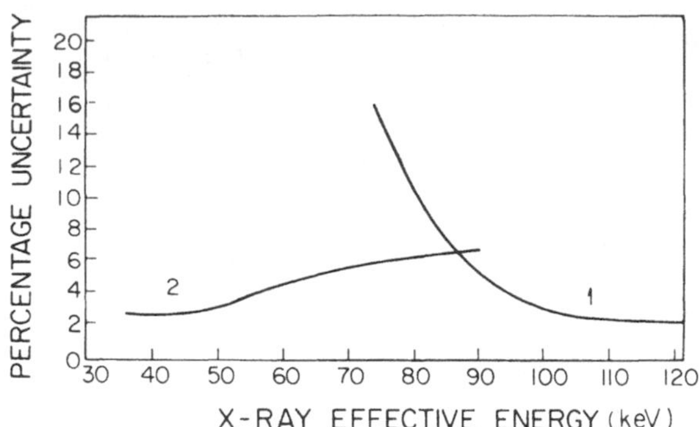

FIGURE 13. The percentage uncertainty in the determination of the effective energy of the incident X-ray beam from TL measurements in water. (1) TL CaF_2/TL LiF at 5-mm depth in water, SSD = 50 cm, field = 100 cm^2. (2) TL LiF 2 cm/TL LiF 10 cm, SSD = 30 cm, field = 36 cm^2. (Adapted from Puite, K. J. and Crebolder, D. L. S. M., *Phys. Med. Biol.*, 19, 341, 1974.)

II. BETA DOSIMETRY

The short range and resulting spatial inhomogeneity of beta rays make it necessary to use thin foils in beta-ray dosimetry, otherwise, the TLD will exhibit severe energy dependence due to the thickness being much greater than the beta particle range. A dosimeter intended for measurement of skin dose should respond in a manner analogous to that of the dose to the sensitive basal skin layer. At a depth of 7 mg cm^{-2} the actual skin dose relative to surface dose will be about 0.9, 0.7, and 0.07 for beta rays with average energies of 0.93, 0.26, and 0.05 MeV, respectively. Skin beta dose should therefore be measured ideally by a thin 5 to 10 mg cm^{-2} tissue-equivalent TLD covered with a 5 mg cm^{-2} entrance tissue-equivalent window.[64] Such a device is, of course, not of universal applicability in view of the range of skin thicknesses in the general population. In situations where, e.g., fingertip irradiations are of concern (skin layer 30 to 50 mg cm^{-2}), a thicker entrance window is mandatory. Of popular use in beta skin dosimetry have been thin microtomed slices of LiF in a polytetrafluoroethylene (PTFE) matrix which are available from several commercial sources. Marshall and Docherty[65] and Gibson et al.[66] describe the use of PTFE discs (30% LiF), 1.27 cm diameter, 8.9 ± 0.3 mg cm^{-2} thick with a window of 3.7 mg cm^{-2} polyethylene. The background reading was 0.4 ± 0.06 mGy corresponding to a lower limit of detectability of approximately 0.12 mGy (2 S.D.) with precision of ±10% (1 S.D.) at 1 mGy. The energy response was equivalent to the dose under 7 mg cm^{-2} of tissue to within 10% for beta rays with maximum energies between 0.15 and 2.27 MeV. A frequently heard complaint, however, is that the thin PTFE TLDs buckle during readout resulting in poor thermal contact. Lowe et al.[67] have pointed out that the microtoming also results in the slicing and scouring of the phosphor powder within the PTFE making it impossible to control the grain size at the surface of the dosimeter. This is important because of the variation of TL efficiency with grain size; moreover, the grains exposed at the surface tend to fall out with repeated use resulting in a significant loss in sensitivity especially for the thinner dosimeters. Charles and Khan[68,69] have described an ultrathin bonded (UTB) dosimeter 6 mg cm^{-2} (30 μm) LiF-PTFE disc thermally bonded to a thick (approximately 0.2 mm) PTFE base to produce greater mechanical rigidity. A 40-μm thick melinex window served as the entrance window. Minimum detectable dose was approximately 170 mR with a

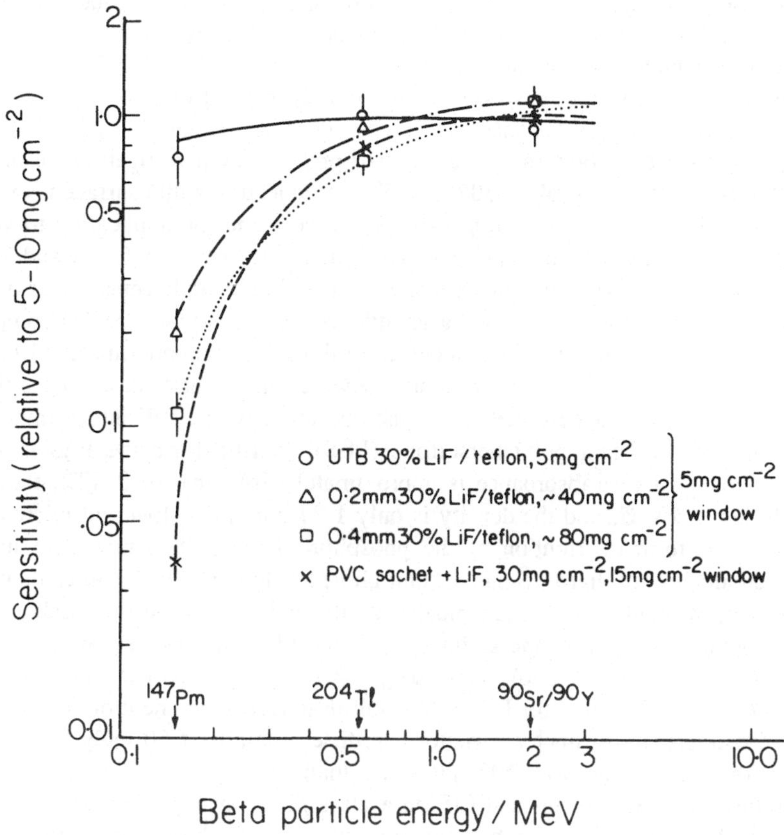

FIGURE 14. The energy response for beta irradiation of various TLDs. (Adapted from Charles, M. W. and Khan, Z. U., *Phys. Med. Biol.*, 23, 972, 1978.)

precision of $\pm 20\%$ (1 S.D.) at 10 mGy and $\pm 3\%$ at 0.1 Gy. The UTB dosimeter provided good agreement with estimated 5 to 10 mg cm^{-2} skin dose down to approximately 60 keV. Figure 14 shows the energy response for beta irradiation of several types of TLDs including the UTB dosimeter. The thicker dosimeters have a response within 30% of the true dose down to endpoint energies of 0.5 MeV. At energies of approximately 0.15 MeV they underrespond by as much as a factor of 5 to 25 depending on their thickness. The UTB dosimeter, on the other hand, is in good agreement with the real 5 to 10 mg cm^{-2} skin dose over the entire energy range from 150 keV to 2 MeV. A limitation of the UTBs, however, is their light sensitivity which gives rise to a large minimum detectable dose, e.g., a 10-min exposure to light in air produces a background signal of approximately 700 mR.

Harvey and Felstead[70] describe attempts to produce LiF dosimeters using thin powdered layers on commercially available self-adhesive tape (Kapton) which will withstand temperatures of up to 400°C. Dosimeters of 5.9 mg cm^{-2} thickness could be prepared with good reproducibility when sieved and washed ground crystalline material was used, however, several problems were encountered: (1) a decrease of sensitivity of approximately 1.3% for each reading cycle; (2) a background equivalent to a radiation dose of 0.70 ± 0.06 and 0.91 ± 0.21 mGy, respectively, for unirradiated and irradiated dosimeters; (3) a strong light sensitivity resulting in a 3.3 ± 0.5 mGy background following 2 hr in daylight in previously irradiated dosimeters; and (4) slightly "tacky" surfaces which tend to pick up dirt readily. Other authors have reported related work[71-73] on a wide range of dosimetry applications based on Kapton tapes; the main advantages are good reproducibility, ease of manufacture, and low cost. Uchrin[74] described LiF powders (a few mg cm^{-2}) cold pressed to aluminum discs which measured the skin dose for $E_{max} > 0.2$ MeV. The loss in sensitivity (TL/unit mass) at 0.2 MeV was, however, approximately a factor of 3 compared to 2 MeV. The lower detection limit was 0.1 mGy and precision of ±10% (1 S.D.) at 10 mGy was achieved. Christensen and Majborn[75] have reported preliminary investigations of the use of a new high-temperature glow peak (~305 to 335°C) produced in a thin surface layer of LiF-TLDs by diffusion of boron. The boron diffusion process and the application to very low energy electron dosimetry (0.1 to 30 keV) were originally described by Lasky and Moran.[76] For personnel dosimetry higher energies are relevant (>60 keV) and deeper diffusion depths are required. Unsolved problems include a greatly reduced sensitivity (~4%) compared to nondiffused dosimeters, which results in a detection threshold of approximately 1 to 2 mGy and the lack of a well-defined diffusion depth. Lowe et al.[67,77] have described a thin-film TLD (25-μm thick) using a new method of incorporation of the LiF into a thin layer of transparent and heat-resistant polyethersulfone (PES). In film form the PES is virtually transparent to visible light (absorbance is approximately 10% that of PTFE), the melting point is well above 400°C, and the density is only 1.37 gm cm^{-2}. Electron microscanning reveals a more uniform distribution of the phosphor in the PES matrix than in PTFE. Additional advantages are that PES can sustain phosphor loadings up to an estimated 80% (30% for PTFE) without losing mechanical stability and there is less phosphor wastage during manufacture. Because of the sulfur content the PES is, however, significantly less tissue equivalent than PTFE; $(\mu_{en}/\rho)_{PES}/(\mu_{en}/\rho)_{tissue} \simeq 2.4$ for gamma-ray energies below 60 keV compared to a value of 1.4 for PTFE, and sulfur activation via neutron irradiation may also occur. Dosimeter-to-dosimeter variations in dose readings at 10 mGy ^{60}Co were 7% for a 25-μm thick dosimeter with 60% phosphor loading prepared by spreading a homogeneous mixture of PES, solvent, and LiF on a glass plate, drying, followed by punching out the various dosimeters. This can be compared with 23% for batch-uncalibrated ultrathin LiF-PTFE dosimeters for the same irradiation. Precision at low dose was not reported, but theoretically at least the higher phosphor loading and decreased light absorbance should result in a significantly decreased threshold dose.

Attempts to improve the energy response of LiF- and $Li_2B_4O_7$:Mn-sintered pellets[78] by artificially reducing light transparency with the addition of graphite suffer from poor TL sensitivity, which makes them unsuitable for low dose dosimetry (4% graphite loading for $Li_2B_4O_7$:Mn yields ±10% (1 S.D.) at 4 mGy. Pradhan and Bhatt[79] partially circumvented this problem by using graphite-loaded $CaSO_4$:Dy PTFE discs accompanied unfortunately by the loss of approximate tissue equivalence. The energy dependence of the graphite-loaded $CaSO_4$:Dy-PTFE discs is shown in Figure 15. At 5% graphite loading the minimum measurable dose was still a relatively high 30 mR. Finally, Lakshmanan et al.[80] argue that due to the very low dose contribution from beta rays of energies less than 100 keV, 0.1-mm thick $CaSO_4$:Dy PTFE TLDs bonded to a 0.7-mm PTFE base (minimum detectable dose 0.13 mGy) can be used in personnel monitoring. As can be seen in Figure 16, however, the energy dependence of these TLDs, even above 100 keV, is still very significant. Aside from the obvious applications to skin dosimetry, PTFE dosimeters have been applied to interface and cavity dose distribution studies[81] and to central axis depth dose and isodose curves.[82]

III. CAVITY THEORY

The purpose of cavity theory is to relate the absorbed dose in a cavity to the absorbed

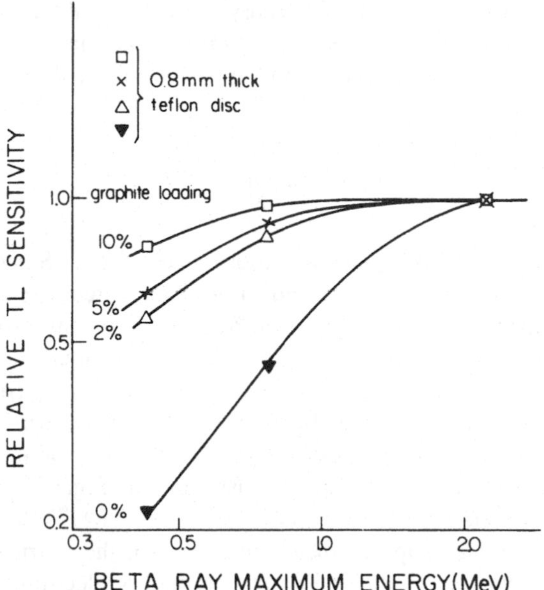

FIGURE 15. Energy response (TL/unit mass) of $CaSO_4$:Dy teflon® dosimeters with 0, 2, 5, and 10% graphite loading relative to ^{90}Sr-^{90}Y beta radiation. (Adapted from Pradhan, A. S. and Bhatt, R. C., *Phys. Med. Biol.*, 22, 873, 1977.)

dose in the medium of different atomic number or composition by means of the equation

$$D_m = \frac{1}{f_{cm}} D_c \tag{6}$$

where D_c is the cavity dose and D_m is the dose when the cavity is filled with medium material. In general, f_{cm} is a function of gamma ray energy, the composition of the cavity and surrounding medium, and the cavity size. When the cavity dimensions are very much smaller than the electron ranges, the electron spectrum within the cavity is determined only by the medium. In this case

$$f_{cm} = \overline{S_m^c} \tag{7}$$

where $\overline{S_m^c}$ is simply the continuous energy loss mass stopping power of the cavity to the medium. The assumptions implicit in Equation 7 are (1) the effects of delta-ray production and bremsstrahlung are negligible, (2) the cavity does not perturb the equilibrium electron distribution created in the medium, and (3) photon interactions in the cavity material are negligible. In the continuous slowing down approximation (CSDA) the electron spectrum in the medium is then simply the reciprocal of the mass stopping power of the medium so that

$$f_{cm} = \overline{S_m^c} = \frac{1}{T_o} \int_0^{T_0} \left(\frac{1}{\rho}\frac{dT}{dx}\right)_{cm} dT \tag{8}$$

where $\overline{T_0}$ is the mean starting energy of secondary electrons in the medium. At the other extreme, when the cavity dimensions are many times larger than the range of the most energetic electron, the electron spectrum within the cavity will be determined by the cavity material so that

$$f_{cm} = \overline{(\mu_{en}/\rho)_{cm}} \tag{9}$$

Figure 17 shows the electron distributions in various size cavities. Since both μ_{en}/ρ and S are energy dependent, accurate application of Equations 7 and 8 requires accurate knowledge of the distribution in energy of the radiation field both in the medium (buffer zone) and in the cavity. For example, Ehrlich[83] measured absorbed dose distributions in CaF_2:Mn TLDs of thicknesses from 0.5 to 3.5 mm irradiated in ^{60}Co parallel beam geometry in media of polystyrene, copper, and lead. Deviations between theory and experiment were attributed to "front-rear asymmetries" (i.e., the effect of the angular distribution of the secondary electrons impinging on the cavity), and also to assumptions regarding the shapes of the secondary electron and primary photon spectrum.

Spencer and Attix[84] derived an approximate expression for the restricted stopping power ratio that takes into account the production of fast secondary electrons (delta rays) and is based on an analysis of the different electron contributions of the medium and cavity

$$f_{cm} = (Z/A)_{cm} \left\{ 1 + \frac{1}{T_0} \left(\int_{\Delta}^{T_0} R_m(T_0,T) \left(\frac{B_c(T)}{B_m(T)} - 1 \right) dT + \Delta R_m(T_0,\Delta) \left(\frac{B_c(\Delta)}{B_m(\Delta)} - 1 \right) \right) \right\} \tag{10}$$

where $R_m(T_0,T) = I_m(T_0,T) \cdot S_m(T)$ and $I_m(T_0,T)$ is the equilibrium electron spectrum in the medium generated by a primary electron with energy T_0, Z, and A are the effective atomic number and weight, respectively, and B is the stopping number equal to S(Z/A). Δ was defined rather arbitrarily as the electron energy whose range was equal to the cavity thickness. Equation 10 neglects the effects of cavity geometry and a cavity source term arising from gamma-ray interactions in the cavity. Δ was restricted to energies greater than the K binding energy and T_0 to energies where the bremsstrahlung contribution is not significant. The functions $R_m(T_0,\Delta)$ and $B_m(T)$ are tabulated[85] in NBS Handbook No. 79. The importance of the Spencer-Attix formulation increases when the cavity and medium are significantly mismatched and delta-ray equilibrium does not exist. In the case of a gross mismatch (e.g., lead and air), the cavity size that determines Δ also becomes important.

A. Burlin General Cavity Theory

For solid-state devices such as TLDs, typical dimensions of the order of 1 mm do not fall into either limiting category. The solution of the intermediate-sized cavity problem would require exactly the solution of an extremely complex electron transport problem which probably can best be handled by Monte Carlo techniques.[86]

Burlin,[87] however, proposed an approximate general cavity theory for photons for all sizes that approaches the Spencer-Attix theory in the small size limit and the ratio of the mass energy absorption coefficients for the large cavities. It takes into consideration a gamma-generated source term in the cavity as well as perturbation of the medium-generated electron spectrum in the cavity. No consideration, however, is given to the effects of electron scattering, which will influence the distribution of absorbed dose within the cavity as well

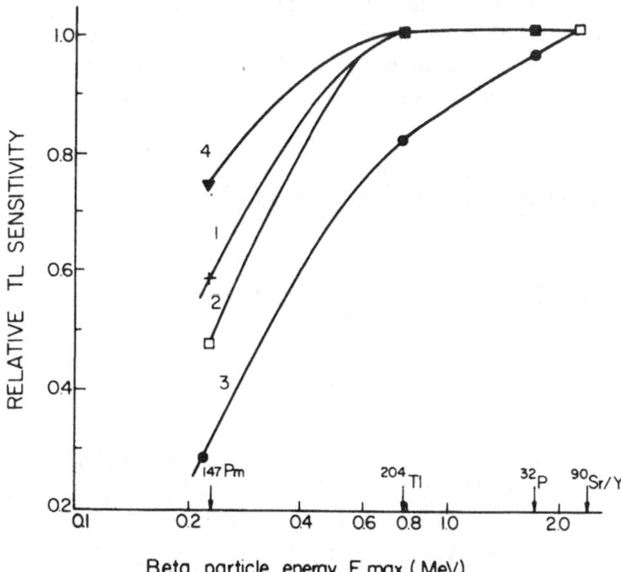

FIGURE 16. Energy dependence of various $CaSO_4$:Dy dosimeters to beta radiation. (1) 0.1-mm $CaSO_4$:Dy teflon® disc bonded to 0.7-mm graphite mixed teflon® base. (2) 0.1-mm $CaSO_4$:Dy teflon® disc bonded to 0.7-mm pure teflon® base. (3) 0.2-mm $CaSO_4$:Dy teflon® disc bonded to 0.6-mm graphite mixed teflon® base. (4) 25-μm LiF teflon® disc as reported by Charles and Khan.[69] (Adapted from Lakshmanan, A. R., Chandra, B., Pradhan, A. S., Kher, R. K., and Bhatt, R. C., *Int. J. Appl. Radiat. Isot.*, 31, 107, 1980.)

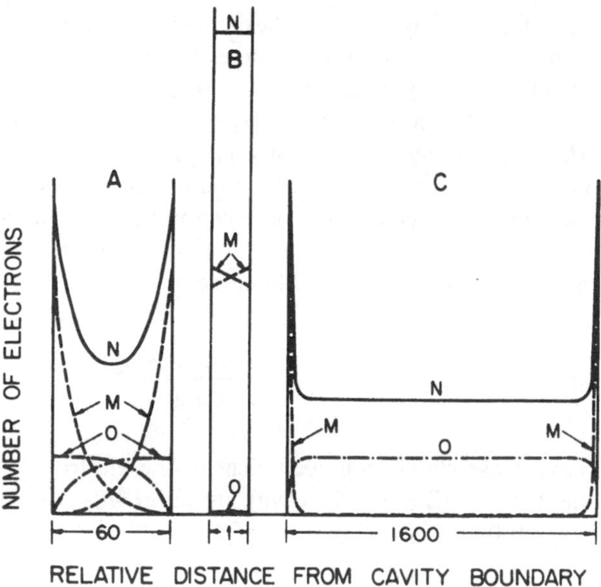

FIGURE 17. Electron distributions in various size cavities. M, electrons entering the cavity from the medium; O, electrons generated by photon interactions in the cavity; and N, total number of electrons. (Adapted from Simons, G. G. and Jule, T. S., *Nucl. Sci. Eng.*, 53, 162, 1974.)

as the angular distribution of secondary electrons impinging on and emerging from the cavity. These effects have been shown to be important[33,88-90] especially when the energy absorbed in the dosimeter arises approximately equally from electrons generated in the medium and in the detector. In the Burlin general cavity theory f_{cm} is given by:

$$f_{cm}(T_0,\Delta) = (Z/A)_{cm}\left\{1 + \frac{d}{T_0}\left[\int_{\Delta}^{T_0} R_m(T_0,T)\left(\frac{B_c(T)}{B_m(T)} - 1\right) dT + \Delta R_m(T_0,\Delta)\left(\frac{B_c(\Delta)}{B_m(\Delta)} - 1\right)\right] + (1 - d)\left[(\mu_{en}/\rho)_{cm}\,(Z/A)_{mc} - 1\right]\right\} \quad (11)$$

which reduces to

$$f_{cm}(T_0,\Delta) = d(Z/A)_{cm}\left\{1 + \frac{1}{T_0}\left[\int_{\Delta}^{T_0} R_m(T_0,T)\left(\frac{B_c(T)}{B_m(T)} - 1\right) dT + \Delta R_m(T_0,\Delta)\left(\frac{B_c(\Delta)}{B_m(\Delta)} - 1\right)\right]\right\} + (1 - d)\,(\mu_{en}/\rho)_{cm} \quad (12)$$

or in a more simple form

$$f_{cm} = d\,\overline{S}_{cm} + (1 - d)\,\overline{(\mu_{en}/\rho)_{cm}} \quad (13)$$

Equations 12 and 13 have an appealing intuitive form wherein the first term represents electrons generated in the medium and the second term is an additional source term representing electrons generated in the cavity via photon interactions in the cavity. Methods for estimating $\overline{S}_{cm}$ have been discussed by Burlin.[87] The continuous slowing down approximation results in an equilibrium electron spectrum that ignores the generation of high energy delta rays but for typical TLD dimensions (e.g., 0.4 × 3 × 3 mm hot pressed LiF-TLD) the influence of delta rays is negligible (for ^{60}Co energies) so that the Spencer-Attix theory agrees with the CSDA theory to within 1%.[91] For very small size cavities (e.g., monolayer TLD samples) the Spencer-Attix theory should be used to calculate $\overline{S}_{cm}$. The factor d is a weighting factor that expresses the reduction of the electron fluence from the medium inside the cavity. If g is the average path length of electrons crossing the cavity, then, on the average, the medium spectrum will be reduced by a factor

$$\int_0^g \exp(-\beta x)dx \Big/ \int_0^g dx = \frac{1 - \exp(-\beta g)}{\beta g} = d \quad (14)$$

where β is the effective mass attenuation coefficient of the electron flux penetrating the cavity material. In the theory of Burlin[87] the cavity spectrum will build up to a fraction of its equilibrium value given by

$$\int_0^g (I - \exp(-\beta x)dx \Big/ \int_0^g dx = 1 - d \quad (15)$$

In the following section a modification developed by Horowitz et al.[89,90] is introduced which takes into account the fact that the average path length for electrons crossing the cavity from outside need not equal the average path length for electrons created within the cavity.

Many experimental studies have indicated that the exponential attenuation of *beta ray spectra* is determined by the maximum energy, E_{max}. Burlin[87] has adopted the formula

$$\beta = 16/(E_{max} - 0.036)^{1.40} \text{ cm}^2 \text{ g}^{-1} \tag{16}$$

whereas others have adopted slightly different approaches (e.g., Eggermont et al.[92] use $\beta = 4.605/R_{ex}(T_0)$ where R_{ex} is the extrapolated range of Tobata et al.[93]). β was measured for beta rays in LiF by Paliwal and Almond[94] with the result

$$\beta_{LiF} = 14/E_{max}^{1.09} \quad \text{for } 0.23 \text{ MeV} < E_{max} < 2.27 \text{ MeV} \tag{17}$$

whereas for higher energy monoenergetic electrons between 8 and 20 MeV

$$\beta_{LiF} = 37.9/E_{max}^{1.61} \tag{18}$$

In fact, monoenergetic electrons do not attenuate exponentially (Section III.D). For photons in the hundreds of keV to MeV range (1.25 MeV, ^{60}Co), the general cavity theory of Burlin has been applied to TLDs by several authors.[5,83,88,91,92,95] Figure 18 shows the variation of 1/f as a function of gamma-ray energy for a 1 × 1 × 6 mm ^{7}LiF dosimeter surrounded by iron using the large, small, and intermediate cavity expressions. The mass energy absorption coefficients of Storm and Israel[96] and stopping powers of Berger and Seltzer[97] were used. The weighting factor d was calculated for $g = 4V/S = 0.243$ g cm^{-2}. Figure 19 shows 1/f for the same TLD surrounded by various materials (intermediate cavity expression). Bertilsson[88] irradiated LiF-teflon® discs of thicknesses ranging from 4×10^{-3} g cm^{-2} to 3×10^{-2} g cm^{-2} in a plane-parallel beam of ^{137}Cs gamma rays and found disagreement with the Burlin theory as shown in Figure 20. The function a(Z,h) is a correction factor which multiplies the restricted stopping power term in Equation 13. The solid curves were calculated using a primary photon beam with 80% of the energy fluence at 662 keV and 20% from photons with an equivalent energy of 400 keV. The dashed curves are calculated using a photon spectrum given by Costrell.[98] The deviations from the Burlin theory appear to be significantly dependent on the choice of the primary photon spectrum, the Z of the surrounding media, and the dosimeter size. As previously mentioned the deviations are most probably due to differences in the electron-scattering properties of the medium and the cavity leading to interface effects, front-back asymmetry, etc. It seems, therefore, that if there is a significant mismatch between the TLD material and the medium, there is considerable uncertainty in the application of 1/f factors.

Although it has been stated[91] that the value of β is generally not critical in comparison to the theoretical and experimental error, recent work[89,90] on a modified general cavity expression indicates a significant dependence of the goodness of fit between theory and experiment on β. Ogunleye et al.[91] applied the Burlin general cavity theory to Compton electron (^{60}Co) spectra, and the theory appeared to apply equally well to Compton electron spectra as to beta-ray spectra. In this experiment, hot-pressed TLD-100 dosimeters (0.4 × 3 × 3 mm) were irradiated by ^{60}Co photons and good agreement was obtained with the experimental results for polystyrene while the theory tended to overestimate the dose in the LiF dosimeters enclosed in Al, Cu, or Pb. The general trend predicted by the Burlin theory

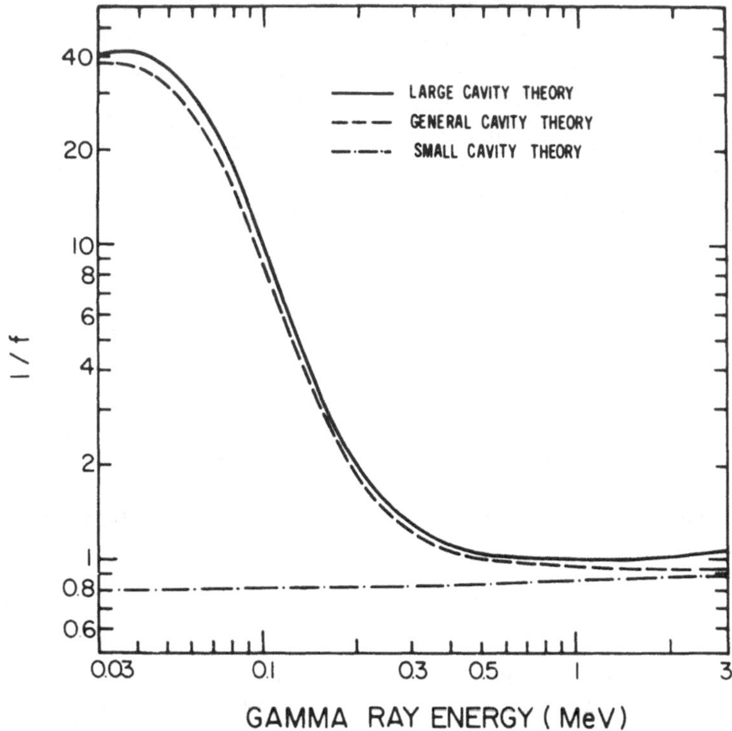

FIGURE 18. Variation of 1/f as a function of gamma-ray energy for a 1 × 1 × 6 mm (g = 4V/S = 0.243 g cm^{-2}) ^{7}LiF dosimeter surrounded by iron, and based on "large cavity", general cavity, and "small cavity" ionization theories. (Adapted from Simons, G. G. and Jule, T. J., *Nucl. Sci. Eng.*, 53, 162, 1974.)

was somewhat steeper relative to cavity size than that observed experimentally. Nonetheless, it may be that the deviations between theory and experiment arise at least in part to nonexponential attenuation of Compton electron spectra. In fact, there is no basis for the assumption that a secondary electron spectrum from gamma-ray interactions will attenuate exponentially. Finally, it should be mentioned that Ogunleye et al.[91] using a ^{60}Co photon beam perpendicular to the plane of rectangular dosimeters take g = 1.2t (t = dosimeter thickness) since the average angle of the Compton electrons with respect to the incident photon beam direction is 34°. This estimate of g is more realistic than g = 4V/S (isotropically impinging radiation) or g = t (dosimeter thickness) sometimes used by other investigators such as Bertilsson.[88]

B. Modified Burlin-Horowitz General Cavity Theory

Horowitz et al.[89,90] modified the Burlin general cavity theory by noting that, in general, the average path length for radiation crossing the cavity, g, need not equal the average path length for radiation created within the cavity as assumed by Burlin and other investigators. This inequality can be illustrated for isotropic radiation in the following manner.

Let x be a point in the connected volume, V, and let R(x,Ω) be the distance from that point to the surface in the direction Ω. Then the average absorption path length $\langle R \rangle$ for isotropic radiation homogeneously created within the volume V is given by

$$\langle R \rangle = \int_V \int_{4\pi} \frac{R(x,\Omega)d\Omega dV}{4\pi V} \tag{19}$$

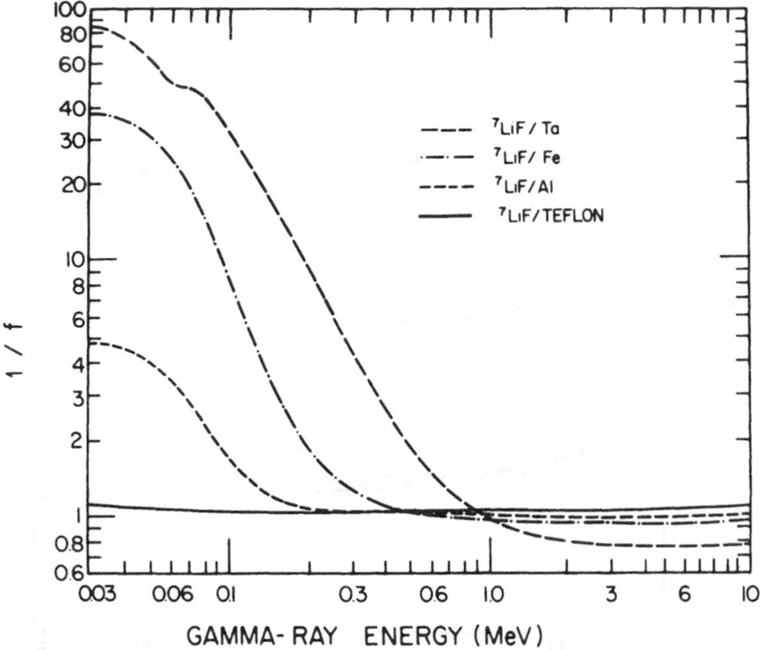

FIGURE 19. Variation of 1/f as a function of gamma-ray energy for a 1 × 1 × 6 mm (g = 4V/S = 0.243 g cm^{-2}) ^{7}LiF dosimeter surrounded by teflon®, Al, Fe, and tantalum. (Adapted from Simons, G. G. and Jule, T. J., *Nucl. Sci. Eng.*, 53, 162, 1974.)

It is easy to misinterpret this average path length $< R >$ as half of the mean chord length through the volume for isotropically impinging flux, $< R' > = 2V/S$. That this is not the case (i.e., $< R > \neq < R' >$) can be illustrated in the following manner (see Figure 21). The volume integral (Equation 19) can be transformed into a surface integration

$$< R > = \int_V \int_{2\pi} \int_0^{R_{max}} \frac{R \, d\Omega ds \, (\vec{\Omega} \cdot \vec{n}) \, dR}{4\pi V} \tag{20}$$

where ds is a surface element, $\vec{n}$ is a unit vector in the direction of the surface element and directed inwards and $\vec{\Omega}$ is the unit direction vector. It follows that

$$\begin{aligned} < R > &= \int\int \frac{(R^2/2) \, d\Omega ds \, (\vec{\Omega} \cdot \vec{n})}{4\pi V} \\ &= \frac{< R'^2 >}{4 < R' >} \end{aligned} \tag{21}$$

Since $< R'^2 > \geqslant 4 < R' >^2$ it follows that

$$< R > \geqslant 2V/S \tag{22}$$

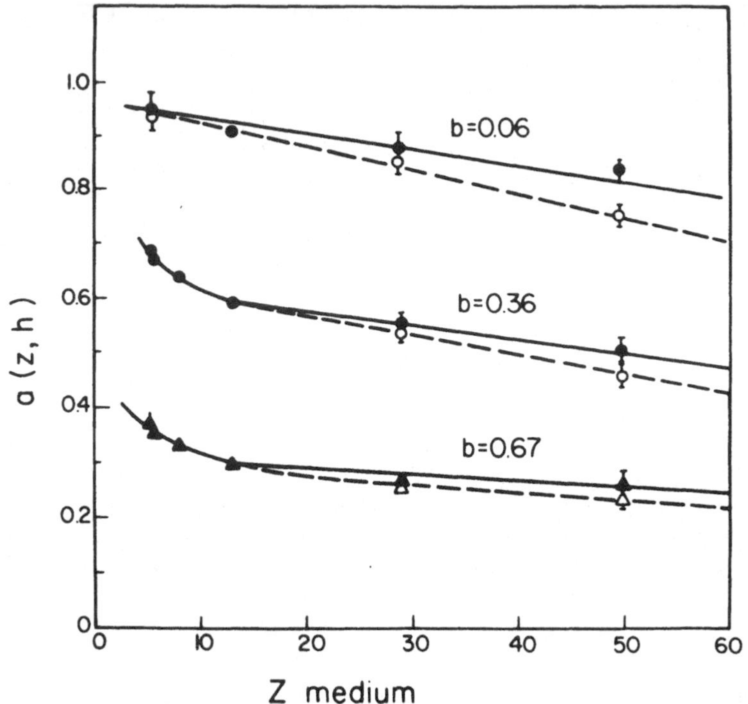

FIGURE 20. Correction factors a(z,h) (a = d and b = 1 − d in cavity theory nomenclature) calculated for three sizes of dosimeters as a function of the atomic number of the medium. The dashed curves are based on calculations assuming a photon spectrum given by Costrell;[98] solid curves are based on calculations assuming a simplified photon spectrum with 80% of the energy fluence at 662 keV and 20% from photons with an equivalent energy of 400 keV (note the strong dependence on the assumed photon energy spectrum). (Adapted from Bertilsson, G., Proc. 4th Int. Conf. Luminescence Dosimetry, Institute of Nuclear Physics, Krakow, 1974, 907.)

Moreover, since $R^2 d\Omega = ds'(\vec{\Omega}\cdot\vec{n})$ where ds′ is the surface element subtended by Ω (Figure 21B)

$$< R > = \int_S \int_S \frac{(\vec{\Omega} \cdot \vec{n})\, ds\, (\vec{\Omega} \cdot \vec{n}')ds'}{8\pi V} \leq \frac{S^2}{8\pi V} \tag{23}$$

Thus

$$\frac{2V}{S} \leq R \leq \frac{S^2}{8\pi V} \tag{24}$$

In summary, it can be seen that the average path length for electrons impinging on the cavity (and responsible for attenuation of the medium spectra within the cavity) need not be equal

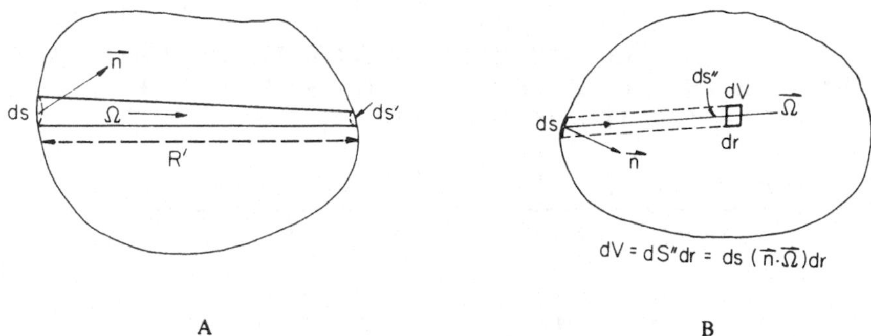

FIGURE 21. (A) Parameters used in the derivation of Equations 17-19. (B) Parameters used in the derivation of Equation 20. (Adapted from Horowitz, Y. S. and Dubi, A., *Phys. Med. Biol.*, 27, 867, 1982.)

to the average path length for the electrons created within the cavity. It follows that the Burlin general cavity theory should be modified to

$$f = \frac{(Z/A)_c}{(Z/A)_m}\left\{1 + d\left[\frac{(Z/A)_m}{(Z/A)_c} S_{cm} - 1\right] + d'\left[\frac{(\mu_{en}/\rho)_c}{(\mu_{en}/\rho)_m}\frac{(Z/A)_m}{(Z/A)_c} - 1\right]\right\} \tag{25}$$

Note that Equation 25 does not revert to the simple form of Equation 13. The symbol d is as previously defined and d' is given by

$$d' = \int_0^{g'} (1 - \exp(-\beta x))dx \Big/ \int_0^{g'} dx \tag{26}$$

where g' is the average absorption path length for electrons created within the cavity by the photon fluence interactions within the cavity. Janssens[250] has argued that the condition of $d + d' = 1$ is fundamental to general cavity theory and should not be violated since it arises from the equilibrium condition that the sum of fluences originating in and outside the cavity must equal the equilibrium fluence. This argument, however, totally ignores electron multiple scattering processes, i.e., electrons originating in the cavity arise from both photon interactions in the cavity (which determines d') and buffer electron-cavity electron multiple scattering processes within the cavity. Kearsley,[251] for example, has recently developed a general cavity expression which treats secondary electron scattering at the cavity medium interfaces and also derives $d + d' \neq 1$. It should be noted that in the limit of very large and very small cavities Equation 25 correctly approaches Equations 7 and 9, respectively. Furthermore as $c \rightarrow m$ we have f approaching unity as required. To illustrate that g need not equal g' we calculate (for typical cavity dimensions of 1.65 × 1.65 × 8.3 mm Harshaw TLD-700 hot-pressed chips) via Monte Carlo for homogeneously created isotropic cavity electrons g' = 1.05 mm whereas g = 4V/S = 1.5 mm. The situation where isotropic conditions do not exist may be considerably more complicated. For example, Ogunleye et al.[91] using a ^{60}Co photon beam perpendicular to the plane of rectangular TLD-100 dosimeters took g = 1.2t (t = dosimeter thickness = 0.38 mm) since the average angle of the Compton electrons with respect to the incident photon beam direction is 34°. In this geometry also, g will clearly not equal g'. Since the cavity electrons are homogeneously created within the cavity volume it follows that to a good approximation g' = g/2. Table 5 and Figure 22

Table 5
A COMPARISON OF BURLIN AND BURLIN-HOROWITZ GENERAL CAVITY THEORIES WITH THE EXPERIMENTAL RESULTS OF OGUNLEYE ET AL[91] FOR TLD-100 DOSIMETERS (g cm²)

	Polystyrene[a]				Aluminum[b]			
t	f_B	f'_{BH}	f''_{BH}	Exptl	f_B	f'_{BH}	f''_{BH}	Exptl
0.101	0.841	0.844	0.846	0.844	1.000	0.994	0.988	0.975
0.201	0.849	0.851	0.851	0.848	0.985	0.979_5	0.979	0.962
0.302	0.853	0.854	0.853	0.854	0.977	0.973_5	0.976	0.966
0.503	0.856	0.857	0.854	0.859	0.970	0.968_5	0.974	0.975
0.704	0.858	0.858	0.855	0.859	0.968	0.966	0.973	0.964
χ^2/n[c]	0.036	0.017	0.047	—	1.25	0.73	0.63	—

	Copper[d]					Lead[e]				
t	f_B	f'_{BH}	f''_{BH}	$f^{\bullet}_{BH}$	Exptl	f_B	f'_{BH}	f''_{BH}	$f^{\bullet}_{BH}$	Exptl
0.101	1.108	1.094	1.079	1.092	1.069	1.202	1.215	1.182	1.180	1.082
0.201	1.068	1.057	1.056_5	1.056	1.030	1.042	1.062	1.061	1.042	0.991
0.302	1.049	1.041	1.048	1.041	1.010	0.968	0.985	1.000	0.974	0.941
0.503	1.033	1.028	1.042	1.028	1.013	0.902	0.915	0.943	0.910	0.891
0.704	1.026	1.022	1.037	1.022	1.010	0.873	0.883	0.917	0.870	0.873
χ^2/n	4.29	2.25	3.29	2.14	—	14.2	20.6	20.6	11.0	—

(β = 16.26 cm²g⁻¹)

	Polystyrene				Aluminum			
t	f_B	f'_{BH}	f''_{BH}	Exptl	f_B	f'_{BH}	f''_{BH}	Exptl
0.101	0.843_5	0.846	0.848	0.844	0.995_5	0.989_5	0.984	0.975
0.201	0.851	0.851_5	0.853	0.848	0.981	0.976_5	0.976	0.962
0.302	0.853	0.855	0.854	0.854	0.977	0.971	0.974	0.966
0.503	0.857	0.858	0.856	0.859	0.969	0.967	0.972	0.975
0.704	0.858	0.859	0.856	0.859	0.967	0.965	0.971	0.964
χ^2/n	0.019	0.022	0.07	—	0.90	0.48	0.37	—

	Copper					Lead				
t	f_B	f'_{BH}	f''_{BH}	$f^{\bullet}_{BH}$	exptl	f_B	f'_{BH}	f''_{BH}	$f^{\bullet}_{BH}$	Exptl
0.101	1.096	1.082	1.068	1.081	1.069	1.155	1.171	1.140	1.136	1.082
0.201	1.058	1.049	1.048	1.048_6	1.030	1.004	1.023	1.023	1.008	0.991
0.302	1.049	1.043	1.041	1.043	1.010	0.965	0.955	0.968	0.947	0.941
0.503	1.028_5	1.024	1.035	1.024	1.013	0.884	0.895	0.919	0.892	0.891
0.704	1.023	1.020	1.032	1.019_7	1.010	0.861	0.869	0.897	0.867	0.873
χ^2/n	2.82	1.51	1.94	1.54	—	4.99	7.18	5.73	2.55	—

Note: f_B (Burlin) calculated using $g = t/\cos\bar{\theta} = 1.2t$ and $g' = g$; f_{BH} (Burlin-Horowitz) calculated using $g = \int_0^{\pi/2} \frac{t\,P(\theta)\,d\theta}{\cos\theta} = 1.539t$ and $g' = g/2$; f''_{BH} calculated using $g = 4V/S$ and $g' = \frac{1}{2}\int \frac{t\,P(\theta)\,d\theta}{\cos\theta} = \frac{1.539t}{2}$; and $f^{\bullet}_{BH}$ calculated with increase of g,g′ due to backscattering at the cavity-medium interface.

Table 5 (continued)
A COMPARISON OF BURLIN AND BURLIN-HOROWITZ GENERAL CAVITY THEORIES WITH THE EXPERIMENTAL RESULTS OF OGUNLEYE ET AL[91] FOR TLD-100 DOSIMETERS (g cm²)

[a] Polystyrene: $\overline{S}_{cm}$ = 0.821, $\overline{(\mu_{en}/\rho)}_{cm}$ = 0.861, $(Z/A)_{cm}$ = 0.858.

[b] Aluminum: $\overline{S}_{cm}$ = 1.04, $\overline{(\mu_{en}/\rho)}_{cm}$ = 0.961, $(Z/A)_{cm}$ = 0.9599.

[c] Chi-squared values calculated assuming an experimental error of 1.5% (1 S.D.).

[d] Copper: $\overline{S}_{cm}$ = 1.209, $\overline{(\mu_{en}/\rho)}_{cm}$ = 1.008, $(Z/A)_{cm}$ = 1.0134.

[e] Lead: $\overline{S}_{cm}$ = 1.610, $\overline{(\mu_{en}/\rho)}_{cm}$ = 0.802, $(Z/A)_{cm}$ = 1.1691.

compare the Burlin and modified Burlin-Horowitz expressions with the experimental results of Ogunleye et al.[91] Following Ogunleye et al. a value of $\beta = 13.4\ cm^2\ g^{-1}$ was used that was obtained from the formula[95] $\beta = 14/E_{max}^{109}$ with E_{max} = 1.04 MeV for 1.25 MeV Compton scattering. f_B(Burlin) was calculated using $g = t/\cos\bar{\theta} = 1.2t$ and $g' = g$ as suggested by Ogunleye et al., whereas f_{BH} (Burlin-Horowitz) was calculated using a more rigorous estimation of the average path length

$$g = \int_0^{\pi/2} \frac{t\ P(\theta)d\theta}{\cos\theta} = 1.539t \quad \text{and} \quad g' = g/2 \tag{27}$$

where $P(\theta)$ is the Klein-Nishina electron angular probability distribution function for 1.25 MeV gamma-ray Compton scattering. Chi-squared values are also tabulated in Table 5 to indicate goodness of fit between theory and experiment. For polystyrene, aluminum, and copper the Burlin-Horowitz expression is in better agreement with experiment as shown by the significantly lower values of χ^2/n. In the case of lead, the Burlin expression yields a better fit although both fits are very poor (χ^2/n = 14.2 and 20.6, respectively), especially for low values of t where the dose in the medium is significantly underestimated. The third column of Table 5 shows f''_{BH} evaluated using g = 4V/S and g′ = 1.539t/2. The justification for the choice of g is that the equilibrium medium-generated spectrum reaching the cavity cannot be expected to have the pure angular distribution of the Compton electrons generated on the medium cavity interface; rather the distribution in angle will have been randomized due to multiple electron scattering in the buffer layer. Actually a more realistic value of g is certainly somewhere between these two extreme values. The χ^2/n is improved for f''_{BH} only in the case of aluminum. In general with g = 4V/S, f approaches $(\mu_{en}/\rho)_{cm}$ in the limit of large t far too slowly. For small t, however, the fit is somewhat improved for Al, Cu, and Pb. Table 5 also shows the Burlin and Burlin-Horowitz expressions recalculated for β = 16.26 based on the formula,[99] $\beta = 17/E_{max}^{1.14}$ (radioisotope attenuation data in aluminum). The χ^2/n values for Al, Cu, and Pb are significantly improved for both theoretical expressions. It follows that the assumption of exponential attenuation for the Compton-generated electrons may significantly influence the agreement between theory and experiment. To summarize up to this point, the experimental data for polystyrene, aluminum, and copper clearly favor the Burlin-Horowitz expression with g = 1.539t and g′ = g/2 and evaluated with Evan's expression for β.

Burlin[87] noted that the relatively good agreement between theory and experiment for the general cavity expression probably arises from cancellation of errors in the treatment applied to the medium- and cavity-generated electrons. This is most likely the reason that the Burlin-Horowitz expression (which more accurately treats the cavity-generated electrons) does not readily result in closer agreement with experiment in the case of lead. In support of this argument we note that inclusion of electron-backscattering effects (which would be expected

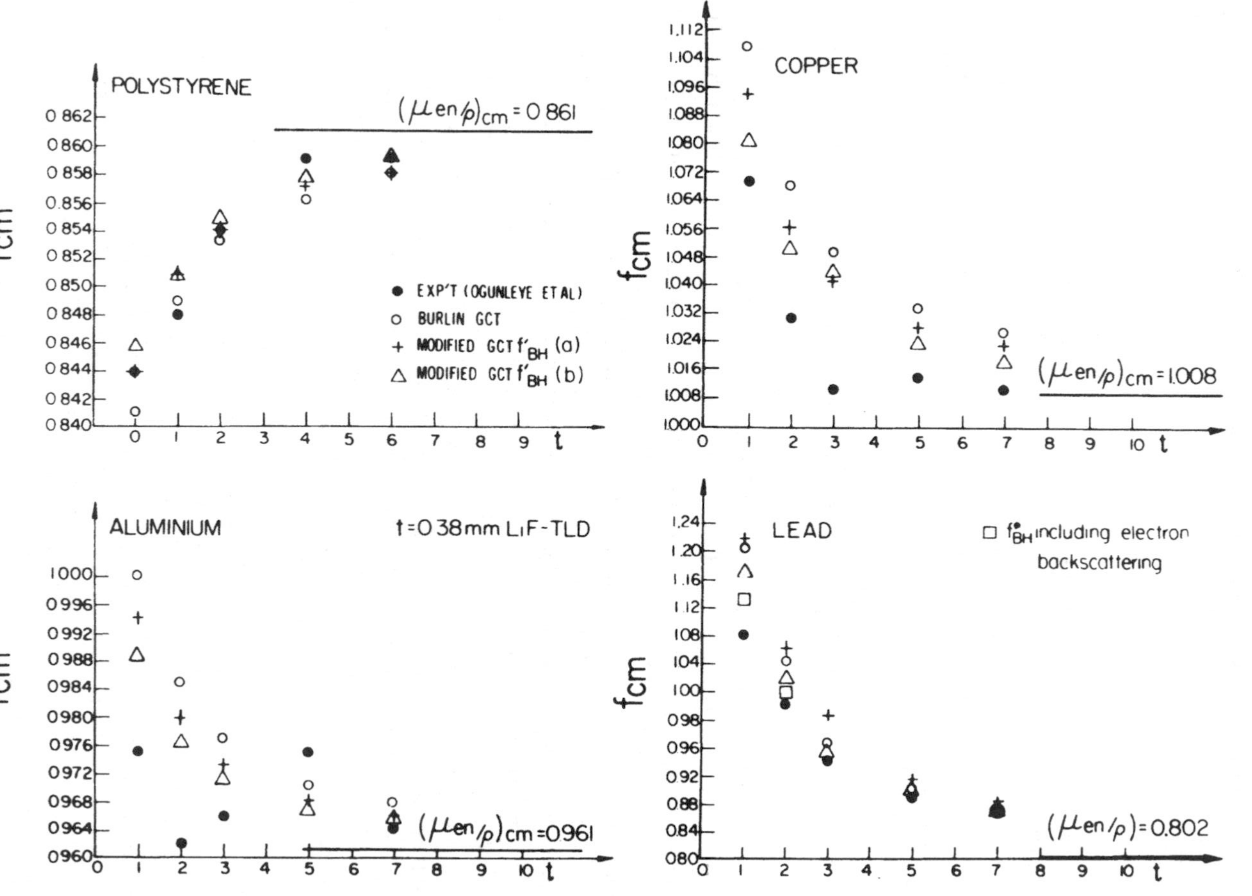

FIGURE 22. A comparison of Burlin and modified Burlin-Horowitz general cavity theory with the experimental results of Ogunleye et al.[91] using TLD-100 chips.

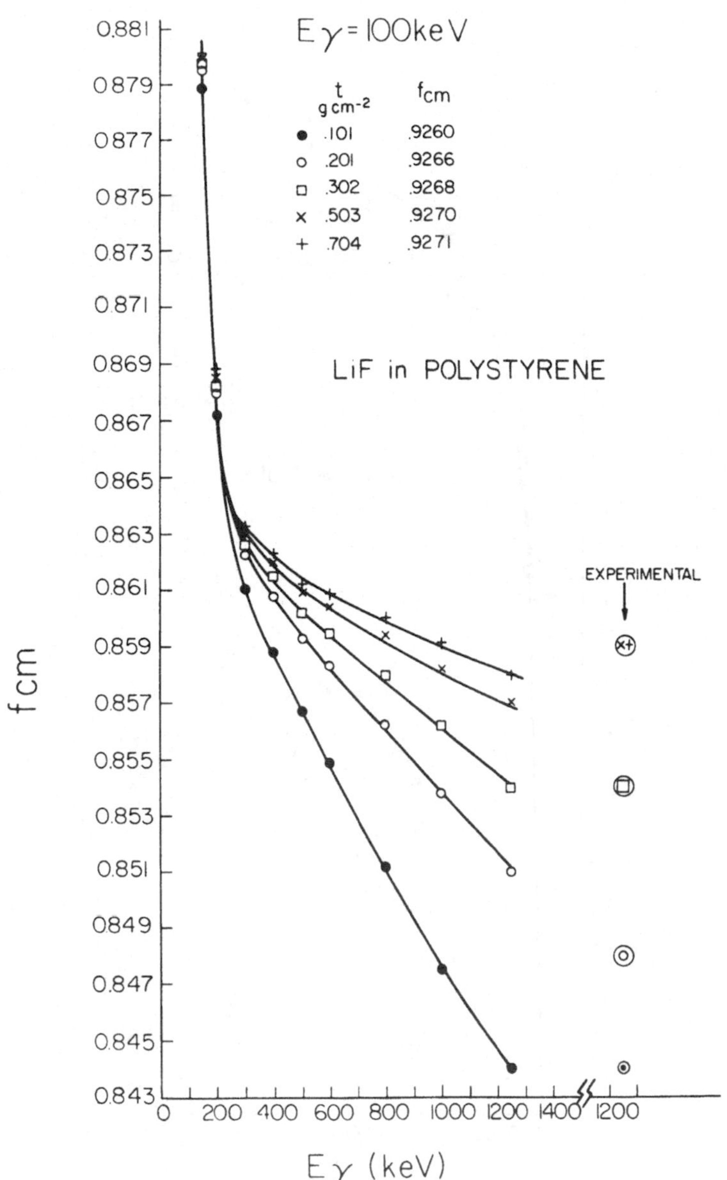

FIGURE 22A. Calculated values for f_{cm} (LIF in polystyrene) as a function of gamma ray energy and cavity thickness using modified Burlin-Horowitz general cavity theory.

to increase f_{cm}) has exactly the opposite effect when introduced via the formalism of general cavity theory. We can demonstrate the expected increase of f_{cm} due to electron backscattering in the following manner. The electron-backscattering coefficients for perpendicular incidence are roughly constant as a function of energy (above approximately 20 keV) and are equal to 0.49, 0.25, 0.1, and 0.05 for Pb, Cu, Al, and LiF, respectively.[100] It follows that approximately 49, 25, and 10% of the cavity-generated electrons that reach the second cavity-medium interface (Pb, Cu, and Al, respectively) will be backscattered into the cavity. At the medium-cavity interface, on the other hand, a relatively insignificant fraction of the

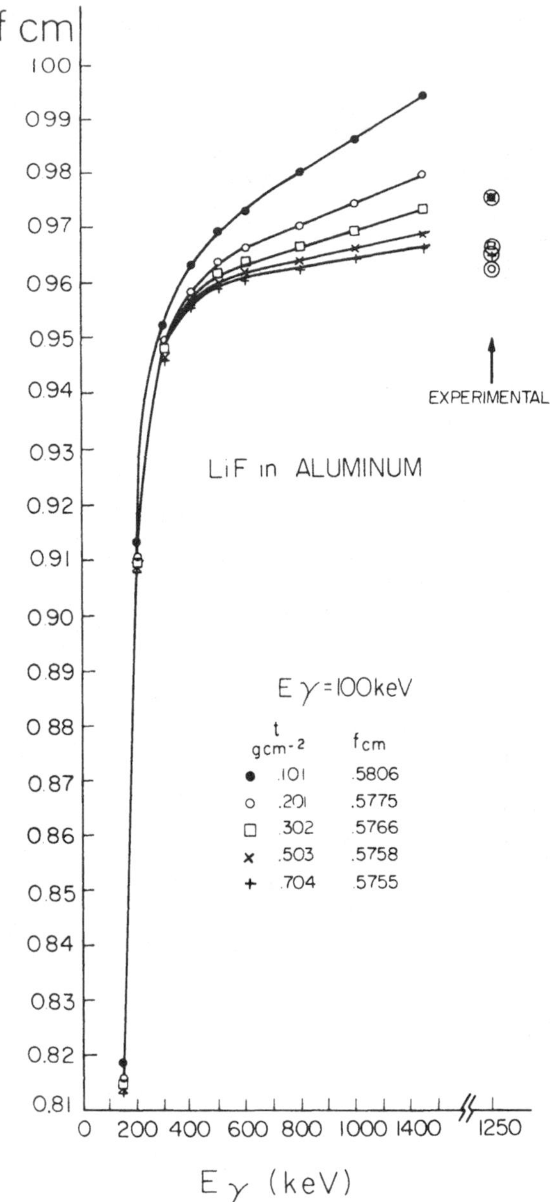

FIGURE 22B. Calculated values for f_{cm} (LIF in aluminum) as a function of gamma ray energy and cavity thickness using modified Burlin-Horowitz general cavity theory.

medium-generated electrons is backscattered out of the cavity because of the very low value of η (the electron-backscattering coefficient). To a first approximation, therefore, the contribution to the cavity dose from cavity-generated electrons can be expected to increase by the factor

$$\frac{n \cdot f \cdot \int_0^{g'} \exp(-\beta x)\, dx}{\int_0^{g'} dx} \tag{28}$$

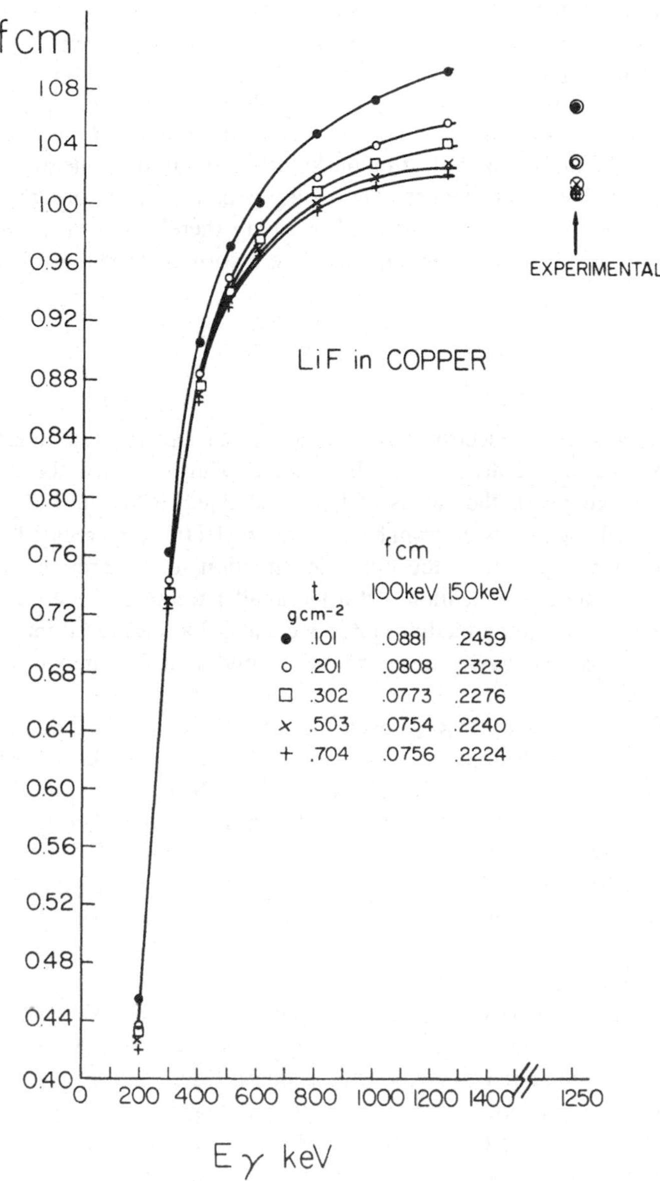

FIGURE 22C. Calculated values for f_{cm} (LIF in Copper) as a function of gamma ray energy and cavity thickness using modified Burlin-Horowitz general cavity theory.

where f is the mean energy of the backscattered electrons expressed as a fraction of the initial energy for normal incidence. To a reasonable approximation, f is independent of energy[101] and is given by (123 + Z)/216 so that f = 0.95, 0.70, and 0.63 for Pb, Cu, and Al, respectively.[102] The average value of $\exp(-\beta x)$ in Equation 28 estimates the fraction of the cavity-generated electrons reaching the cavity-medium interface. For t = 0.101 g cm^{-2} Equation 28 yields 0.29, 0.11, and 0.04 for Pb, Cu, and Al, respectively. Thus inclusion of electron backscattering must increase f_{cm} thereby increasing the discrepancy between theory and experiment as tabulated in Table 5. The percentage increase in f_{cm} will be given by 0.29, 0.11, and 0.04 multiplied by the relative contribution of cavity-generated electrons to the total cavity dose. These considerations are in qualitative agreement with the dose distribution results of Ogunleye et al. for TLD stacks, which show increased dose due to electron backscattering in the extremity dosimeters close to the cavity-medium interface. In these experiments the dose in the TLD closest to the cavity-medium interface is approximately 45% (Pb), 15% (Cu), and 5% (Al) greater than the dose in the middle dosimeter. On the other hand, the inclusion of electron backscattering into the general cavity formalism *decreases* f_{cm} in the following manner. The cavity-generated electrons reaching the LiF-medium interface are backscattered into the LiF cavity thereby increasing g′. If we adopt $g_{bs} = 1.539t/2 + 1.539t$, it is possible to calculate a corrected value of d′ given by

$$d'^{*} = (1 - I_{bs})d' + I_{bs}d'_{bs} \tag{29}$$

where I_{bs} is the fraction of the electron fluence reaching the LiF-medium interface. Adopting this approach, which admittedly ignores the angular distribution of the cavity-generated backscattered electrons yields the values of f^{*}_{BH} as tabulated in Table 5. Thus the increase in g′ for that part of the dose arising from backscattered electrons decreases f_{cm} and improves the agreement with experiment in the opposite direction to the previously demonstrated expected behavior. It deserves mention that a far smaller fraction of the medium-generated electrons also reaches the cavity-medium interface and is backscattered into the cavity with similar effect. That is, the cavity dose increases, but inclusion into the general cavity formalism decreases f_{cm}.

In conclusion, general cavity theory, even in its modified form[89,90] does not yield satisfactory agreement ($\chi^2/n < 1$) in cases where there is a very significant mismatch between the cavity and the medium (Pb and Cu in this example). Nonetheless, the Burlin-Horowitz expression with $g = 1.539t$ and $g' = g/2$ yields significantly better agreement with experimental data for polystyrene, aluminum, and copper than the Burlin expression. In the case of Pb, the smallest χ^2/n value of 2.55 is achieved for f'^{*}_{BH}, which includes the effect of electron backscattering in the estimation of g and g′. If $g = 1.539t$ is inserted in the Burlin expression instead of $g = 1.2t$ as suggested by Ogunleye et al.,[91] then the two expressions yield very similar results at 1.25 MeV. However, the Burlin-Horowitz expression is preferred due to its more realistic treatment of the cavity-generated electrons. Future experimental and theoretical studies as dicussed in the following can be expected to yield even better agreement due to the relaxation of the constraint that d + d equal unity. Figures 22A, 22B, and 22C illustrate the application of modified general cavity theory to gamma ray energies from 0.1 MeV to 1.25 MeV in the Ogunleye et al.[91] geometry for LiF-TLDs in polystyrene, Al, and Cu, respectively. Expression 27 was used to calculate $g' = g/2$ as a function of gamma ray energy. Mass energy absorption coefficients were taken from Hubbell[18] and Evans,[99] collision mass stopping powers from ICRU Report 21[252] and values of $\overline{T}_0$ from NBS Handbook 64.[253] The simplicity of general cavity theory and its evident ability to give good agreement with experimental results (when the cavity and medium are not extremely mismatched — polystyrene and aluminum in this example) make it obligatory to continue its development via further theoretical and experimental studies. For example, the significant dependence of the theoretical results on β (especially in the case of strong mismatch between cavity and medium) is strong motivation to study the attenuation of Compton-generated electron spectra to determine the validity of the assumption of exponential attenuation. The electron spectrum rather than being a pure Compton spectrum can be expected to possess a strong low energy component due to multiple scattering in the buffer layer surrounding the cavity. If the attenuation is faster than exponential due to this low energy component, the agreement between theory and experiment will improve. Experiments of this nature are currently being designed in the Radiation Physics Laboratory of the Ben Gurion University of the Negev.

C. Extension of General Cavity Theory to Electrons

For application to electrons, Burlin et al.[103] omitted the photon-generated cavity term to obtain

$$f = d\,\overline{S} \tag{30}$$

and qualified their treatment to situations where the electrons crossing the cavity were in a state of quasidiffusion, this in order to justify the assumption of exponential attenuation and the choice of g = 4V/S (the mean chord length for isotropically incident radiation). Equation 30 suffers from the flaw that as the cavity material approaches the medium material, f approaches d rather than unity as it should. Moreover, we shall see in the following that Equation 30 does not agree with experimental data or Monte Carlo calculations. Almond and McCray[104] and Paliwal and Almond[105] introduced a cavity source term analogous to the photon case and arrived at the expression

$$f = d\,\overline{S_{cm}} + (1 - d)\,(Z/A)_{cm} \tag{31}$$

where the second term was interpreted as an electron-generated cavity source term. Equation 31 is also conceptually troublesome since in the electron case *there are no high energy primary electrons being generated in the cavity* (except, of course, those due to bremsstrahlung photons, which are relatively insignificant). Almond and McCray[106] refer to "many secondary electrons generated in the cavity which will contribute significantly to the dose". These secondary electrons due to ionization following electron-electron interactions, however, are already included in the S_{cm} ratio in the Burlin expression (Equation 30). Burlin[107] argued that the large cavity limit of Equation 31 is incorrect, however, Almond and McCray[106] maintained that for very large cavities the limit $f \rightarrow (Z/A)_{cm}$ is acceptable since one would expect to a first approximation that the absorbed dose be proportional to the ratio of the electron densities. It deserves mention that as $c \rightarrow m$, $f \rightarrow 1$ as required in Equation 31. Both expressions (Equations 30 and 31) have been criticized by O'Brien[86] because they neglect the secondary photon flux produced both in the cavity and the medium. However, even at 40 MeV the absorbed dose due to bremsstrahlung radiation is only approximately 5% of the maximum electron-induced absorbed dose[108] and since the cavity effect is of the order of 10%[86,109] the maximum error in f_{cm} due to neglect of radiative processes should not exceed 0.5%. A more important and astonishing fault in Equation 31, however, lies in the fact that the minimum value of $f^{E}_{^{60}Co}$ as calculated from the Almond and McCray expression, and using correct values for $\overline{S_{cm}}$ is 0.965 as has been pointed out by Shiragai[110] and Lowe.[111] This can be seen by a trivial examination of Equation 31 after numerical values have been inserted for polystyrene as the medium and LiF as the cavity. Lowe[111] traced the discrepancy to the $\overline{S_{cm}}$ (Spencer-Attix) values used by Almond and McCray: 3.55 MeV (0.676), 5.59 MeV (0.721), 7.31 MeV (0.736), 10.51 MeV (0.751), 13.34 MeV (0.756), and 25.04 MeV (0.769) compared to his own values of 0.905, 0.897, 0.892, 0.889, 0.888, and 0.885, respectively. Returning to Equation 31, the values of $\overline{S_{cm}}$ are 0.833 and 0.828 for 30 MeV and 3 MeV electrons, respectively,[108] so that $\overline{S_{cm}}$(CSDA) = 0.83 to very good accuracy and $(Z/A)_{cm}$ = 0.858. It follows that

$$f^{E}_{^{60}Co} = \frac{(0.83\,d + (1 - d)\,0.858)_E}{(0.821\,d + (1 - d)0.861)_{^{60}Co}} \tag{32}$$

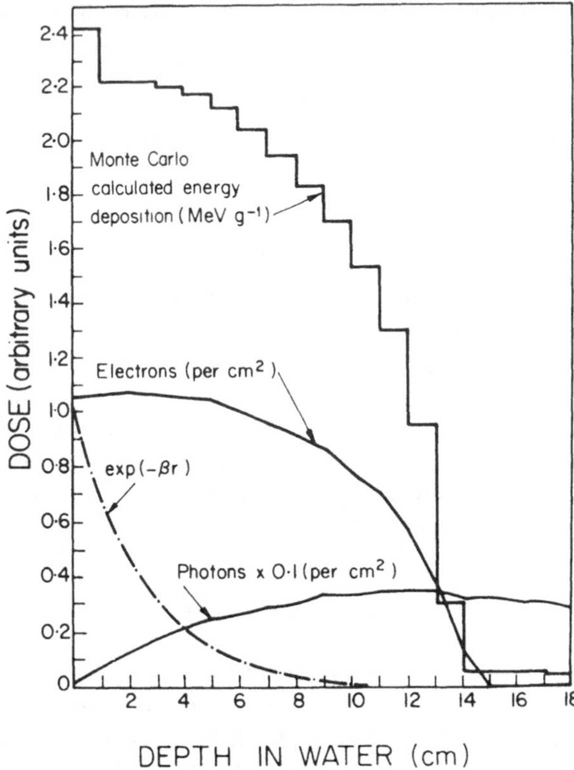

FIGURE 23. Monte Carlo calculations of energy deposition in water by 28-MeV electrons as well as calculations of the secondary photon flux. (Adapted from O'Brien, K., *Phys. Med. Biol.*, 22, 836, 1977.)

where $\overline{S_{cm}} = 0.821$ and $(\mu_{en}/\rho)_{cm} = 0.861$ for ^{60}Co photons. Inspection of Equation 32 yields

$$0.965 \leqslant f^{E}_{60_{Co}} \leqslant 1.045 \tag{33}$$

so that the Almond and McCray[104] and Paliwal and Almond[105] calculation of $f^{E}_{60_{Co}}$ are incorrect as shown by Shiragai.[110] Furthermore the Almond and McCray[104,106] expression cannot predict the Almond and McCray[104,106] experimental result $f^{E}_{60_{Co}} \simeq 0.90$ as claimed by Paliwal and Almond.[105]

We have further applied Equations 30 and 31 to the experiments of Ogunleye and Paliwal[109] using Equation 34 to estimate g_E and assuming exponential attenuation. Substitution of $E_{max} = 22$ MeV in Equation 17 yields $\beta = 0.261$ cm^2 g^{-1} and $d = 0.986$. The Burlin expression yields $f_{cm} = 0.974$ and the Almond and McCray expression $f_{cm} = 0.987$, both in contradiction with the experimentally observed value of 0.91 ± 0.02.

In the application of general cavity theory the calculation of the parameter d could be of critical importance since it determines the relative contribution to the dose from electrons generated in the medium. Convincing evidence has been presented[85,86,107,112] that the assumption of exponential attenuation is not valid for high energy electrons when the cavity is near the surface and/or surrounded by low atomic number material (water, perspex, polystyrene, LiF). Figure 23 shows the Monte Carlo depth dose calculations of O'Brien[86] for 28-MeV electrons slowing down in water. Clearly, the assumption of continuous and uniform energy loss which could lead to exponential attenuation is very crude. When the cavity is surrounded by high Z material (e.g., Pb) the electrons are so multiply scattered in the high atomic number material that they reach a state of quasidiffusion in which their attenuation is more closely exponential. Paliwal and Almond[94] attempted to illustrate that high energy electrons were also attenuated exponentially, however, their transmission experiments did not correctly simulate the required geometry for a justification of the use of exponential attenuation. What is more correctly required is an electron beam entering two adjacent media of different atomic number.

Additional serious error in the calculation of d may arise from incorrect estimation of g (the path length across the cavity). For high energy electrons, the choice of $g = 4V/S$ is equivalent to the assumption of quasidiffusion since $g = 4V/S$ is correct for isotropically incident radiation. Eggermont et al.[92] have calculated d for plane parallel and spherical cavities using the average chord length $g = 4V/S$ and also by integrating the exponential attenuation over the chord length distribution. The result was percentage deviations in d of the order of 1 to 5%. The choice of $g = 4V/S$ as used by Almond and McCray[104] and Paliwal and Almond[105] for high energy electrons slowing down in polystyrene when the TLD was near the surface of the medium is unjustified since the electrons will surely be very significantly biased in the forward direction. These considerations should make it clear that in the case of a surface exposed to high energy electrons, or in other situations where charged particle equilibrium is in doubt such as transition layers and beam edges, the angular distribution of the electrons incident on the cavity is of great importance in determining the cavity effect. The tracks of the electrons are scattered more and more with increasing depth. At the surface where the electron tracks are essentially straight and parallel a can or pillbox shape will result in minimum perturbation with the electrons impinging vertically on the flat surface. At greater depths where the flux is more nearly isotropic a pillbox would be a mistake because electrons could then enter more or less parallel to the long side of the cavity and in this case for minimum perturbation a small spherical cavity is recommended.

Holt et al.[113] calculated g realistically for high energy electrons by relating the mean path length to the mean square scattering angle, $\bar{\theta}_{RMS}$, through which the electrons will have been scattered before reaching the cavity. The mean path length was then taken to be the rod diameter multiplied by $\sec(\bar{\theta}_{RMS})$ where

$$\bar{\theta}_{RMS} = \frac{180}{\pi}\frac{E_s}{E_0}(\ell/X_0)^{1/2} \tag{34}$$

$X_0 = 37.1$ cm for water and is the radiation length in the scattering material[108] and E_s has a constant value of 21 MeV. More precise formulations for $\bar{\theta}_{RMS}$ are given by Rossi.[114] For TLD dimensions of $3 \times 3 \times 0.4$ mm exposed to 20 MeV electrons 2 cm beneath the surface (the geometry of Ogunleye and Paliwal[109]) we calculate via Equation 34, $g = 0.41$ mm whereas $g = 4V/S = 0.63$ mm. However, Holt et al.[113] in their semiempirical model for f_{cm} given by

$$f_{cm} = S_{mc}\{1 + (S_{mc} - 1)\}\frac{<x>}{R_{max}} \tag{35}$$

artificially generated their experimentally observed reduced response at low electron energies (3 to 6 MeV) by assuming that the change in energy fluence is given by the change in mean energy and attributed the reduction in the cavity dose to increased multiple scattering at lower energies. O'Brien,[86] however, correctly pointed out that the product of flux and stopping power are approximately independent of energy and that any increase in electron path length in the cavity that results from multiple scattering will tend to increase the cavity dose rather than decrease it. Moreover, Holt et al. assumed that their TLDs behaved in the small cavity limit for ^{60}Co radiation which is not accurate.

Fregene[115] has proposed a phenomenological theory for high energy electron radiation based on results from experimental measurements of the build-up of secondary electrons in different media, and depth dose data of high energy electrons which led to the formulation that the variation of mass stopping power with thickness of cavity material would be linear for depths much less than the equilibrium depth (R_e) of the secondary electrons. R_e is taken as the depth of maximum dose in water and where there is a plateau in the percentage of depth dose distribution, the midpoint of the plateau is equated to R_e.

$$\frac{1}{f_{cm}} = \frac{1}{S_{cm}} + \left(1 - \frac{1}{S_{cm}}\right)\frac{< x >}{R_e} \qquad (36)$$

for dosimeter thicknesses less than R_e.

D. High Energy Electron Response: Experimental Results

The definite existence of a cavity effect for high energy electrons and photons has been convincingly illustrated by both O'Brien[86] and Ogunleye and Paliwal.[109] The former carried out Monte Carlo calculations in the Paliwal and Almond[105] and Holt et al.[113] geometries and obtained $f^{E}_{^{60}Co}$ = 0.906 and 0.910, respectively. Ogunleye and Paliwal illustrated experimentally the presence of a cavity effect for the interaction of 6- to 18-MeV electrons and 22-MeV photons with hot-pressed TLD-100 (3 × 3 × 0.4 mm). The dosimeters were irradiated in LiF, polystyrene, and water phantoms, and a decreased response of approximately 8% was observed for the latter two materials but was not observed in the homogeneous irradiation in the LiF phantom. The decreased response has been observed by many other groups (Table 6) in TLD-100, -700, and other unidentified LiF materials in various configurations (powder, extruded rods, chips, single crystals) of various dimensions ranging from the thin (0.4 mm) to the thick (5 mm) and light (approximately 10 mg) to the very heavy (174 mg). On the other hand, the cavity effect has not been observed by an equally impressive list of investigations (Table 7) in TLD-100, -600, -700, LiF(IAO), LiF-NTL-50p in a similar variety of configurations (powder, plaques, extruded rods), dimensions (0.5 to 6 mm), and mass (approximately 10 to 100 mg). The experiments of Holt et al.[113] and to a lesser extent Liu and Bagne[138] form, in fact, a third subset of findings in which there is a decreased response at low energies (3 to 6 MeV) with essentially no decrease in response at the higher energies. Ogunleye and Paliwal[109] also observed a statistically significant lesser response for TLD-100 in water (0.88 ± 0.01) at 6 MeV compared to 0.92 ± 0.01 at the higher electron energies. At the present time the results shown in Tables 6 and 7 are impossible to reconcile and imply the existence of unknown experimental variables which affect the TL sensitivity of LiF-TLD when exposed to high energy electrons and photons.

Ogunleye[117] carried out cavity measurements with TLD-100 (3 × 3 × 0.3 mm) in stacks (0.09 to 0.63 g cm^{-2} thick) in LiF, polystyrene, aluminum, copper, and lead media exposed to an 18-MeV electron beam. Relative experimental values of average absorbed dose for different dosimeter thicknesses were compared to the various cavity theories. The agreement of Burlin's expression with experiment was quite poor for LiF dosimeters in polystyrene, aluminum, and copper; only in lead were they approximately comparable. It follows, therefore, that Burlin's expression is more valid when the electrons are in a quasidiffused state and the exponential treatment describes the attenuation of the electrons within the cavity medium with sufficient accuracy. Almond and McCray's[104,106] expression shows good agreement with experiment only in aluminum and copper, while it is poor for polystyrene and worse for lead. Another noticeable trend, although small in aluminum, is the progressive divergence of their theory from experiment with increasing cavity size in all media. The agreement observed with aluminum and copper is probably somewhat fortuitous due to the closeness in stopping power and electron density ratios of the media and LiF. The empirical expression suggested by Holt et al.[113] is in agreement with experimental results only for LiF in Al, and is poor in describing the results in all the other media. The agreement in the case of Al is again probably fortuitous due to the closeness of the mass collision stopping power ratio of Al and LiF. The modifying term in the expression is therefore close to unity. It is further observed that the slope of the curve (with increasing cavity thickness) resulting from the Holt et al. expression is always opposite to that of the experiment. Agreement of the Fregene expression with experiment is very good for polystyrene and aluminum, adequate (within 1 S.D. of all the experimental points) for copper, but again very poor in the case of lead although it does predict the correct experimental trend with increasing cavity size. The Fregene expression provides the best fit to the experimental points in low Z media (polystyrene and aluminum in the Ogunleye experiment).

Some additional comments are necessary to clarify further the experimental situation and discourage attempts to trivialize the serious discrepancies outlined in Tables 6 and 7.

1. Almond and McCray[104] and O'Brien[86] attempted to dismiss the experimental discrepancies by referring only to a subset of experiments (Suntharalingam and Cameron[116] and Antoku et al.[139]) in Table 7. The results of the former were explained as due to the use of a fine powder so that the cavity effect would not be expected to be observed; the latter results were dismissed because of the relatively poor precision. In fact, Suntharalingam and Cameron[116] did not report their cavity dimensions and a later effort studied the high energy response for TLD-100 dosimeter thicknesses from 1 to 6 mm[118,119] with identical results, i.e., no decrease in sensitivity for 10- to 50-MeV electrons and 4 to 45 MV X-rays.
2. Gantchew and Toushlekova[120] observed different response to high energy electrons in two LiF materials produced by Chemiewerk GDR (LiF-200-Ti-80T and LiF-200-Mg-S). The former material showed an average relative response of 0.93 and the latter 1.01 for 5.7 to 36.9-MeV electrons. Slight batch-dependent effects have also been reported by Binks[121] and Almond et al.,[122] however, these were barely statistically significant. An exchange of letters[123,124] did not alter the possibility of the suggestion by Gantchew and Toushlekova[120] that impurity composition may play an important role in the divergence of the results via an impurity composition material-dependent ionization density effect. In our opinion, the two LiF materials studied by Gantchew and Toushlekova represent extremes in material defect composition which therefore do not significantly reflect on the discrepancies in Tables 6 and 7. The 200-Mg-S is a radiation-sensitized material with no added Ti dopant, whereas the 200-Ti-80T material is doped with both Mg and Ti. On the other hand, most of the electron response studies have been carried out on LiF (Harshaw) of nominally similar if not identical doping characteristics. It should be mentioned further that the batch-dependent variations in relative TL response reported by Horowitz et al.[125-127] at high ionization densities were not observed between different batches of TLD-100 or TLD-600 but rather between TLD-100 and TLD-600 and TLD-700.
3. Most previous investigations have not reported important details of the experimental procedures (peak height or integral TL, T_8, dose rate, etc.). It should be recalled that dose rate independence in LiF has not been established to better than approximately 5%, and a possible dose rate dependence at higher dose rates has been reported.[128,129]

Table 6
EXPERIMENTS WHICH REPORT S(E) ≃ 0.9 FOR HIGH ENERGY ELECTRONS IN LiF-TLD

Material	Geometry	Irradiation depth (cm)	Response at energy, E						Ref.
Type N	Flat capsules	0.55 (^{60}Co)	10—15 MeV						121
Type 7	Polystyrene	1.6 (e^-)	0.888 ± 0.017						
(Conrad)	[20 mm (ID) 1-mm deep]		0.917 ± 0.03						
TLD-700	Tubes (174 mg)	—	(e^-, 6—18 MeV) 0.935 ± 0.021 (X-rays, 22 MV_p) 0.929 ± 0.014						133
TLD-700	Capsules (~75 mg)	—	(e^-, 10 MeV) 0.899 ± 0.021 (e^-, 15 MeV) 0.916 ± 0.018						131
TLD-700	Teflon® capsules [174 mg 25 mm (ID) 5-mm deep]	1.5 (e^-) 3 (18.5 MV_p) 4 (22 MV_p)	(e^-, 6—18 MeV) 0.938 ± 0.02 (18.5-, 22 MV_p X-rays) 0.93 ± 0.02						104, 122
TLD-700	Rods (1 × 1 × 6 mm)	6—9	35-MV photons, 0.899 ± 0.007 50-MV photons, 0.907 ± 0.009 65-MV photons, 0.907 ± 0.011						134
TLD-100	Teflon® tube [8-mm long, 1 mm (ID)]	—	(e^-, 15—33 MeV) 0.92 ± 0.03						135
			X-rays						
			6 MeV	9 MeV	12 MeV	15 MeV	18 MeV	22 MeV	
TLD-100	0.125 × 0.125 ×	LiF phantom	0.98 ± 0.03	0.98 ± 0.02	0.99 ± 0.02	0.99 ± 0.02	0.98 ± 0.01	1.01 ± 0.02	109
	0.015 in.	polystyrene	0.89 ± 0.02	0.91 ± 0.02	0.91 ± 0.01	0.91 ± 0.02	0.90 ± 0.02	0.91 ± 0.02	
	hot pressed	water	0.88 ± 0.01	0.93 ± 0.01	0.93 ± 0.01	0.92 ± 0.01	0.91 ± 0.03	0.93 ± 0.01	

LiF (Harshaw)	Extruded rods — (1 mm in diameter, 6 mm in length, ~12 mg)		(e^-, 5—39 MeV) 0.945 ± 0.015	112
LiF	Chips (3.2 × 3.2 × 0.9 mm)	0.5—2(e^-) 6 (X-rays)	(e^-, 5—40 MeV) 0.930 ± 0.02 (X-rays, 42 MV_p) 0.940 ± 0.02	136
LiF crystals	9 × 11 × 1.7—1.9 mm	1.5	(e^-, 7—28 MeV) 0.878 ± 0.013	105

$\sigma = \pm 2\%$

	7 MeV	9 MeV	11 MeV	15 MeV	18 MeV	28 MeV	
As in Paliwal et al.[105] via Monte Carlo[a]	0.925	0.901	0.903	0.912	0.903	0.893	88

Note: Some of the values of S(E) are normalized to the same G_{Fe}^{3+} values [15.70 ions/100 eV for electrons and 15.45 ions/100 eV for ^{60}Co photons (ICRU, 1972)] and following Fregene.[123] Variations in the values of G_{Fe} used in the various investigations are far too small to be able to explain the observed discrepancies.

[a] Monte Carlo calculations were also carried out simulating the two geometries used by Holt et al.[113] The experimental dependence on energy was not reproduced via the Monte Carlo calculations which were consistently ~7% lower than the experimental values.

Table 7
EXPERIMENTS WHICH REPORT S(E) = 1 FOR HIGH ENERGY ELECTRONS IN LiF-TLD

Material	Geometry	Irradiation depth (cm)	Response at energy, E						Ref.
Type N	Loose powder	5 (8, 15 MV)	8 MV	15 MV	33 MV				140
Type 7	Polythene capsules	10 (33 MV)	1.005 ± 0.02	0.982 ± 0.02	0.96 ± 0.02				
	(4 mm (ID) 6 mm in length)		1.007 ± 0.015	0.99 ± 0.02	0.997 ± 0.015				
LiF-IAO	Powder, 100—150 mesh lucite capsule (4 mm (ID) 16 mm in length)	2	10 MeV	15 MeV	20 MeV	25 MeV	30 MeV		141
			1.00 ± 0.01	1.01 ± 0.03	1.02 ± 0.03	1.03 ± 0.02	1.02 ± 0.03		
Aloka-NTL 50-p	Polyethylene (10 × 10 × 1 mm, 60 mg)	'1.5	0.99 ± 0.03	1.00 ± 0.04	0.99 ± 0.02	1.00 ± 0.03	0.99 ± 0.04		110
200-Mg-S	Powder in	1—3	5.7 MeV	9.4 MeV	19.4 MeV	28.1 MeV	36.9 MeV		120
200-Ti-80T	teflon® (4 mm (ID)		1.01	1.01	1.01	1.01	1.00		
	5 mm in length)		0.92	0.93	0.94	0.94	0.95		
TLD-700	Extruded rods (1 × 6 × 1 mm)	Depth of maximum buildup	3 MeV	5 MeV	10 MeV	15 MeV	20 MeV		115
			0.945 ± 0.013	0.964 ± 0.012	0.959 ± 0.013	0.999 ± 0.013	0.994 ± 0.014		
			25 MeV	30 MeV	35 MeV	40 MeV	45 MeV		
			1.00 ± 0.013	1.012 ± 0.013	0.987 ± 0.014	1.003 ± 0.013	0.996 ± 0.013		
			6 MV (1.001 ± 0.022), 33 MV (1.001 ± 0.013), 45 MV (0.996 ± 0.013)						
TLD-700	Rods (1 mm (ID) 6 mm in length)	1.5[a]	3 MeV	6 MeV	8 MeV	10 MeV	15 MeV	20 MeV	113
			0.87	0.97	0.97	0.984	0.99	1.003	
		Height	10 MeV	20 MeV	30 MeV				
TLD-100	Loose powder	0.5 mm	0.99 ± 0.05	1.00 ± 0.03	1.00 ± 0.03				139
TLD-600	polystyrene [10 mm	1.0 mm	1.00 ± 0.06	1.00 ± 0.04	1.00 ± 0.04				
TLD-700	(ID)]	2.0 mm	0.99 ± 0.03	1.00 ± 0.05	1.00 ± 0.04				
		4.0 mm	0.99 ± 0.03	0.99 ± 0.02	1.00 ± 0.03				
TLD-100	Loose powder, thin paper sachets		15 MeV, 25 MeV, 33 MeV consistent with unity ±3%						138

TLD-100	Plaques, gelatine capsules [4.5 mm (ID) 10 mm in length]	1.5	15 MeV electrons: after a 2—4% correction using general cavity theory, $\eta(E) = 1$ for the plaques and 1.04 for the powder, consistent with $S(E) \simeq 1$, overall experimental precision ~5%	142
TLD-100	Loose powder, gelatine capsules (thickness from 1—6 mm)		10 MeV 0.991; 20 MeV 1.003; 30 MeV 0.996; 40 MeV 1.012; 50 MeV 1.004 4 MV(1.002), 32 MV(0.994), 45 MV(0.997) 3% precision throughout	118,119

Note: Rank et al. (as reported in Ehrlich[142]), Rossow and Streeter,[143] Degner et al.[144] and Pinkerton[145] also report no decrease in response within the precision of their measurements.

[a] An additional experiment was carried out at 20 MeV and varying depths of irradiation with similar results.

Pinkerton et al.[130,131] observed a significant difference in glow curve shape between ^{60}Co and 15-MeV electron irradiations with peaks 4 and 5 shifted by approximately 20°C to lower temperatures (this could arise via a preferential population of peak 4 for 15-MeV electrons). Pinkerton et al. suggested that this slight dissimilarity (?) probably arose from variations in N_2 flow rate or phosphor disposition in the planchet; however, if this were true then the shift in temperature should have occurred randomly for the two types of radiation. The glow curve shape is very often a reliable signature indicating the absence or presence of batch effects, ionization density effects, or other experimental anomalies. It is highly recommended that further studies pay careful attention to these details. Any difference in glow curve shape between the various radiations should be carefully investigated and reported.

4. Batch and material variations in η for densely ionizing radiations (low energy X-rays, heavy charged particles, etc.) are conceptually possible on microdosimetric grounds because of the well-known dependence of LiF supralinearity on ionization density and batch defect composition. On the other hand, both Suntharalingam and Cameron[116] and Eggermont et al.[132] report f(D) essentially identical for ^{60}Co photons and high energy electrons. Further studies should include measurement of f(D) for ^{60}Co and higher energy electrons so as to provide further information on possible ionization density effects.
5. Inspection of published details on beam parameters, depth of irradiation, etc. reveals no apparent correlation with decrease in response contrary to the suggestion of Liu et al.[138] that an overabundance of lower energy scattered electrons near the surface of the medium may be a contributing factor. Moreover, very few investigators have clearly observed a decreased response at lower energies.

1. Conclusions

The nature of the various experimental results (small standard deviation for each major subset of data) suggests that the discrepancies are not due to a large number of small systematic errors, but rather to a small number (perhaps even one) of large errors. Due to inadequate attention to and/or publication of relevant experimental details, it is impossible at the present time to trace the source of the reported discrepancies in high energy electron interactions with LiF-TLD. The discrepancies indicate the presence of hidden experimental variables which affect the TL response at the 10% level. The translation of the TL signal to absorbed dose in tissue or water may therefore be compromised at the 10% level. Since the cavity effect has been experimentally established in certain TLD-beam configurations,[86,109] corollary dosimetry experiments following Ogunleye and Paliwal[109] should be carried out to experimentally determine the cavity effect if the use of LiF-TLD is contemplated in high energy electron radiation dosimetry or high energy photon dosimetry. Finally, none of the existing cavity theories for high energy electrons as proposed by Burlin et al.,[103] Almond and McCray,[104] Holt et al.,[113] or Fregene[115] is reliable in both high Z and low Z media. The expression suggested by Fregene seems to be the most reliable of the expressions (at least from the point of view of describing the experimental behavior as a function of cavity thickness) in low Z media (LiF in polystyrene and aluminum).

IV. NEUTRON DOSIMETRY

Many studies have been carried out on the characteristics of the neutron relative TL response of LiF (TLD-100: 7.4% ^{6}Li, 92.6% ^{7}Li; TLD-600: 95.62% ^{6}Li, 4.38% ^{7}Li; TLD-700: 0.007% ^{6}Li, 99.993% ^{7}Li) as well as many other TL materials. The motivation for these studies comes from the dosimetric application of TL materials in various mixed-field n − γ environments (e.g., accident dosimetry, personnel dosimetry, certain types of in vivo dosimetry; 14-MeV collimated neutron beams produced by d-T generators are also finding increased use in certain radiotherapy applications). Becker,[146] Furuta and Tanaka,[147] and Griffith et al.[148] have reviewed the use of TL in personnel dosimetry, specifically, albedo techniques (which are based on the moderating and backscattering properties of the human thorax) and various efforts to increase TL fast neutron sensitivities (external hydrogenous radiators, internal hydrogen doping, etc.). As Becker[149] has pointed out it has not been possible to overcome the intrinsic problems in the application of TL to fast neutron dosimetry which include a rapid loss in sensitivity for neutron energies in excess of approximately 10 keV, strong directional dependence of response (albedo detectors), and strong influence of other geometrical factors (e.g., distance between the body surface and the detector). At the present time, fast neutron TLD should therefore be considered only as a semiquantitative supplement to more convincing fast neutron dosimetry techniques especially in situations where a large intermediate energy contribution to the total dose is expected. The main successful applications of TLD in neutron dosimetry have thus been in thermal neutron detection and in the estimation of γ-ray dose in n − γ mixed radiation fields. The following sections are oriented towards a review of the basic factors affecting the sensitivity of "bare" TLDs and their use in these latter applications.

A. Thermal Neutron Sensitivities

The factors which determine the thermal neutron sensitivity, S_i (the TL neutron sensitivity is defined as the exposure dose in ^{60}Co R required to give the same TL signal as an impinging neutron fluence of 10^{10} neutrons cm^{-2}) can be derived in the following manner. The relative TL response for the heavy charged particles liberated by the neutron interactions is given by

$$\bar{\eta}_{HCP,\gamma} = \frac{\text{kerma in TL sample due to } S_i \text{ R } (^{60}\text{C})}{\text{HCP kerma in sample due to an impinging-neutron fluence of } 10^{10} \text{ n cm}^{-2}} = S_i(\rho V)C_i/\bar{\Sigma\phi}\ V\ \bar{E}_{HCP}\ (1.6 \times 10^{-8}) \qquad (37)$$

It follows that

$$S_i = \frac{\bar{\eta}_{HCP,\gamma}\ \bar{\Sigma}\ \bar{\phi}\ V\ \bar{E}_{HCP}\ (1.6 \times 10^{-8})}{(\rho V)\ C_i} = \frac{\bar{\eta}_{HCP,\gamma}\ \bar{\Sigma}\ (\bar{\phi}/\phi)\ \phi\ E_{HCP}\ (1.6 \times 10^{-8})}{\rho C_i} \qquad (38)$$

where $\bar{\Sigma}$ is the neutron macroscopic cross section given by $\sigma N_0 \rho \omega / A$ (σ is the microscopic cross section, N_0 is Avogadro's number, A is the atomic number, and ω is the percentage isotopic content), $(\bar{\phi}/\phi)$ is the self-shielding factor, ϕ is the neutron fluence equal to 10^{10} n cm^{-2}, E_{HCP} is the kinetic energy of the HCP liberated by neutron interaction, 1.6×10^{-8} converts the HCP energy in MeV to ergs, and C_i is the conversion factor from ^{60}Co R to ergs/g. It is important to note that the geometry of the fluence-dosimeter sample only enters through the self-shielding factor. This assumes that all the kinetic energy of the HCP is absorbed in the TLD. Thermal neutron TL sensitivities are usually measured using a Maxwellian spectrum of thermal neutrons at T = 293.6 K. In this case, $\bar{\Sigma}$ is an average over the Maxwellian spectrum. Table 8 lists the experimentally observed thermal neutron sen-

Table 8
EXPERIMENTALLY MEASURED THERMAL NEUTRON SENSITIVITIES OF VARIOUS TL MATERIALS

^{60}Co equivalent R/10^{10} n cm^{-2}

TL material	Minimum sensitivity		Maximum sensitivity	Ref.
TLD-100	65		535	156
				157
TLD-600	870		2400[a]	158
				153
TLD-700	0.19		2.5	153
				159
LiF (Conrad)		165		160
^{6}LiF (Conrad)		2650		159
^{7}LiF (Conrad)		23!		159
$Li_2B_4O_7$[b]	230		670	157
				161
BeO	0.13 ± 0.08		0.45	162
				163
CaF_2:Mn[c]	0.07 ± 0.01		1.05 ± 0.08	164
				165
CaF_2:Dy[c]	0.33 ± 0.021		0.65	166
				157
$CaSO_4$:Dy	0.38		0.63	163
				167
$CaSO_4$:Tm	0.21		0.23	163
				168
Mg_2SiO_4:Tb		0.21		163
Al_2O_3:Si,Ti	0.05[d]		0.4	169

[a] Corrected for self-shielding.
[b] It is interesting to note that the value $\bar{\eta}_{HCP,\gamma}$ = 1.1 reported by Horowitz et al.[154] is equivalent to a sensitivity of approximately 400 ^{60}Co R per 10^{10} n cm^{-2}.
[c] Horowitz et al.[166] have shown that the thermal neutron sensitivity of these phosphors is directly proportional to the Mn or Dy content.
[d] Measured for a dosimeter thickness of 25 mg cm^{-2}.

sitivities of most of the common TL materials. As can be seen, the various measurements often vary by as much as an order of magnitude. This variation in sensitivity is due to many factors including self-shielding in the strongly absorbing TL materials, variations in ^{6}Li content in the highly enriched TLD-700 and ^{7}LiF dosimeters (Ayyangar et al.[150] reported ^{6}Li content varying between 0.034% and 0.087% in TLD-700), variations in Mn or Dy content in the CaF_2 and $CaSO_4$ TLDs, and batch and material variations leading to changes in η and others. In the following sections these factors will be discussed one by one.

1. Self-Shielding Factors

Equation 38 shows that the thermal neutron sensitivity is directly proportional to the self-shielding factor. For strongly absorbing TLDs (e.g., TLD-100 and TLD-600) this proportionality may be of crucial importance since the TLDs are usually calibrated using known Maxwellian thermal neutron fluences (reactor cores or moderated fast neutron fluences). In both cases (but especially the latter) care must be taken to ensure that the values of $\bar{\phi}/\phi$ are identical for the calibration field and the radiation field in which dosimetric measurements are to be carried out. This latter requirement can be difficult to fulfill since $\bar{\phi}/\phi$ can be significantly dependent on the angular distribution and the distribution in energy of the thermal neutrons. The most obvious method to avoid problems of self-shielding is to use ultrathin dosimeters (e.g., Lowe[151] used LiF-PES TLDs of 3 mg cm^{-2} thickness for which $\bar{\phi}/\phi$ equals essentially unity).

Horowitz et al.[152-155] have calculated via Monte Carlo the absorption probability (or equivalently the self-shielding factor) for various cylindrical and rectangular TLDs in iso- and anisotropic thermal neutron fluences. The self-shielding factor and the absorption probability are connected by the following relation when the neutron fluence is isotropic

$$\phi A\chi/4 = \bar{\Sigma}\,\bar{\phi}\,V \text{ or } (\bar{\phi}/\phi) = A\chi/4\bar{\Sigma}V \tag{39}$$

where ϕ is the unperturbed thermal neutron fluence in neutrons cm^{-2}, $\bar{\phi}$ is the average thermal neutron fluence within the TL sample, χ is the absorption probability, A is the surface area of the TLD, V is the volume of the TLD and the factor 1/4 arises from the fact that for an isotropic fluence of ϕ neutrons cm^{-2}, the number of neutrons impinging on a surface of unit area is given by

$$\int_0^{2\pi}\int_0^{\pi/2} (\phi/4\pi)\cos\theta\sin\theta\, d\theta\, d\phi = \phi/4 \tag{40}$$

For ^6Li, $\bar{\sigma}$ is the $^6Li(n,\alpha)\tau$ microscopic cross section averaged over the Maxwellian distribution. Current best value at E_n = 25.3 meV[170] is 942.4 × 10^{-24} cm^2. $\bar{\sigma} = \sigma(25.3\text{ meV})/0.886$. The factor 1/0.886 arises from the fact that the average neutron velocity is given by $(2/\pi^{1/2})v_t$ where v_t is the neutron velocity at E_n = 25.3 meV.

The Monte Carlo calculations are carried out in the following manner. The neutron energy is sampled over the Maxwellian distribution using the inverse equation method. The neutron flux is energy distributed according to

$$\frac{\phi(E)dE}{\phi} = \exp^{-(E/E_T)}\frac{E}{E_T}\frac{dE}{E_T} \tag{41}$$

where E_T = kT = 0.0253 eV and E is the energy of the neutron. Letting $x = E/E_T$ we have

$$f(x)dx = x\exp^{-x}dx \tag{42}$$

so that

$$\zeta = \int_0^E f(X)dx = \int_0^E x\,e^{-x}\,dx = -(x+1)e^{-x}\Big|_0^E \tag{43}$$

or

$$\zeta = -(1+E)\,e^{-E} + 1 \tag{44}$$

Now since $1 - \zeta$ is the same random number as ζ, the Monte Carlo sampling is carried out by solving the transcendental equation

$$\zeta = e^{-E}(1+E) \tag{45}$$

where ζ is a randomly generated number between 0 and 1. Once E is known, Σ can be calculated easily from the equation $\Sigma = \Sigma_T (E_T/E)^{1/2}$ where Σ_T is the macroscopic cross section for neutrons of energy E_T. Both E_T and Σ_T are input data. Now since

$$\phi(r) = \phi(r - \ell)e^{-\ell\Sigma} \tag{46}$$

$$\chi = \bar{\Sigma} V \phi_0 \int \frac{e^{-\ell\Sigma} f(\Omega) dV}{V} \tag{47}$$

where $f(\Omega)\phi_0$ is the impinging flux, $f(\Omega)$ being the angular distribution of the flux. Thus

$$\chi = \bar{\Sigma} V \bar{\phi} = \bar{\Sigma} V \phi_0 \int \frac{e^{-\ell\Sigma} f(\Omega) dV}{V} \tag{48}$$

or

$$\bar{\phi}/\phi_0 = \int \frac{e^{-\ell\Sigma} f(\Omega) dV}{V} \tag{49}$$

The Monte Carlo program randomly generates a point P(x,y,z) in the TL volume, which is the point of absorption of an incoming neutron. The direction cosines of the path of the neutron (w1,w2,w3) are then randomly generated and their negative values are taken in order to calculate the neutron path length. In the case of nonisotropicity the path direction can be weighted by the angular distribution of the impinging thermal neutrons. A line with direction cosines (-w1,-w2,-w3) is drawn from P and the intersection points with all the planes in the geometry of the crystal are found. The distance from these points to P is calculated and the minimal length, L, is taken as the length of the neutron path in the crystal. The factor $e^{-\ell\Sigma}$ is then calculated using a value of the macroscopic cross section evaluated from Equation 45 and the process is repeated from the beginning. The values $e^{-\ell\Sigma}$ are accumulated and eventually, when the variance is below a required value (e.g., 2%) this sum is divided by the number of generated neutrons to obtain the self-shielding factor. Tables 9 and 10 give calculated self-shielding factors for TLD-100 and TLD-600 in various cylindrical and rectangular geometries.

The importance of accurate calculation of the self-shielding factor can best be illustrated by illustrative examples:

1. Ayyangar et al.[163] used a very thin sample (approximately 0.15 mm) of TLD-100 and TLD-600 in an attempt to obtain experimental results free of self-shielding. Monte Carlo calculations for a cylinder of radius 5 mm and height 0.15 mm lead to self-shielding factors of 0.72 and 0.48 for TLD-100 and TLD-600, respectively. This would indicate that the thermal neutron sensitivities reported by Ayyangar et al. are approximately 40 and 110% too low for TLD-100 and TLD-600, respectively.
2. Tanaka and Furuta[171] described the use of ^{6}LiF TLDs as thermal neutron detectors of high sensitivity. Their estimate of $\bar{\phi}/\phi = 0.233$ contributed to a discrepancy between theory and experiment which was corrected by Horowitz and Dubi[172] who calculated 0.156 for the TLD geometry employed by Tanaka and Furuta.
3. As shown in Table 8, the thermal neutron sensitivity of TLD-100 as reported in the literature has varied between 65 to 535 ^{60}Co R/10^{10} n cm^{-2} and for TLD-600 has varied between 870 and 2400 ^{60}Co R/10^{10} n cm^{-2}. Horowitz et al.[153] have shown how self-shielding is a major contributor to this wide variation in the reported results.

Table 9
SELF-SHIELDING FACTORS IN RECTANGULAR GEOMETRIES

Dimensions of crystal (cm)	Type	Density (g cm^{-3})	Self-shielding factor
0.3175 × 0.175 × 0.0889	TLD-600	2.15	0.22 (± 0.99%)
0.3175 × 0.3175 × 0.0889	TLD-100	2.15	0.79 (± 0.66%)
0.3175 × 0.3175 × 0.0254	TLD-600	2.15	0.43 (± 0.95%)
0.3175 × 0.3175 × 0.0254	TLD-100	2.15	0.90 (± 0.37%)
0.1 × 0.01 × 0.6	TLD-600	2.15	0.26 (± 0.98%)
0.1 × 0.1 × 0.6	TLD-100	2.15	0.83 (± 0.52%)
0.165 × 0.165 × 0.83	TLD-600	2.15	0.17 (± 0.98%)
0.165 × 0.165 × 0.83	TLD-100	2.15	0.75 (± 0.78%)
0.5 × 0.5 × 0.1	TLD-600	2.65	0.71 (± 0.96%)
0.5 × 0.5 × 0.1	TLD-100	2.65	0.15 (± 0.97%)
1.0 × 1.0 × 0.15	TLD-600	2.65	0.96 (± 0.99%)
1.0 × 1.0 × 0.15	TLD-100	2.65	0.60 (± 0.97%)
1.0 × 1.0 × 0.1	TLD-600	2.65	0.13 (± 0.99%)
1.0 × 1.0 × 0.1	TLD-100	2.65	0.66 (± 0.82%)

Table 10
SELF-SHIELDING FACTORS IN CYLINDRICAL GEOMETRIES

Dimensions of crystal		Self-shielding factor[a]	
Radius (cm)	Height (ρ = 1.25 g cm^{-3})	TLD-100	TLD-600
0.1	0.1	0.85	0.33
0.1	0.5	0.73	0.056
0.1	1.0	0.86	0.20
0.1	1.5	0.92	0.20
0.2	0.1	0.82	0.27
0.2	0.5	0.70	0.14
0.2	1.0	0.71	0.12
0.2	1.5	0.72	0.11
0.5	0.1	0.77	0.23
0.5	0.5	0.57	0.080
0.5	1.0	0.50	0.060
0.5	1.5	0.48	0.054
1.0	0.1	0.75	0.21
1.0	0.5	0.49	0.059
1.0	1.0	0.39	0.039
1.0	1.5	0.21	0.034

[a] Error in the self-shielding factor is less than 1% (1 S.D.) for all cases.

4. The assumption of isotropicity in thermal neutron radiation fields is not always valid.[154] Measurements were carried out at two points (in water and in air) 11 cm distant from a 3 Ci Am-Be source situated in a 1.5 × 1.5 × 1.5 m water tank. The angular distribution of the thermal neutrons was measured using the Cd difference technique with indium foils. Cadmium-backed indium foils were rotated and integral measure-

Table 11
THERMAL NEUTRON ABSORPTION PARAMETERS OF TLD-100, -600, AND -700

Reaction	2200 m sec^{-1} cross section (10^{-24} cm^2)	Macroscopic cross section Σ (cm^{-1})		
		TLD-100	TLD-600	TLD-700
$^6Li(n,\alpha)\tau$[a]	942.4[b]	4.33	57.14	4.03×10^{-3}
$^6Li(n,\gamma)$	40×10^{-3}	1.8×10^{-4}	2.43×10^{-3}	1.71×10^{-7}
$^7Li(n,\gamma)$	36×10^{-3}	2.1×10^{-3}	1.01×10^{-4}	2.2×10^{-3}
$^{19}F(n,\gamma)$	9×10^{-3}	5.5×10^{-4}	5.72×10^{-4}	5.5×10^{-4}

[a] The thermal neutron sensitivity is not influenced by the kerma liberated by the tritium decay because of the long half-life (12.5 years) and low beta-ray energies (E_{max} = 18 keV). The tritium beta decay does, however, result in a time-dependent build-up of the zero dose reading after neutron preexposure.[176]

[b] Vlasov et al.[170] quote a confidence level of 0.5%. Below 10 keV, σ follows the l/v law to within 1%.

ments of the foil activation were carried out every 20°. Following Beckurtz and Wirtz[173] the Cd correction factor was determined in order to subtract the contribution to the foil activation due to epithermal neutrons. Each angular distribution point is an integral measurement over 180°, but again, following Beckurtz and Wirtz, it is possible to expand the vector flux F(r,Ω) as a series of Legendre polynomials in cosθ where θ is the angle to the field axis. Assuming that F(r) does not change appreciably along the foil surface, Horowitz et al. measured:

$$F(\Omega) = P_0\cos\theta + (0.114\pm0.008)P_1(\cos\theta) + (0.126\pm0.029)P_2(\cos\theta) + (0.062\pm0.02)P_3(\cos\theta) + (0.112\pm0.037)P_4(\cos\theta) \quad (50)$$

The resulting self-shielding factor was 0.164 instead of 0.17 for f(Ω) = 1. In a mixed thermal neutron-gamma ray radiation field essentially free of epithermal or fast neutrons, where the gamma dose can be fairly accurately estimated using TLD-700, this 3.5% decrease in the self-shielding factor is not significant. However, in a mixed radiation environment not limited to thermal neutrons where the TLD-700 responds to gamma rays and fast neutrons and the thermal neutron induced signal in TLD-600 may be far greater than the fast neutron or gamma ray-induced TL signals, Horowitz et al.[154] showed that the 3.5% change in $\bar{\phi}/\phi$ due to the anisotropy can result in a 70% error in the gamma dose if the anisotropy is not taken into account.

2. *The Dependence of S_i on the Relative TL Response, $\eta_{HCP,\gamma}$*

The thermal neutron absorption parameters of the three isotopically different LiF dosimeters (TLD-100, TLD-600, TLD-700) are shown in Table 11. The very large thermal neutron sensitivity of TLD-100 and TLD-600 is thus due to the large cross section and large positive Q value of the $^6Li(n,\alpha)\tau$ reaction which yields a 2.06 MeV alpha particle and a 2.72 MeV triton for every absorbed neutron. The macroscopic cross section for the $^6Li(n,\alpha)\tau$ reaction is at least 10^4 times greater than any of the competing (n,γ) reactions. In TLD-700, the total macroscopic cross section is less than 7×10^{-3} cm^{-1}, i.e., 3 to 4 orders of magnitude less than the total macroscopic cross section in TLD-100 or TLD-600, which explains why the TLD-700 thermal neutron sensitivity is negligible compared to the TLD-100 or TLD-600 thermal neutron sensitivity (cf. Table 8). The contribution to the macroscopic cross section from various impurity concentrations typical to LiF-TLD can also be expected to be negligible.[153]

These considerations form the basis for the paired TLD-600, TLD-700 technique for the separation of thermal neutron dose and gamma dose in mixed thermal neutron-gamma ray radiation fields. The TLD-700 is usually assumed to measure the gamma dose only and subtraction from the TLD-600 or TLD-100 signal gives a measure of the thermal neutron-absorbed dose.

The thermal neutron sensitivity, S_i, is proportional to the product $\bar{\eta}_{HCP,\gamma}E_{HCP}$ which for TLD-100 and TLD-600 can be written

$$\bar{\eta}_{HCP,\gamma}\,\bar{E}_{HCP} = \eta_{\tau\gamma}E_{\tau} + \eta_{\alpha\gamma}E_{\alpha} = 2.72\,\eta_{\tau\gamma} + 2.06\,\eta_{\alpha\gamma} \tag{51}$$

No direct experimental data exist for 2.72-MeV tritons but $\eta_{\alpha\gamma}$ for 2-MeV alpha particles can be expected to lie between 0.1 and 0.5 (Section V). The nonuniversality in $\eta_{HCP,\gamma}$ due to batch variations in impurity and/or defect compositions is the other major factor (aside from the variations in the self-shielding factor) responsible for the large variations reported in the thermal neutron sensitivities of TLD-100, TLD-600, and the other TL materials. The nonuniversality in $\eta_{HCP,\gamma}$ was established beyond reasonable doubt by a series of definitive experiments carried out by Horowitz et al.[125-127]. In these experiments 13.8- and 81.0 meV monoenergetic neutrons from a Kandi-II diffractometer in parallel beam geometry were used to irradiate TLD-100, TLD-600, and TLD-800 ($Li_2B_4O_7$:Mn,Si — Harshaw). The meV neutron irradiations (in comparison to direct HCP irradiation) are especially useful because they deliver particulate high ionization density radiation throughout the volume of the dosimeter and are therefore not sensitive to various surface effects (contamination, degradation, etc.) which might be present for direct surface HCP irradiation. Furthermore, Horowitz et al. used bulb dosimeters for the meV neutron irradiations with an inviolate inert gas atmosphere within the bulb. In a parallel beam configuration estimation of the self-shielding factor does not require Monte Carlo calculations since

$$\bar{\phi}/\phi = \frac{1 - \exp(-\Sigma d)}{\Sigma d} \tag{52}$$

and ϕ was measured using thin gold foils ($\bar{\phi}/\phi = 1$) and was monitored using a calibrated fission monitor. The glow curve was integrated to approximately 300°C with the result

$$\begin{aligned} \bar{\eta}_{(\tau+\alpha),\gamma} &= 0.340 \pm 0.007 \text{ (TLD-100)} \\ \bar{\eta}_{(\tau+\alpha),\gamma} &= 0.420 \pm 0.007 \text{ (TLD-600)} \end{aligned} \tag{53}$$

i.e., a difference in η of approximately 20% where the values of $\bar{\eta}_{\tau+\alpha,\gamma}$ are separated by more than 10 standard deviations. Since no physical reason exists which could correlate η with isotopic composition of Li in LiF, Horowitz et al. suggested the likelihood that ppm variations in the impurity composition or that variations in Ti concentration were responsible for the variations in η. DC spectrochemical analyses were carried out on the batches of TLD-100, 600, and 700. In all cases the Mg,Ti concentration was found to be 540 and 7 ppm/w of Mg and Ti, respectively, where in the latter case (because of the very low concentration) the accuracy of the analysis was estimated to be 20%. Other elemental concentrations in which no significant variations were observed were Fe(20),Si(100), Cu(125), V (25), Ca (40), and B (25). Consistently significant variations in Ba, Cr, and Al concentrations were discovered between TLD-100, TLD-600, and TLD-700. These are summarized

Table 12
CONCENTRATION OF ELEMENTS (PPM BY WEIGHT OF Li BASE)

Batch no.	TLD type	Ba	Cr	Al
1	TLD-100	30	25	80
2	TLD-600	5	100	40
3	TLD-700	$<1^a$	50	50

[a] Lower limit of detection.

Table 13
THERMAL NEUTRON ABSORPTION CROSS SECTIONS FOR $Li_2B_4O_7$

	Cross section	
Reaction	Microscopic (cm^2) E_n = 0.0253 eV	Macroscopic (cm^{-1})
$^6Li(n,\alpha)\tau$	942×10^{-24}	847×10^{-4}
$^7Li(n,\gamma)$	0.037×10^{-24}	0.4×10^{-4}
$^{10}B(n,\alpha)^7Li$	$3{,}837 \times 10^{-24}$	$18{,}430 \times 10^{-4}$
$^{10}B(n,\gamma)$	0.5×10^{-24}	2.4×10^{-4}
$^{11}B(n,\gamma)$	0.005×10^{-24}	0.11×10^{-4}
$^{16}O(n,\gamma)$	0.0002×10^{-24}	0.008×10^{-4}
$^{nat}Si(n,\gamma)$	0.16×10^{-24}	0.0085×10^{-4}
$^{55}Mn(n,\gamma)$	13.3×10^{-24}	0.124×10^{-4}

in Table 12. It is worthwhile mentioning that both Ba and Al are well known as potential coactivators of the LiF TL process.[174]

Other measurements of $\bar{\eta}_{\tau+\alpha,\gamma}$ in LiF[14,153,175] have yielded values as low as 0.13 and as high as 0.41. The value of 0.13 obtained by Horowitz et al.[153] were on the identical batches in which the later studies yielded $\eta_{\tau+\alpha\gamma}$ = 0.42, however, the later studies were carried out using inert N_2 gas annealing compared to air annealing in the earlier studies. As discussed in Volume II, Chapter 3, experiments in the Radiation Physics Laboratory of the Ben Gurion University of the Negev have also revealed that the high temperature annealing environment drastically affects $f_\gamma(D)$. In the framework of TST, changes in $f_\gamma(D)$ are correlated with changes in $\eta_{HCP,\gamma}$. Horowitz et al.[125-127] have emphasized that the later studies (Equation 53) were carried out with extreme care to ensure identical measuring systems and experimental conditions.

Similar experiments were carried out by Horowitz et al. for two batches of $Li_2B_4O_7$:Mn (TLD-800, Harshaw). The elemental cross section parameters of TLD-800 are shown in Table 13. It is clear that the only kerma producing reactions of significance are $^{10}B(n,\alpha)^7Li$ and $^6Li(n,\alpha)\tau$. The former reaction yields a 1.015 MeV ^7Li ion and a 1.775-MeV alpha particle with 6.1% probability and a 0.84-MeV ^7Li ion and 1.47-MeV alpha particle with 93.9% probability. The average charged particle energy liberated per absorbed thermal neutron in $Li_2B_4O_7$ is thus 2.73 MeV. The relative TL response was again found to be batch dependent

$$\eta^1_{\tau+\alpha+{}^7Li} = 0.775 \pm 0.007 \text{ (1 S.D.)}$$
$$\eta^2_{\tau+\alpha+{}^7Li} = 1.105 \pm 0.017 \text{ (1.S.D.)} \qquad (54)$$

Since only the relative values of η are of significance, the quoted errors are statistical only. The total statistical and systematic error was approximately 7%. Most previous measurements of $\eta_{HCP,\gamma}$ in $Li_2B_4O_7$ have been via HCP surface irradiation and are, therefore, suspect because of the well-known hygroscopicity of this material. Lakshmanan and Ayyangar[177] used very thin TLD samples and found η for 4.78 MeV alpha particles to be 0.37. In this case the irradiations were throughout the entire volume, but the grain size employed was 1 to 8 μm and Wallace et al. reported a strong grain size effect in $Li_2B_4O_7$. Thermal neutron sensitivity measurements (cf. Table 8) have not yielded values of η because of unknown self-shielding factors. Spectrochemical analysis of the two batches of TLD-800 used by Horowitz et al. revealed significant variations in the Al and Si impurity concentration level. In the case of Al, batch 1 and batch 2 contained 100 and 450 ppm/w, respectively, and in the case of Si, 300 and 1800 ppm/w, respectively. The results for LiF and $Li_2B_4O_7$ establish that at the ppm level elemental concentrations of impurities vary greatly from batch to batch even if purchased from the same supplier.

These results for $\eta_{HCP,\gamma}$ unequivocally establish that the ionization density dependence of the HCP relative TL response in LiF and $Li_2B_4O_7$ is batch and material dependent at the high ionization densities produced by low energy (~1 MeV/amu) heavy charged particles. Horowitz et al. concluded, therefore, that the batch variations in the relative TL response are due to variations in the batch impurity composition at the ppm level, or to variations in the mode of incorporation of these impurities or to other unknown chemical or physical characteristics. The nonuniversality in $\eta_{HCP,\gamma}$ leads directly to the conclusion that fast neutron TL sensitivities are similarly nonuniversal as a function of neutron energy. This conclusion is unassailable since the fast neutron response (as well as the thermal neutron response) arises mainly from recoiling high ionization density particles. It is more than likely that the nonuniversality in $\eta_{HCP,\gamma}$ extends to all TL materials and to a certain extent to other regions of ionization density. Recall that the results for $\eta_{x,\gamma}$ in the literature also indicate the possible role of batch effects in relative TL response ionization density studies. These results are further substantiated on theoretical grounds in Volume II, Chapter 3, where we have shown via a modified track structure theory analysis that $\eta_{HCP,\gamma}$ is strongly correlated with $f_\gamma(D)$. Batch and material variations in $f_\gamma(D)$ are very well documented and established beyond doubt (Chapter 1, Section I.A). An extreme example illustrating the dependence of η on $f_\gamma(D)$ can be seen in the work of Goldstein et al.[178] who obtained $\eta_{HCP,\gamma} = 1$ for a 450°C peak in LiF (Isomet 1) which saturated at 10^6 Gy instead of the usual γ saturation at 10^3 to 10^4 Gy which leads to the values of $\eta_{HCP,\gamma}$ between 0.1 and 0.4 observed for most commercial TLD materials.

Because of the important practical and theoretical implications, Horowitz et al.[127] investigated further the possibility that the nonuniversality in $\eta_{HCP,\gamma}$ arises from batch variations in the relative intensity of the high temperature TL in LiF and $Li_2B_4O_7$. If this were the case, the nonuniversality in η could have been hypothesized as due to variations of Ba, Cr, Al or possibly Ti in LiF (Al, Si in $Li_2B_4O_7$) affecting the TL efficiency of peaks 6 to 8 but leaving unaffected the TL efficiency of peaks 4 and 5. This would then imply a universal value of η^{4+5} where η^{4+5} is the partial relative TL response corresponding to peaks 4+5. A further implication of the possible universal behavior of η^{4+5} would then have been that measurement of the TL signal corresponding to peaks 4+5 via the peak-height method or via a relatively low temperature cut-off would have hopefully been free of the nonuniversal contaminant behavior of peaks 6 to 8.

In $Li_2B_4O_7$ (TLD-800, Harshaw) no changes in the glow curve structure were observed (Figure 24) either as a function of batch or type of radiation so that the two batch values of $\eta_{HCP,\gamma}$ (Equation 54) remain unchanged. As usual the results for LiF were considerably more complex. No significant glow curve variations were observed between dosimeters of the same batch, however, $\epsilon_{T>}$, (the ratio of the TL efficiency of peaks 6 to 8 divided by

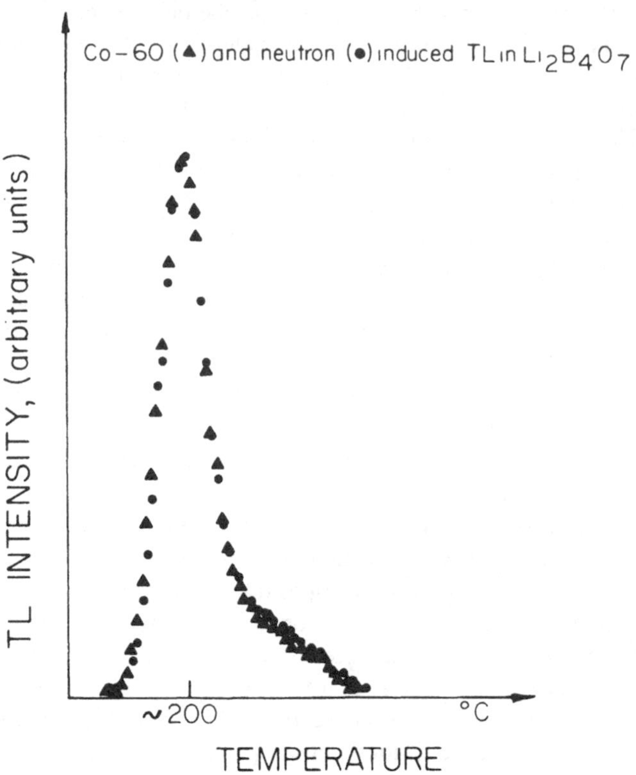

FIGURE 24. ^{60}Co and meV neutron-induced TL in $Li_2B_4O_7$:Mn,Si (TLD-800). The TL glow curves are indistinguishable and no batch effect was observed in $\epsilon_{T>}$. (Adapted from Horowitz, Y. S., Kalef-Ezra, J., Moscovitch, M., and Pinto, H., *Nucl. Instrum. Methods*, 172, 479, 1980.)

the total TL intensity of peaks 4 to 8) was found to be both batch- and radiation-type dependent. The results are summarized in Table 14 and illustrated in Figure 25. The minimum value of $\epsilon_{T>}$ recorded was 0.035 ± 0.01 following ^{60}Co gamma irradiation of TLD-100 and the maximum value was 0.355 ± 0.02 following alpha particle irradiation of TLD-700. As expected, the values of $\epsilon_{T>}$ obtained for meV neutron irradiation and alpha irradiation were almost identical because of the very similar ionization densities of the two radiations. Separation of peaks 6 to 8 from peak 5 was carried out using an analytic peak shape fitting procedure following Podgorsak et al.[179] illustrated in Figure 26

$$\frac{I(T - T_m)}{I_m} = \exp\left\{1 + \frac{(T - T_m)y}{T_m} - \exp\frac{(T - T_m)y}{T_m}\right\} \tag{55}$$

where I is the glow peak intensity at temperature T; I_m is the maximum glow peak intensity at temperature T_m; and $y = E/kT_m$ where E is the trap activation energy and k is the Boltzmann constant. Equation 55 is accurate for $y >> 1$. The batch ratios of the partial relative TL response, η^{4+5}, are quite insensitive to the specific values of y used in the separation procedure. Combining the values of $\epsilon_{T>}$ with the values for η^{total} yields the values for η^{4+5} and η^{6-8} displayed in Table 15. The batch variations in η^{4+5} for thermal neutrons are slightly reduced but still separated by over 5 S.D., and a proportionally greater increase

Table 14
RELATIVE HIGH TEMPERATURE TL INTENSITY IN LiF, $\epsilon_{T>}$

	Radiation type		
TLD type	**Gamma, ^{60}Co**	**13.54-meV neutrons**	**3.8-MeV alpha particles**
Batch 1			
TLD-100	0.035 ± 0.01	0.206 ± 0.02	0.205 ± 0.02
Batch 2			
TLD-600	0.102 ± 0.015	0.288 ± 0.02	0.316 ± 0.02
Batch 3			
TLD-700	0.123 ± 0.015	—	0.355 ± 0.02

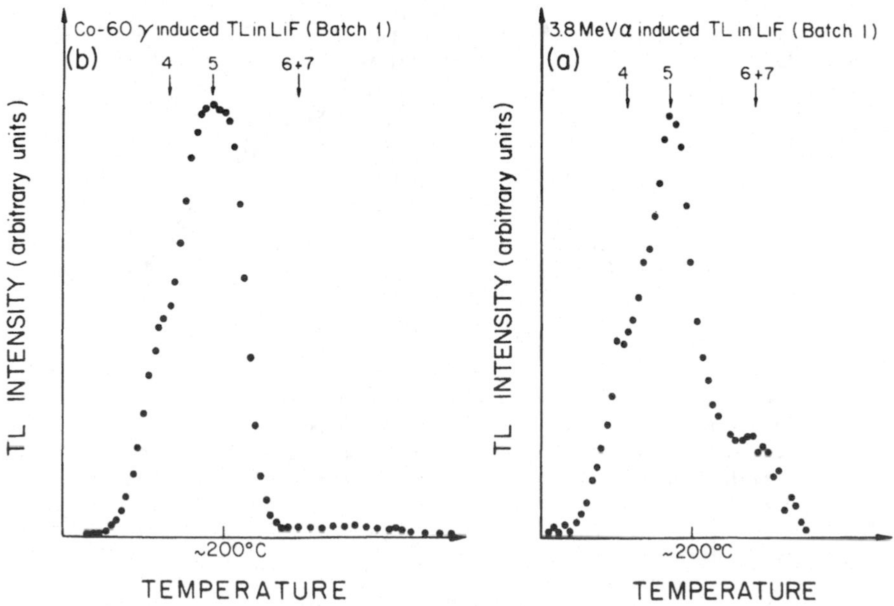

FIGURE 25. (a) 3.8 MeV alpha-induced TL in LiF(TLD-100). (b) ^{60}Co gamma ray-induced TL in LiF (TLD-100). Note the very weak population of peaks 6 and 7. (Adapted from Horowitz, Y. S., Kalef-Ezra, J., Moscovitch, M., and Pinto, H., *Nucl. Instrum. Methods,* 172, 479, 1980.)

in the batch variations of η^{6-8} is also observed. Experimental results for 3.8-MeV alpha particles reveal even greater batch variations (~50%) between TLD-100 and TLD-700.

a. Conclusions

The relative TL response, $\eta_{HCP,\gamma}$, is batch dependent due to variations in impurity composition and/or defect structure variations. The nonuniversality in $\eta_{HCP,\gamma}$ is not due to batch-dependent variations in high temperature TL in either LiF or $Li_2B_4O_7$. In Volume II, Chapter 3, $\eta_{HCP,\gamma}$ is shown via modified track structure theory to be the convolution of $f_\gamma(D)$ with $D_{HCP}(r)$ so that the nonuniversality in $\eta_{HCP,\gamma}$ can be interpreted to arise from the nonuniversality of $f_\gamma(D)$ which has been extremely well documented. The nonuniversality in $\eta_{HCP,\gamma}$ results in nonuniversality of both thermal and fast neutron sensitivities since the fast neutron response as well as the thermal neutron response arise from recoiling heavy charged particles

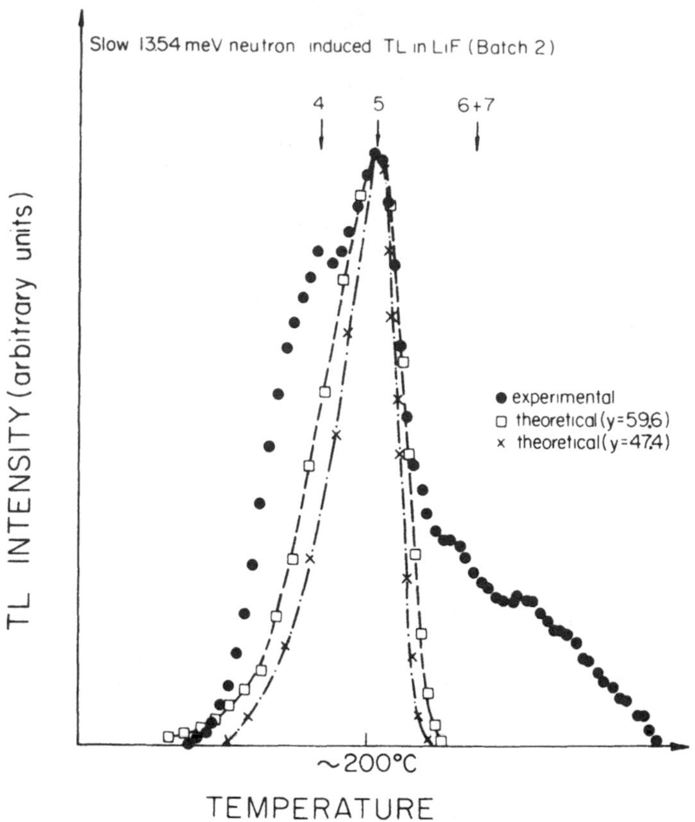

FIGURE 26. Slow 13.54 meV neutron-induced TL in TLD-600. The solid lines indicate two "theoretical" fits to the shape of peak 5 with maximum reasonable variation in the parameter y. (Adapted from Horowitz, Y. S., Kalef-Ezra, J., Moscovitch, M., and Pinto, H., *Nucl. Instrum. Methods*, 172, 479, 1980.)

Table 15
RELATIVE PARTIAL TL RESPONSE IN LiF

	Neutrons		Alpha particles	
	η^{4+5}	η^{6-8}	η^{4+5}	η^{6-8}
Batch 1				
TLD-100	0.288 ± 0.008	2.06 ± 0.3	0.140 ± 0.025	0.996 ± 0.3
Batch 2				
TLD-700	0.333 ± 0.008	1.19 ± 0.2	0.160 ± 0.01	0.65 ± 0.1
Batch 3				
TLD-700	—	—	0.213 ± 0.015	0.836 ± 0.1

induced by neutron interactions. There are, however, exceptions to this statement. For example, the thermal neutron response of Al_2O_3(Si,Ti) arises in its entirety from the beta decay of ^{28}Al so that the thermal neutron sensitivity of this material will not be subject to variations in $\eta_{HCP,\gamma}$.

Table 16
THERMAL NEUTRON-INDUCED KERMA IN TLD-700

Reaction	Charged particle species	Kerma (MeV/absorbed n_{th})	(Kerma)$\bar{\Sigma}$ (MeV cm^{-1})
$^6Li(n,\alpha)\tau$	α	2.02	8.14×10^{-3}
	τ	2.76	11.12×10^{-3}
$^8Li \xrightarrow{\beta} {}^8Be$	e^-	~0.6	~1.3×10^{-3}
$^8Be \rightarrow 2\alpha$	α	~3	~7×10^{-3}
$^{20}F \xrightarrow{\beta} {}^{20}Ne$	e^-	~0.7	~0.4×10^{-3}

3. Thermal Neutron Sensitivity of TLD-700

Attix[180] and others have pointed out that the thermal neutron-induced TL signal in TLD-700 is not necessarily negligible in mixed $n_{th} - \gamma$ radiation fields. Improved accuracy of separation requires, therefore, separate measurement of the thermal neutron sensitivity of TLD-700 which has been reported to vary between 0.19[153] to 2.5 ^{60}Co R/10^{10} n cm^{-2} for TLD-700[159] and as high as 23 ^{60}Co R/10^{10} n cm^{-2} for ^{7}LiF (Conrad).[159] Becker[146] has pointed out that negligible thermal neutron absorption arises from typical values of ppm trace impurities such as Al (20), Ca (6), Mg (300), Si (40), Ti (10), Dy (1.5×10^{-3}), Eu (7×10^{-5}), and Mn (1.5×10^{-2}) but Ayyangar et al.[150] reported variations of ^{6}Li content in TLD-700 (Harshaw) varying from 0.034 to 0.087%. Horowitz[181] has pointed out that the widespread results can also be partly due to variations in sample size and batch origin. The various components of kerma induced in TLD-700 per absorbed thermal neutron are shown in Table 16. The last column of Table 16, the product of macroscopic cross section and kerma, illustrates the relative contribution to the total TL signal. The total beta-induced kerma is approximately 6% of the (α,τ) induced kerma. Since the electron-induced kerma is expected to be approximately five times more efficient than the (α,τ)-induced kerma in producing TL, it follows that approximately 30% of the neutron-induced TL signal in TLD-700 arises from electrons in the beta decay of ^{8}Li and ^{20}F. The effect of sample size arises from the fact that the maximum electron energies in these beta decays are 14 and 5.4 MeV, respectively. These energies correspond to ranges of approximately 1 and 2.5 cm, respectively, in LiF ($\rho = 2.64$ g cm^{-3}). It follows that the thermal neutron sensitivity will increase with increasing sample size due to increasing electron absorption up to very large samples indeed. Similar behavior has been observed for Al_2O_3:Si,Ti where the thermal neutron sensitivity arises entirely from the ^{28}Al beta decay (E_{max} = 2.87 MeV). Mehta and Sengupta[169] measured S_i = 0.05 ^{60}Co R per 10^{10} n cm^{-2} and 0.4 ^{60}Co R per 10^{10} n cm^{-2} for two samples, 25 mg cm^{-2} (50 mg spread over 2 cm^2) and a far thicker sample (100 mg in a 10 × 2 mm diameter capsule). In the latter sample all the beta-ray energy is absorbed in the TL sample, whereas in the former only approximately 10% of the beta-ray energy is absorbed. The effect of batch origin arises from the dependence of $\eta_{HCP,\gamma}$ on ionization density as previously discussed. Aside from the 2.06-MeV alpha particle and 2.72-MeV triton produced in the ^{6}Li capture reaction, the breakup of ^{8}Be produces a broad alpha particle spectrum with a most probable energy of 1.6 MeV.

The maximum thermal neutron sensitivity for TLD-700 may be calculated (assuming charged particle total absorption) as the sum over the various charged particle species of the expression $\phi\, \bar{\Sigma}\, \bar{E}\, (1.6 \times 10^{-6})\, \eta_{HCP,\gamma}/\rho C$. Assuming $\eta^{max}_{HCP,\gamma} = 0.4$ for the combined alpha and triton kerma and $\eta_{e,\gamma} = 1$ for the electron kerma results in a maximum thermal neutron sensitivity for TLD-700 of 1.6 ^{60}Co R per 10^{10} n cm^{-2} (assuming 0.007% ^{6}Li). In conclusion the measured thermal neutron sensitivities of TLD-700 cannot be expected to be even roughly identical and individual measurements are of significance only for the specific dosimeter geometry and batch under investigation. The practice of comparing thermal neutron sensitivities of different batch origin and different dosimeter geometries is of questionable scientific or practical significance.

4. Thermal Neutron Sensitivity of Other TL Materials

The thermal neutron response of both CaF_2:Mn and CaF_2:Dy has been experimentally shown[166] to be due to both a prompt gamma dose (~20%) arising from thermal neutron capture in Ca,F and either Mn or Dy and an electron dose (~80%) arising from the beta decay of ^{56}Mn ($\tau_{1/2}$ = 139.2 min, E_{max} = 1.3 MeV) and ^{165}Dy ($\tau_{1/2}$ = 154.56 min, E_{max} = 2.85 MeV). The relative importance of the two doses is sample size dependent; similarly, the sensitivity to thermal neutrons is directly dependent on the percentage of Mn and Dy content and is also dosimeter size dependent because of the high energy electrons liberated in the prompt gamma and delayed beta decay. The prompt gamma dose arises from the ${}^{A}_{Z}X(n,\gamma){}^{A+1}_{Z}X$ capture reaction (X = Ca, F, Dy, or Mn). The spectrum of gamma rays emitted in the various decays is extremely complex but has been tabulated in detail.[182] In addition, the results of calculations are available[183] which give the gamma fluence in photons cm^{-2} required to produce 1 R as a function of photon energy. The maximum TL sensitivity due to prompt gamma rays arising from each elemental constituent (j) can then be estimated in the following manner.

The source strength, q_j, of the prompt gamma rays is given by

$$q_j = \phi_n \, \Sigma_j \, \overline{N}_\gamma \text{ (gamma cm}^{-3}\text{)} \tag{56}$$

where $\overline{N}_\gamma$ is the average number of gamma rays per radiative capture in the j-th element. The resulting photon fluence is not necessarily uniform throughout the TL volume; however, the average photon fluence $\bar{\phi}_\gamma$ can be estimated by

$$\bar{\phi}_\gamma = \iiint_0^{R_{max}} \frac{(q_{ds})}{4\pi} \frac{d\Omega dV}{V} \tag{57}$$

$$= q \int\int^{R_{max}} \frac{d\Omega dV}{4\pi V} \tag{58}$$

$$= q < R > \tag{59}$$

where < R > is the average path length for isotropically emitted radiation homogeneously created within the TLD volume.

For typical TLD dimensions (e.g., 0.165 × 0.165 × 0.83 cm) Horowitz et al.[153] calculate < R > = 1.05 mm via Monte Carlo. The average distance of 1.05 mm corresponds to a range of 0.9 MeV electrons in CaF_2 (ρ = 3.05 g cm^{-3}). Since a large fraction of the prompt gamma rays has energy greater than even twice this energy this electron irradiation cannot fulfill charged particle equilibrium requirements so that it gives an estimate only of the maximum prompt gamma-induced dose. The average photon fluence $\phi_n \Sigma_j \overline{N}_\gamma < R >$ gives rise to

$$\phi_n \, \Sigma_j \, \overline{N}_\gamma < R > \frac{\Sigma I_i / X_i}{\Sigma I_i} \text{ (Roentgen)} \tag{60}$$

Table 17
MAXIMUM PROMPT GAMMA CONTRIBUTION TO THE TL SENSITIVITY[a]

	^{60}Co R per 10^{10} n cm^{-2}				
	Ca(n,γ)	F(n,γ)	Mn(n,γ)	Dy(n,γ)	Total
CaF_2:Mn (2%)	0.042	0.0014	0.028	—	0.07
CaF_2:Dy (0.35%)	0.042	0.0014	—	0.124	0.168

[a] Calculated for a Maxwellian distribution of thermal neutrons at T = 293.6 K for a TL sample with < R > = 1.05 mm.

Table 18
AVERAGE ELECTRON ENERGY IN ^{56}Mn AND ^{165}Dy BETA DECAYS

	E_{max} (MeV)	$\bar{E}_e/E_{max}$[a]	$\bar{E}_e$ (MeV)	Branching ratio[b]	$\bar{E}_e$ (absorbed — MeV)
Mn	2.85	0.43	1.22	0.53	
	1.0	0.35	0.35	0.3	0.8
	0.75	0.33	0.25	0.16	
Dy	1.3	0.34	0.44	0.83	
	1.2	0.33	0.40	0.15	0.43

[a] James et al.[184]
[b] Lederer et al.[185]

where X_i is the photon fluence required to give one Roentgen at the energy of the i-th gamma ray and I_i is the relative intensity of the i-th gamma ray. The results of the calculation for < R > = 1.05 mm are listed in Tables 17 and 18 for CaF_2:Mn (2%) and CaF_2:Dy (0.35%). The maximum delayed beta contribution to the TL sensitivity is given by

$$\frac{\phi\, \Sigma\, (1.6 \times 10^{-6})\, \bar{E}_e}{\rho C} \qquad (61)$$

where the only contributors of significance are the ^{55}Mn(n,γ)^{56}Mn and ^{164}Dy(n,γ)^{165}Dy reactions. C is the amount of energy liberated by 1 R in 1 g of CaF_2 (= 84.7 erg) and $\bar{E}_e$ is the average energy loss in MeV of the electrons in the dosimeter per beta decay (Table 18). Equation 61 yields values of 0.4 and 0.7 R per 10^{10} n cm^{-2} for CaF_2:Mn (2%) and CaF_2:Dy (0.35%) respectively for a Maxwellian distribution at T = 293.6 K. Prokic[165] has calculated 0.77 R per 10^{10} n cm^{-2} for CaF_2:Mn (3%) in good agreement with the above calculation and measured 1.05 ± 0.08 R per 10^{10} n cm^{-2} also in good agreement with these calculations considering the very large sample (200 mg) which would ensure maximum electron absorption and a greater prompt gamma response than that listed in Table 17. Horowitz et al.[166] measured the sensitivity for CaF_2:Dy (0.35%) and CaF_2:Mn (2%) extruded bulb dosimeters (0.165 × 0.165 × 0.83 cm) as 0.33 ^{60}Co R per 10^{10} n cm^{-2} and 0.14 ± 0.01 ^{60}Co R per 10^{10} n cm^{-2}, respectively. It is difficult to comment precisely on the measurements of CaF_2:Mn reported in the literature and reviewed by Horowitz et al.[166] since only two other authors[164,165] have reported the percentage Mn content of their sample. As would be expected the result reported by Prokic[165] is the highest because of the very large TL sample employed. On the other hand, the result reported by Puite[164] for CaF_2:Mn (4.1%) of 0.07 ^{60}Co R per 10^{10} n cm^{-2} would appear to be too low by far unless extremely small samples were employed. Ayyangar et al.[186] have studied the thermal neutron sensitivity and gamma-ray sensitivity of $CaSO_4$:Dy as a function of Dy concentration. $CaSO_4$:Dy(0.05%) has high gamma ray sensitivity coupled with low thermal neutron sensitivity (0.38 ^{60}Co R per 10^{10} n cm^{-2} for a 5 g sample) and hence could be useful for gamma ray dosimetry in mixed $n_{th} - \gamma$ fields where the gamma energy dependence of $CaSO_4$:Dy does not pose a serious problem. Attix[180] discussed the feasibility of using rem-equivalent TL dosimeters for mixed neutron ($E_n < 10$ keV) - gamma ray fields, and Ayyangar et al.[163] suggested TLD-700 (5.2 wt % TLD-100) to achieve the rem-equivalent response of 18.6 ^{60}Co R per 10^{10} n cm^{-2}. In view of the nonuniversality of $\eta_{HCP,\gamma}$ at high ionization density, the figure of 5.2% obviously has little universal significance. In fact, Attix[180] on the basis of sensitivity data at his disposal suggested TLD-700 (2 wt % TLD-100). Obviously, the mixing percentage will also be dependent on the self-shielding factor for thermal neutrons so that the use of very thin dosimeters ($\bar{\phi}/\phi \simeq 1$) is mandatory in rem-equivalent dosimetry.

5. TL Dosimetry in Mixed n_{th} — γ-Ray Radiation Fields

The most popular method for separation of the thermal neutron and gamma-ray dose in mixed $n_{th} - \gamma$-radiation fields is via the double TLD technique using TLD-700 and either TLD-600 or TLD-100. In this method

$$N_7 = \phi^n S_7^n + D_7^\gamma S_7^\gamma \tag{62}$$

$$N_6 = \phi^n S_6^n + D_6^\gamma S_6^\gamma \tag{63}$$

where N_i is the number of TL photons (intensity of the TL signal) recorded by the TLD, D_i^γ is the gamma ray-induced dose and S_i^n and S_i^γ are the sensitivities to thermal neutrons and gamma rays, respectively. Because of the very low sensitivity of TLD-700 to thermal neutrons, the approximation

$$O \simeq \phi^n S_7^n \ll D_7^\gamma S_7^\gamma \tag{64}$$

is often invoked and it is assumed that TLD-700 measures the gamma dose only. A further assumption ($D_7^\gamma = D_6^\gamma$) then allows the estimation of ϕ^n. The following points, however, should be remembered.

1. $\phi^n S_7^n$ is not always necessarily much smaller (say less than 1%) of $D_7^\gamma S_7^\gamma$. For example, if $\phi^n = 10^8$ n cm^{-2}, $S_n \simeq 0.5$ ^{60}Co R per 10^{10} n cm^{-2} then $\phi^n S_7^n = 5$ mR. Application of the approximation (Equation 64) at the 1% level would then require that the gamma-induced dose in TLD-700 be greater than 0.5 R.
2. Since S(E), $\eta_{x,\gamma}(E)$ at low gamma-ray energies (less than 150 keV) may be batch dependent at the 10 to 20% level, the assumption $D_7^\gamma \simeq D_6^\gamma$ may be compromised at the 10% level if a significant proportion of the gamma ray-induced dose arises from gamma-ray energies less than 150 keV.
3. Very thin dosimeters ($\bar{\phi}/\phi \simeq 1$) are preferable in order to eliminate errors due to different self-shielding in the calibration and measurement radiation fields. Thermal neutron radiation fields are not always isotropic so that the assumption $(\bar{\phi}/\phi)_{calibration}$ equal to $(\bar{\phi}/\phi)_{measurement}$ is not always applicable.

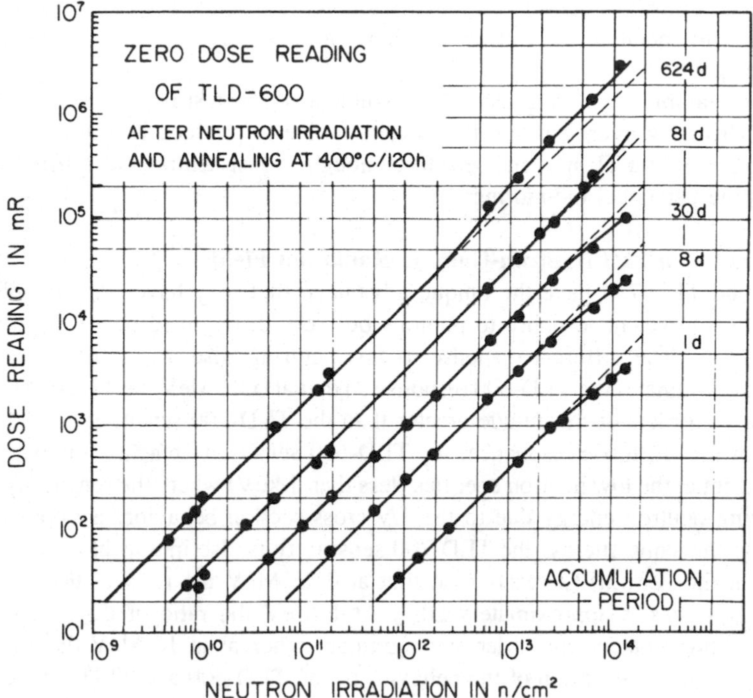

FIGURE 27. Zero dose reading of TLD-600 after thermal neutron preexposure and different "tritium decay" accumulation periods. (Adapted from Piesch, E., Burgkhardt, B., and Sayed, A. M., *Nucl. Instrum. Methods*, 157, 179, 1978.)

4. Tritium production[176,187] may introduce (following neutron preexposure) a time-dependent build-up of the zero dose reading (Figure 27). For example, a preexposure of 10^{10} n cm^{-2} followed by 400°C/120 hr annealing and an accumulation period of 81 days results in a 30 mR-induced signal from the tritium beta decay.

An alternative to the double TLD technique is to use a single TLD-600 dosimeter in which the neutron dose is estimated via the enhanced population of peaks 6 to 8 at high ionization density. Unfortunately, the batch variations in $\epsilon_{T>}$ reported in the literature are even greater than the variations listed in Table 14. For ^{60}Co irradiation $\epsilon_{T>}$ lies between 0.02 and 0.2, between 0.2 and 0.5 for slow neutron irradiation, and between 0.3 and 0.4 for alpha particle irradiation. Many attempts have been made to use this dependence of $\epsilon_{T>}$ on ionization density as an $n-\gamma$ discriminator,[77,188-194] however, the reduced TL efficiency of peaks 6 to 8 relative to peak 5, the erratic batch dependence of $\epsilon_{T>}$, the composite nature of the high temperature glow peak component, the difficulty of accurate separation from peak 5, the strong deviation of $\eta_{x\gamma}$ from unity for the high temperature peaks, and the different behavior of f(D) for the high and low temperature components renders this technique somewhat difficult to apply and of fairly limited accuracy and reliability. Nevertheless, because of the reduction by a factor of 2 in the required number of dosimeters, the single TLD-600 technique is in use in personnel dosimetry in several laboratories (e.g., Naval Research Laboratory, U.S.A. since January 1976). Nash and Johnson[192] have compared the accuracy of the two techniques for various mixed $n_{th}-\gamma$ radiation fields. In general, the performance of the single detector does not equal the performance of the pair because of the lower efficiency and hence lower precision of measurement of the 250°C component. A similar glow peak dependence on ionization density has been observed for CaF_2:Tm (TLD-300, Harshaw) where the high temperature peak at 250°C shows a greater response to fast neutrons than the 150°C peak.[148]

Morata and Nambi[195] have suggested improvement of the fast neutron sensitivity of this phosphor by diffusing hydrogen into the fluorite lattice to produce U centers and thereby, hopefully, to arrive at a TL material sensitive enough to fast neutrons to fulfill the requirements of routine personnel monitoring.

B. TL Dosimetry in Fast Neutron-Gamma Radiation Fields

TLD-600 and TLD-700 have the unique advantage that they have, in principle, almost identical tissue-equivalent response to photons but very different response to both thermal and fast neutrons. The different response to fast neutrons also arises from the ^{6}Li(n,α)τ reaction which dominates the TLD-600 response so that at 0.01 MeV the TLD-600 sensitivity is more than two orders of magnitude greater than the TLD-700 sensitivity. At 0.25 MeV, the location of the ^{6}Li(n,α)τ resonance, the TLD-600 sensitivity reaches a maximum value unequaled except at the low neutron energies (less than 1 keV) where the sensitivity increases with decreasing neutron energy due to the 1/v cross-section behavior. By coincidence, at almost exactly the same energy, the TLD-700 sensitivity is also in a local maximum due to a resonance in the ^{7}Li(n,n)^{7}Li reaction so that at 0.25 MeV the ratio of the TLD-600-to-TLD-700 sensitivities is approximately 20:1. At 1 MeV the ratio of the sensitivities has decreased to approximately one order of magnitude, whereas at 10 MeV they are roughly equal. For a detailed description of the behavior of the TLD-600 and TLD-700 sensitivities as a function of neutron energy see Horowitz and Freeman.[196] The one to two orders of magnitude of greater sensitivity of TLD-600 to fast neutrons has led to the suggestion that paired TLD-600 and TLD-700 probes could be used to separate fast neutron and gamma-induced dose in mixed fast neutron-gamma ray fields as well as in mixed thermal neutron-gamma ray fields.[197-200]

The methodology suggested by these authors[197-200] is based on the assumption that the fast neutron sensitivities of both TLD-600 and TLD-700 are universal quantities, i.e., they do not vary significantly from batch to batch or from manufacturer to manufacturer. Unfortunately, since the neutron sensitivity arises from recoiling high ionization density heavy charged particles induced by neutron interactions, the nonuniversality in $\eta_{HCP,\gamma}$ previously discussed establishes that fast neutron sensitivities are nonuniversal. Supportive evidence arises from direct experimental measurements of the neutron sensitivity of 14 to 15-MeV neutrons which reveal large variations between the various measurements[167] listed in Table 19. In the Radiation Physics Laboratory of the Ben Gurion University of the Negev our measurements for TLD-600 lie within the range of values listed in Table 19, but for TLD-700 we have measured a relative TL response of 0.07 at 14 MeV. It follows that fast neutron sensitivities cannot be assumed to be even approximately constant for all types of LiF so that the use of the paired LiF-TLD technique necessitates independent and accurate experimental measurement of the fast and thermal neutron sensitivities of the TL dosimeters to be employed. A second important limitation involves the effect of thermal neutrons. If the thermal neutron fluence is not insignificant compared to the fast neutron fluence, then the neutron energy spectrum cannot be characterized by a single parameter, e.g., 1/E spectrum, fission spectrum, monoenergetic spectrum, etc. Moreover, the following two factors, i.e., (1) the very high macroscopic cross section of ^{6}LiF to thermal neutrons ($\Sigma = 57.13\ cm^{-1}$) renders even very thin ^{6}LiF dosimeters totally black to thermal neutrons, and (2) the large positive Q value of the ^{6}Li(n,α)τ reaction, which results in a relatively high absorbed dose per impinging thermal neutron, implies that the thermal neutron absorbed dose in ^{6}LiF is most probably not insignificant in most of the $n - \gamma$ radiation fields under investigation. This limitation certainly does not exist to the same degree in other commonly used techniques

Table 19
RELATIVE RESPONSE OF TL MATERIALS TO 14= to 15-MeV NEUTRONS

Material	Neutron energy	$\eta_{n\gamma}$	Ref.
^{7}LiF	14.7	0.113	167
	14.1	0.16	208
	15	0.14	11, 43
	14.7	0.13	201
	14.7	0.186	202
	14.4	0.11	199
	14.7	0.105	203
	14.1	0.07	196
$Li_2B_4O_7$:Mn	14.7	0.14	167
	14.7	0.156	201
CaF_2:Mn	14.7	0.114	167
	14.0	0.22	204
	14.7	0.21	201
	14	0.135	205
	15.5	0.23	205
$CaSO_4$:Dy	14.7	0.12	167
	14.7	0.096	203
	14.4	0.114	199

such as paired ion chambers,[206,207] film,[208] Geiger counter,[209] or tissue-equivalent proportional counters[210] where the very large 940 barn thermal neutron cross section of the ^{6}Li(n,α)T reaction has no parallel. Thus the large thermal neutron-induced dose in TLD-600 (or TLD-100) implies that the paired LiF-TLD technique yields two equations in three unknowns (the thermal neutron, fast neutron, and gamma-ray dose). Consequently, a third independent measurement is required, e.g., a measurement of the thermal neutron fluence via foils. The thermal neutron absorbed dose can then be calculated and the resultant TL signal subtracted from the TLD-600 and TLD-700 signals. Horowitz et al.[154] illustrated the importance of these considerations by measurements in an Am-Be fast neutron field. Measurements were carried out at two points (in water and in air) 11-cm distance from a 3 Ci Am-Be source situated in a 1.5 × 1.5 × 1.5 m water tank. For the point in water, Horowitz et al. found via indium foil activation ϕ_{th} = 1290 ± 4.5% (1 S.D.) n/cm^{-2}/sec^{-1} and ϕ_f = 800 ± 21% (1 S.D.) as measured by the paired TLD technique after subtraction of the thermal neutron-induced signal in TLD-600. These numbers translated to 8700 and 2300 TL photon counts per hour of irradiation due to thermal and fast neutrons, respectively. At the same point, the gamma fluence resulted in 410 ± 15% (1 S.D.) TL photon counts hr^{-1} of irradiation. Two very significant points arise: (1) the thermal neutron-induced signal is by far greater than either of the fast neutron or gamma ray-induced signals and the error in its estimation therefore introduces a comparatively far greater percentage error in the estimation of the fast neutron and gamma-ray dose and (2) as previously discussed, the 3.5% reduction in the self-shielding factor due to the angular distribution of the thermal neutrons translates to approximately a 70% error in the photon dose if the angular distribution is not taken into account in the calculation of the self-shielding factor. The situation is similar for the measurements at the point in air, 11 cm from the source, which yielded 2800, 200, and 325 TL photon counts hr^{-1} of irradiation due to thermal neutrons, fast neutrons, and gamma rays, respectively.

In conclusion, then, the paired LiF-TLD technique for the separation of fast neutron and gamma-ray dose cannot be assumed to be free of very significant complications in certain mixed radiation fields due to the very large thermal neutron-induced signal in TLD-600 or TLD-100. The thermal neutron fluence must be separately measured to high accuracy and the induced TL signal calculated and subtracted from the TLD-600 signal. The thermal neutron-induced TL signal is geometry dependent due to self-shielding (unless very thin TL samples are employed). In cases where $\phi_{th}/\phi_f > 0.05$, even the angular distribution of the thermal neutrons must be measured to determine $\bar{\phi}/\phi$ to sufficient accuracy so as to avoid errors of the order of 10% in the estimation of the gamma dose. In addition, the paired TLD technique requires independent measurement of the TLD sensitivities as a function of neutron energy since it has been demonstrated both experimentally and theoretically that these may vary significantly from batch to batch.

1. Theoretical Calculations of Fast Neutron Sensitivities

The main application of TLD to fast neutron dosimetry has been in the estimation of gamma dose via neutron-insensitive dosimeters (usually TLD-700 or CaF_2:Mn). Obviously, there would exist considerable advantage in accurate knowledge of the TL sensitivity of these dosimeters as a function of neutron energy. Even crude knowledge of the neutron energy spectrum would then allow significantly more accurate estimation of the gamma-induced dose. Unfortunately, the experimental measurement of the fast neutron sensitivity is difficult (it is, in itself, a mixed field problem) due to the presence of photons and scattered neutrons. Moreover, accurate theoretical calculations are equally difficult. Various authors[198,199,211-214] have calculated the fast neutron sensitivity of TL materials (^{6}LiF, ^{7}LiF, CaF_2, etc.) by assuming that the neutron sensitivity is closely related to the kerma. Some calculations[165,199,214] have attempted to take into account varying efficiencies of different ions to induce TL by assuming a universal TL-linear energy transfer (LET) curve. Even this latter approach, however, should be regarded as a crude approximation for the following reasons:

1. As discussed in the previous section, $\eta_{HCP,\gamma}$ is batch and material dependent to a very significant extent. This leads directly to the conclusion that the fast neutron sensitivities are similarly nonuniversal as a function of neutron energy. This conclusion is unassailable since the fast neutron response arises almost exclusively from recoiling high ionization density heavy charged particles.
2. Significant errors may arise even in the kerma calculations due to uncertainties in the nuclear reaction cross sections and angular distribution data. For example, in LiF, at 15 MeV, the neutron-induced reactions of significance are ^{6}Li(n,n)^{6}Li, ^{6}Li(n,p)^{6}He, ^{6}Li(n,α)τ, ^{6}Li(n,n')^{6}Li, ^{6}Li(n,2n)^{5}Li, ^{19}F(n,n)^{19}F, ^{19}F(n,p)^{19}O, ^{19}F(n,α)^{16}O, ^{7}Li(n,n)^{7}Li, ^{7}Li(n,n')^{7}Li, ^{7}Li(n,2n)^{6}Li with further reactions like the ^{7}Li(n,n'α), ^{7}Li(n,2nα), and ^{7}Li (n,d) also contributing approximately 1% to the total kerma. Kerma calculations by Horowitz and Freeman[196] agreed with Tanaka and Furuta[198] and Spurny et al.[212] at low neutron energy, but above 5 MeV (where the number of open reaction channels begins to increase significantly) the results reported by Spurny et al. were approximately 15% higher. The discrepancy almost certainly arises from difference in the data files used by the various authors at higher neutron energies. To illustrate the effect of variations in $\eta_{HCP,\gamma}$ on fast neutron sensitivities, Horowitz and Freeman[196] calculated the ^{6}LiF and ^{7}LiF fast neutron sensitivity from 1 keV to 14 MeV via a differential kerma approach as a function of angle using two body kinematics and angular distribution data whenever available. The charged particle kerma were then folded in with different TL-LET curves as measured by Patrick et al.,[215] Jahnert,[216] and Tochilin et al.[11] The results yielded very different fast neutron sensitivities as a function of neutron energy. For example, at 15 MeV for ^{6}LiF, the Patrick et al.-Jahnert data yield ~5

^{60}Co R per 10^{10} n cm^{-2} and the Tochilin et al. data yield ~10 ^{60}Co R per 10^{10} n cm^{-2}, and for ^{7}LiF the respective sensitivities were calculated as ~3 and 8 ^{60}Co R per 10^{10} n cm^{-2}, respectively. For purposes of comparison, Rinard and Simons[214] calculated 4 ^{60}Co R per 10^{10} n cm^{-2} for ^{7}LiF and Tanaka and Furuta[198] measured 10 ^{60}Co R per 10^{10} n cm^{-2} for both ^{6}LiF and ^{7}LiF. Prior to the work of Horowitz and collaborators, Attix et al.[206] also commented on the dangers of selecting published values of fast neutron sensitivities: measurement of the gamma-ray component of the NRL fast neutron beam with CaF_2:Mn and ^{7}LiF TLDs yielded negative values of D_γ using published fast neutron sensitivities. Blum et al.[205] have also used CaF_2:Mn to measure the gamma dose in the fast neutron beam of the MRC cyclotron. Exposure in pairs (one in polythene, the other in lead) allowed the extraction of the gamma dose (under the assumption that the response of the phosphor in lead represents the effect of the gamma-ray contamination in the fast neutron beam). The angular dependence (arising from the short range of the recoil protons) severely limits the clinical applications of this technique; furthermore, the claimed advantage that there is no interaction between neutron and gamma response with CaF_2[164,217] now appears true for LiF also (see following section).

3. The final inaccuracy in the theoretical calculations of fast neutron TL sensitivities arises from the assumption that LET is a good parameter for the calculation of $\eta_{HCP,\gamma}$. In fact, track structure theory (see the following chapter) predicts that protons, alpha particles, and heavy ions of the same LET can possess very different values of $\eta_{HCP,\gamma}$. Calculation of the relative TL response of the various ions liberated by fast neutrons must therefore be treated in a three-dimensional model which takes into account the electron dose distribution as a function of radial distance from the path of the primary ion folded in with the correct electron-induced f(D) TL dose-response function.

In conclusion, the calculation of fast neutron TL sensitivities even via the more sophisticated kerma-LET approach should be regarded as a first approximation only. Any agreement with a particular set of fast neutron sensitivities should be regarded as fortuitous and overlook the fact that at 14 to 15 MeV, e.g., experimental measurements of fast neutron sensitivities for ^{6}LiF and ^{7}LiF vary between 3 and 10 ^{60}Co R per 10^{10} n cm^{-2} and for CaF_2:Mn between 2 and 15 ^{60}Co R per 10^{10} n cm^{-2}.[165,201,214]

2. *Nonadditivity of n − γ-Induced TL Yields*

a. Fast Neutrons

Oltman et al.[218] reported that fast neutrons from 0.1 to 1 MeV tended to release the stored gamma-induced signal from a previous irradiation (polyethylene-sealed TLD-700 dosimeters were employed) or simultaneous gamma irradiation. A fast neutron fluence of 4 × 10^8 n cm^{-2} was observed to decrease the gamma TL signal by ~10%. The gamma dose was ~3 R (^{60}Co) and the percentage decrease was independent of neutron dose (0.1 to 0.6 mGy) or neutron energy. Wallace and Ziemer[161] also observed similar effects (15 to 20% loss in polyethylene-sealed LiF-Harshaw dosimeters) with 4 × 10^9 n cm^{-2} of Pu-Be neutrons following a 5 R ^{137}Cs gamma-ray exposure and a 13.9% loss in TL signal for polyethylene-sealed $Li_2B_4O_7$:Mn (Harshaw). Mason et al.[219] reported (without giving experimental details) a 6% reduction with 2.8- and 14-MeV neutrons irradiating Conrad ^{7}LiF and in a parallel publication[189] discussed the irradiation of LiF (Conrad) in polythene containers irradiated by thermal neutrons. It is fairly certain that the Mason et al. work with fast neutrons was also with LiF sealed in polythene containers. Finally, Kastner et al. (as reported by Wallace et al.[220]) observed a fast neutron-induced decrease for unencapsulated ^{7}LiF microrods. Wallace et al.[220] reexamined the question of n − γ additivity and found that their earlier results could be explained as arising from a spurious broad high temperature peak induced in LiF via vapor released from the polyethylene tubing during the process of heat sealing the phosphor powders into capsules. No loss in TL signal was observed for glass-encapsulated ^{7}LiF or $Li_2B_4O_7$:Mn or polyethylene-encapsulated $Li_2B_4O_7$:Mn. Goldstein et al.[201] using peak height estimation of the TL signal observed no damage or transfer effects ($\pm 3\%$) for 14.5-MeV neutrons irradiating aluminum or gelatin-encapsulated TLD-100, CaF_2:Mn, BeO, or $Li_2B_4O_7$:Mn. Finally, Bloch and Weber[221] reported no changes in the gamma sensitivity of TLD-700 following 16 Gy of fast neutron-induced dose, and Pearson and Moran[222] observed no irreversible sensitivity changes after exposure of LiF (Harshaw) to fast or thermal neutrons.

b. Thermal Neutrons

Mason[189] and Dua et al.[159] both reported sensitivity enhancement of ^{7}LiF (Conrad) and ^{6}LiF (Conrad) following thermal neutron irradiation. The former work was with encapsulated LiF powders in thin-walled polythene containers, and the latter used planchets covered with a polyimide resin. In lieu of the findings regarding fast neutrons, these experimental details themselves are sufficient to cast serious doubt on the reliability of the results. The onset of thermal neutron-induced supralinearity at very low dose (0.1 Gy as reported by Mason and 2×10^9 n cm^{-2} equivalent to approximately 1 Gy as reported by Dua et al.) also suggests characteristics peculiar to the materials used by these authors. In the Radiation Physics Laboratory of the Ben Gurion University of the Negev we observe no memory or interaction effects in TLD-100, -600, or -700 following thermal neutron irradiations up to approximately 1.5×10^{10} n cm^{-2}. Similar negative results for thermal neutrons have been reported for LiF,[192] $Li_2B_4O_7$:Mn,[223] $CaSO_4$:Dy,[224] and CaF_2:Mn.[164,205,217] It is interesting to note, however, that Puite[164] did find that the fast neutron sensitivity of CaF_2:Mn was a function of neutron dose (no observable threshold) in which decreasing sensitivity to neutrons as a function of neutron dose was hypothesized to arise from competing traps created via lattice recoils.

In conclusion, then, both the fast neutron "erasure" effect and the thermal neutron "enhancement" effect in LiF-TLD can be dismissed as due to spurious phenomena arising from contact poisoning of the TLDs. The LiF (Conrad) materials used by Mason and Dua et al. show peculiar characteristics (enhanced thermal neutron supralinearity, possible sensitivity enhancement).

3. LiF-TLD Fast Neutron Glow Curves

Various groups[189,208] have reported that $\epsilon_{T>}$ for fast neutrons is not increased over the value of $\epsilon_{T>}$ obtained for gamma rays at low dose. Other authors[220] have occasionally published n + γ glow curves which, at first glance, tend to support this observation. Since the enhanced population of the high temperature peaks has been conclusively correlated with increase in ionization density, it would be unusual if fast neutron irradiation did not similarly result in increased values of $\epsilon_{T>}$ associated with high ionization density. This is due to the fact that most of the fast neutron-induced kerma arise via high ionization density recoiling Li and F ions. In our own work with fast neutrons we observe $\epsilon_{T>}$ comparable to the value observed for thermal neutrons. In the case of Wallace et al.,[220] the glow curve (Figure 2 of their work) is given for ^{7}LiF (Harshaw) following 0.5 R of ^{226}Ra gamma rays and 10^8 n cm^{-2} (0.26 MeV). This fluence of fast neutrons deposits approximately 2×10^{-4} Gy and since $\eta_{n\gamma} \simeq 0.2$ for fast neutrons it follows that the fast neutron TL signal is only approximately 1% of the total signal so that obviously $\epsilon_{T>}(n+\gamma)$ will be indistinguishable from the value of $\epsilon_{T>}$ observed for gamma rays alone. A similar explanation most probably also applies to the work reported by Wingate et al.[200] and Mason.[189]

Table 20
FAST NEUTRON ACTIVATION TL DOSIMETRY

Reaction	Threshold	Cross section at 14 MeV (mb)	Decay products	Maximum energy (MeV)	$\tau_{1/2}$ (minutes)
$^{19}F(n,2n)^{18}F$	11	50	β^+ (97%) e^- (3%)	0.64	110
$^{32}S(n,p)^{32}P$	2.5	250	β^-	1.7	14 days
$^{24}Mg(n,p)^{24}Na$	7	200	β^-	2.8	15 hr

4. TL Fast Neutron Activation Dosimetry

This technique consists of exposing a TLD to the mixed radiation field. Only the neutrons are capable of inducing nuclear transformations because of the very high threshold for gamma ray induced nuclear reactions (~20 MeV). After exposure the dosimeter is annealed to eliminate the TL produced during irradiation and then stored to undergo internal or "self"-irradiation from the radioactive products. A final reading gives the TL resulting from the "self"-irradiation which can be related to the neutron fluence. TLD fast neutron activation thus offers the important advantage of complete discrimination against gamma rays. On the other hand, activation techniques suffer from their dependence on a rather large number of other factors. The measured activity of the radioactive product is given by

$$A_d = F(t_1,t_2,t_3)\, N_p \int_0^{\infty} \sigma(E)\, \phi(E)\, dE \qquad (65)$$

where t_1 = exposure time, t_2 = the time lapse until measurement, t_3 = the half-life of the radioactive product, N_p is the number of parent nuclei, $\sigma(E)$ is the energy-dependent cross section, and $\phi(E)$ the energy-dependent neutron fluence. The TL yield is thus proportional to the $\int \sigma\phi dE$ so that the possibility of extraction of accurate information regarding $\phi(E)$ is limited. The TL yield will be dependent on a number of additional factors related to the energy of the radioactive products (usually electrons) and their relative ability to produce TL. Since the beta ranges are not negligible compared to typical dosimeter dimensions, the TL yield will also be size and shape dependent. Further complications and sensitivity changes may arise from the requirement to heat erase the "prompt" dose TL which is usually two to three orders of magnitude greater than the delayed-dose TL from the subsequent decay of the neutron-induced activity. Since the delayed-dose TL is quite small, typically 10^{-5} Gy for a neutron fluence which delivers tens of mGy to tissue, the TL reader must be highly sensitive and have a highly reproducible heating system. In mixed $n-\gamma$ radiation fields when the gamma dose is greater than 10 Gy the possibility of a sensitivity change due to the gamma-ray dose must also be considered. For example, Pradhan et al.[225] found, in reactor $n-\gamma$ measurements using $CaSO_4$:Dy, a 55% enhancement in neutron sensitivity due to 10^3 Gy of gamma rays. An easy method of avoiding this problem is to give a high temperature postirradiation anneal to remove the sensitization. At very high dose, however, even this treatment may be unable to entirely remove the sensitization or damage. The useful fast neutron reactions for the common phosphors are shown in Table 20. Other possible reactions leading to ^{16}N ($\tau_{1/2}$ = 7.2 sec), ^{19}O ($\tau_{1/2}$ = 29 sec), or ^3H ($\tau_{1/2}$ = 12.3 years) are not useful due to the very long or very short half-lives. The thermal neutron capture reaction $^{164}Dy(n,\gamma)^{165}Dy$ (σ_{th} = 2800 barns, $E_{max}(\beta^-)$ = 1.3 MeV, and $\tau_{1/2}$ = 140 min) complicates the situation for CaF_2:Dy and $CaSO_4$:Dy, but the 14-day half-life of the ^{32}P decay allows separate estimation of the thermal neutron- and fast neutron-induced TL signals in the case of $CaSO_4$:Dy. Mayhugh and Watanabe[226] estimated that 10^{-3} mGy in the first day of self-irradiation for $CaSO_4$:Dy requires an original fluence of $\sim 1.5 \times 10^8$ n cm^{-2}, and Pradhan et al.[225] found that a minimum measurable fast neutron dose of 0.12 mGy (16-day postirradiation interval for TL accumulation) was achieved for $CaSO_4$:Dy and sulfur mixture (1%) compared to a minimum fast neutron dose by conventional sulfur activation detection methods of approximately 10^{-2} Gy with an accuracy of 10%. Thus the low minimum measurable dose and reasonable TL accumulation period makes the $CaSO_4$:Dy mixed sulfur pellets an attractive possibility for fast neutron (>2.5 MeV) personnel monitoring. Pearson and Moran[222] have compared the usefulness of the various common phosphors (LiF, CaF_2:Dy, $CaSO_4$:Dy, Mg_2SiO_4:Tb, $MgSO_4$). For 14-MeV neutrons from the D-T reaction all of the dosimeters had adequate sensitivity for measurements out to the 5% dose contours (10^9 n cm^{-2}). Since the body is a large moderator, scattered neutrons down to thermal energy are usually present in clinical applications. The high threshold of the fluorine reaction gives optimal discrimination vs. these scattered neutrons so that LiF is the best choice for 14-MeV clinical applications. Detailed investigations using both fast and thermal neutron activation TLD have been reported extensively elsewhere.[166,176,222,225-229]

V. HEAVY CHARGED PARTICLE RELATIVE TL RESPONSE: EXPERIMENTAL RESULTS

Extensive efforts are currently being invested in the investigation of the use of various thermoluminescent dosimeters in exotic radiation fields, such as mixed n $-$ γ and various heavy charged particle radiation fields. Because of the relevance to accurate intercalibration between these various radiation fields, Tanaka and Furuta[197] and many other investigators attempted, e.g., to establish the existence of a universal $\eta_{HCP,\gamma}$ vs. $1/\rho\ \overline{LET}_\infty$ dependence based on the assumption that the specific stopping power of HCPs is the only or dominant factor that influences the relative TL response and furthermore that there is a unique $1/\rho\ \overline{LET}_\infty$ dependence common to many TL materials (e.g., LiF, CaF_2:Mn, $CaSO_4$:T_m, etc.). Unfortunately, the collected data on relative HCP TL response in Figures 28 to 31 establish that even relative TL properties apparently depend on a formidable array of experimental variables, and, of equal or even greater importance, unequivocally illustrate the convincing lack of a unique correlation between $\eta_{HCP,\gamma}$ and $1/\rho\ \overline{LET}_\infty$. A similar lack of correlation is revealed when Z_{eff}^2/β^2 is used instead of $1/\rho\ \overline{LET}_\infty$ (β is the relative HCP velocity and Z_{eff} the HCP effective charge stopping in the TL material).

Detailed inspection of the data in Figures 28 to 31 reveals the following important features:

1. Although there is a general trend to decreasing values of η, η' (η' is the relative TL response measured via peak height estimation of the TL signal) at $1/\rho\ \overline{LET}_\infty > 10^2$ to 10^3 MeV g^{-1} cm^2 in most materials, the large spread of values for $1/\rho\ \overline{LET}_\infty > 10^3$ MeV g^{-1} cm^2 clearly lays to rest all conjecture regarding the possibility of universal behavior. In LiF-TLD (Harshaw), e.g., many groups have measured η, η' for 4.0-MeV alpha particles stopping in the dosimeter (Table 21) and the results for $\eta_{\alpha\gamma}$ vary from as low as 0.045 $\pm$ 0.003 (Montret-Brugerolle[246]) to as high as 0.27 $\pm$ 0.025 (Horowitz et al.[125]).
2. The large spread of values in η, η' reported in the literature arises from many experimental and material-dependent factors. To investigate the importance of material-dependent factors, Horowitz et al.[125-127] carried out extensive studies of various batches of LiF and $Li_2B_4O_7$ Harshaw TLDs using 13.54-meV neutrons and 3.8-MeV alpha particles. Great care was exercised to ensure identical experimental procedures so that the differences in η shown in Table 15 can only arise from material-dependent factors. Direct current spectrochemical analysis revealed large variations in Ba, Cr, and Al in

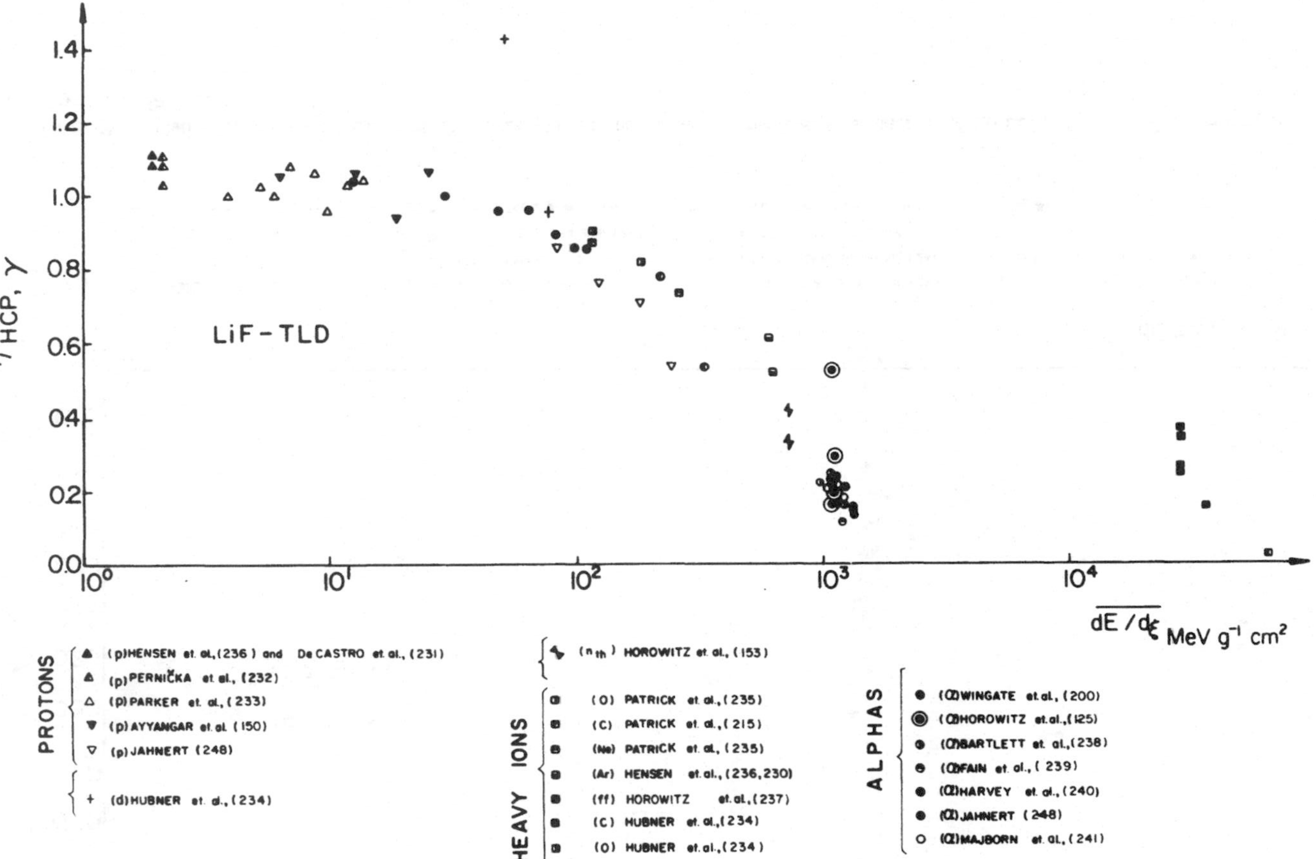

FIGURE 28. Relative TL response of LiF-TLDs to heavy charged particles as a function of $1/\rho$ LET_{∞}. (Adapted from Horowitz, Y. S., *Phys. Med. Biol.*, 26, 765, 1981.)

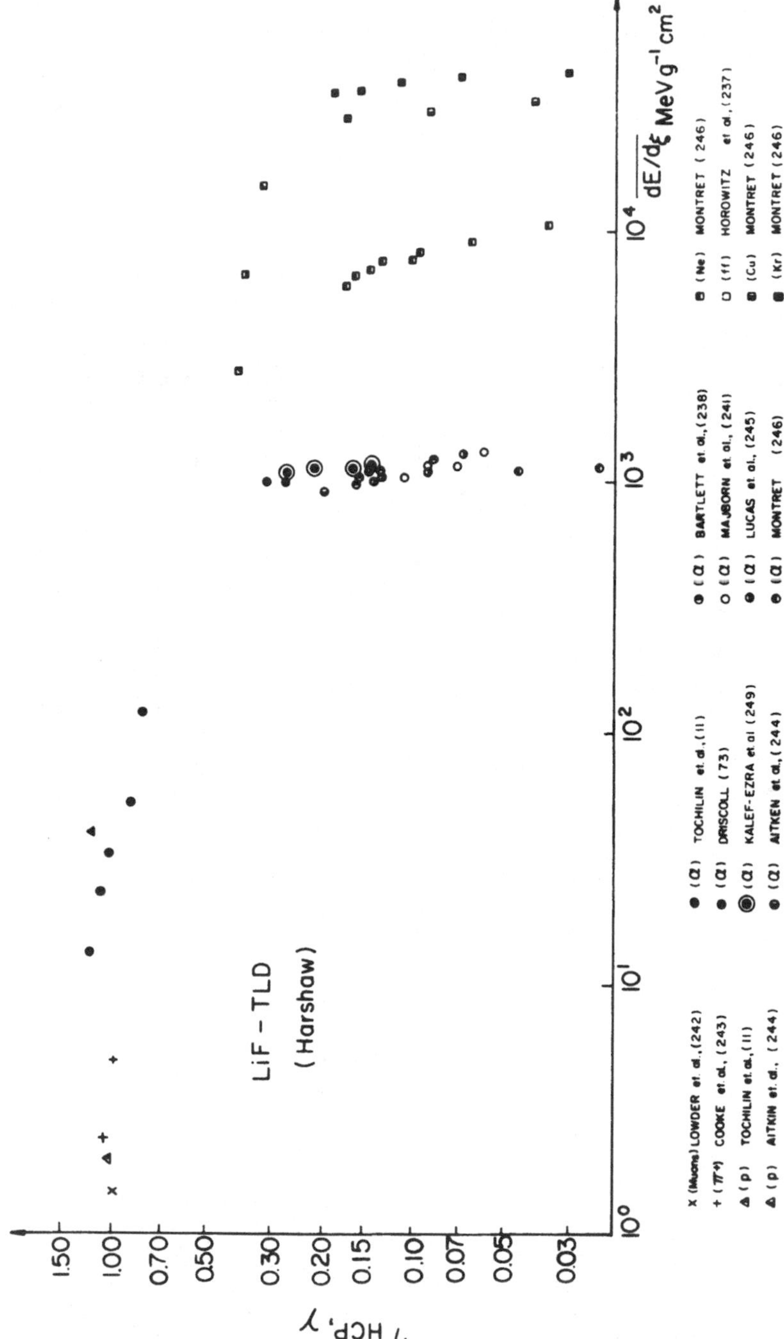

FIGURE 29. Relative TL response (peak-5 height) of LiF-TLD to heavy charged particles as a function of 1/ρ LET_{∞}. (Adapted from Horowitz, Y. S., *Phys. Med. Biol.*, 26, 765, 1981.)

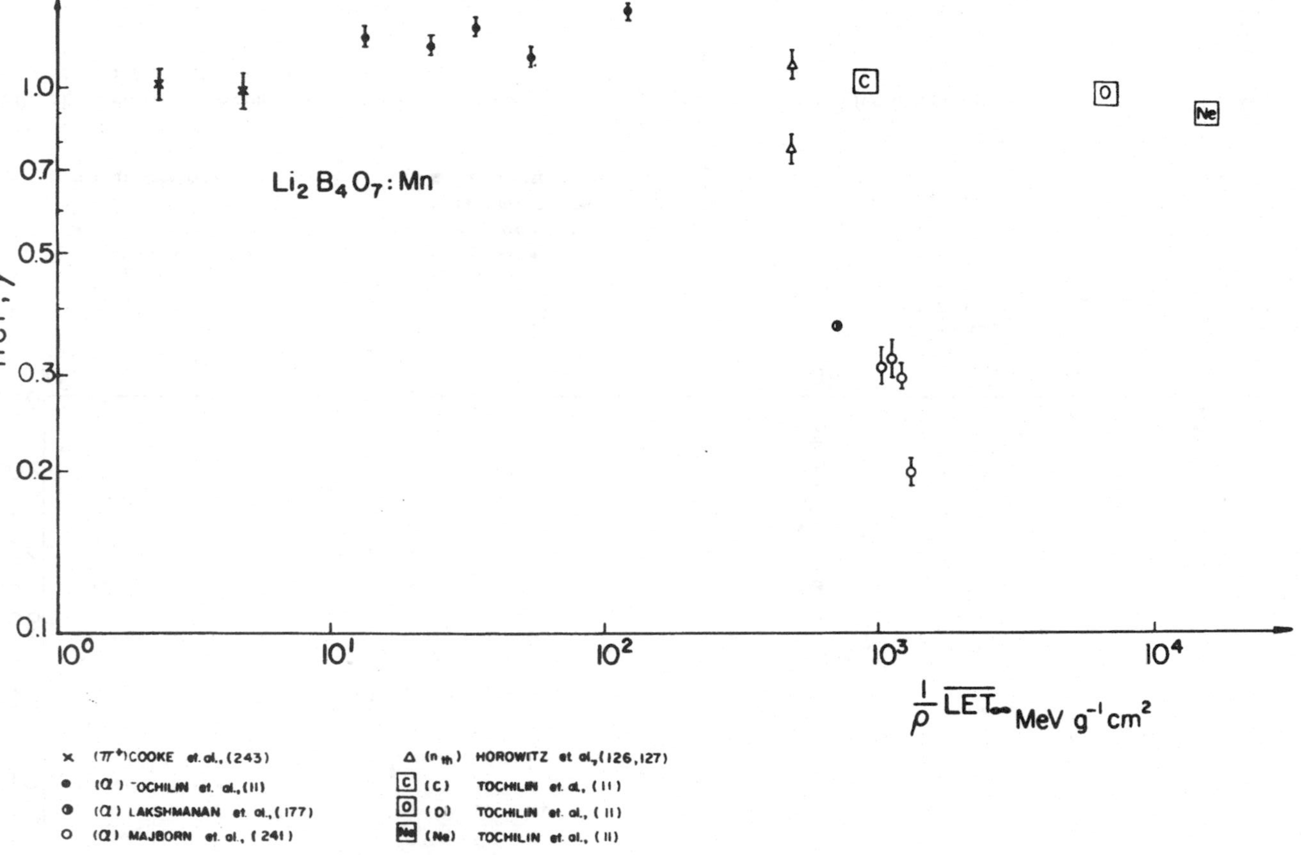

FIGURE 30. Relative TL response (peak height) of $Li_2B_4O_7$:Mn to heavy charged particles as a function of $1/\rho\ LET_\infty$. (Adapted from Horowitz, Y. S., *Phys. Med. Biol.*, 26, 765, 1981.)

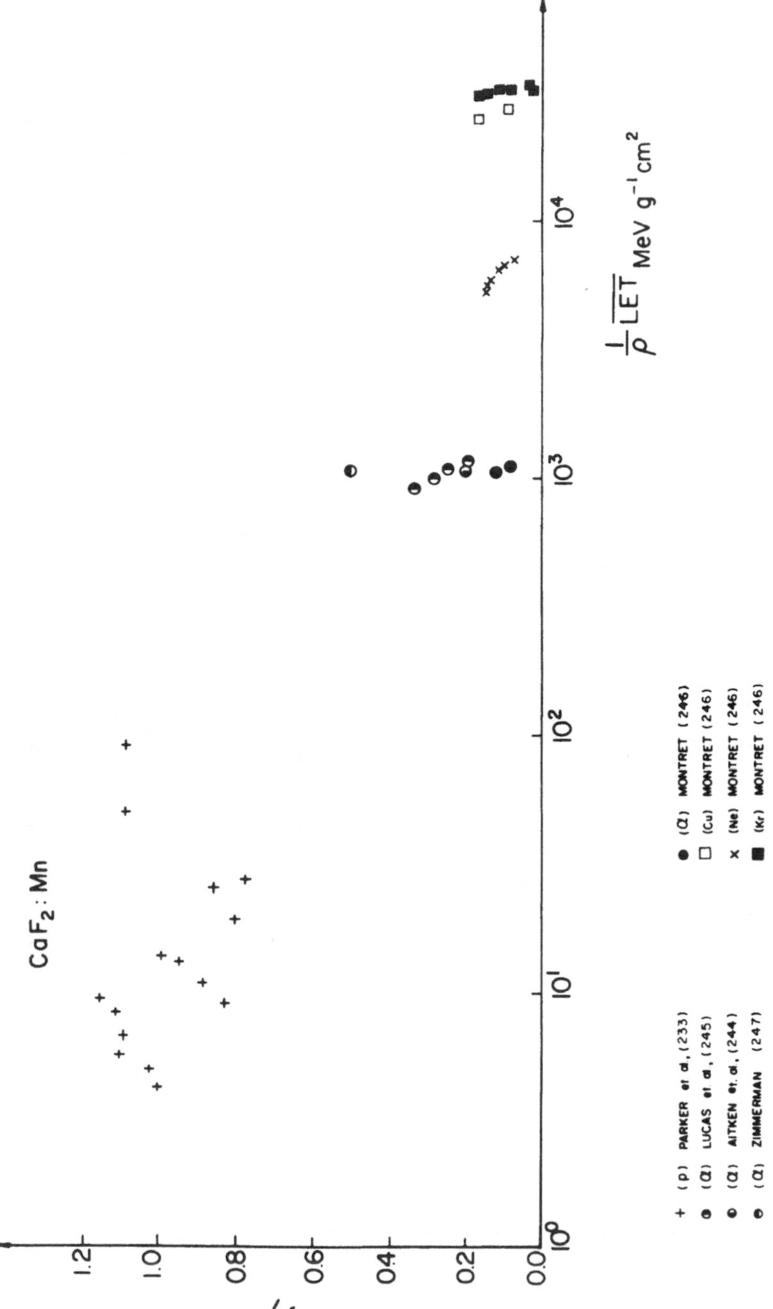

FIGURE 31. Relative TL response (main peak height) of CaF_2:Mn to heavy charged particles as a function of $1/\rho$ $LET_\times$. (Adapted from Horowitz, Y. S., *Phys. Med. Biol.*, 26, 765, 1981.)

Table 21
RELATIVE TL RESPONSE OF LOW ENERGY ALPHA PARTICLES

E (MeV)	$\eta_{\alpha\gamma}$	$\eta'_{\alpha\gamma}$	Ref.
	TLD-700		
5.5	0.314 ± 0.074	0.313 ± 0.080	73
5.7	0.232 ± 0.013	0.158 ± 0.008	238
4.95	0.248 ± 0.013	0.151 ± 0.008	238
4.3	0.219 ± 0.014	0.129 ± 0.007	238
2.5	0.167 ± 0.014	0.087 ± 0.007	238
4.7	0.211 ± 0.003	0.108 ± 0.001	241
3.8	0.219 ± 0.002	0.090 ± 0.001	241
2.8	0.183 ± 0.002	0.072 ± 0.001	241
1.7—3.7	0.165 ± 0.007	—	216
3.8	0.295 ± 0.038	0.213 ± 0.032	125
	TLD-600		
5.5	0.348 ± 0.083	0.313 ± 0.077	73
3.8	0.209 ± 0.026	0.160 ± 0.022	125
	TLD-100		
4.0		0.045 ± 0.003	246
3.2		0.024 ± 0.001	246
5.47		0.135	245
4.48		0.125	245
3.38		0.090	245
2.09		0.066	245
4.48	0.235 ± 0.01		240
3.86	0.245 ± 0.01		240
2.85	0.210 ± 0.01		240
3.8	0.167 ± 0.024		125

the three batches of LiF and variations in Al, Si in the two batches of $Li_2B_4O_7$. Accuracy of the spectrochemical analysis was 10 to 20% depending on the impurity concentration level. The obvious conclusion is that η, $\eta'_{HCP,\gamma}$ are sensitive to ppm concentration of various other TL coactivators, to variations in Mg,Ti concentration at the 10 to 20% level, or to other unknown physical or chemical characteristics. Purchase of LiF-TLD or $Li_2B_4O_7$-TLD from the same supplier with the same nominal major dopant concentrations does not imply similar behavior of the relative TL response. It would be by no means a drastic extrapolation to infer that these considerations apply to most or all TL materials.

3. Additional experiments in the Radiation Physics Laboratory of the Ben Gurion University of the Negev[237] have revealed that $\eta_{HCP,\gamma}$ is strongly dependent on the atmosphere (air or N_2) and the cooling rate employed after the high temperature preirradiation anneal (T_1,t_1); these results are listed in Table 22.
4. A similar dependence of f(D) on these parameters was also observed,[237,249] and in this context it is important to point out that large variations in the TL dose response of LiF-TLD (Harshaw) have also been reported by many investigators.[38] Indeed, we will show in the following chapter that track structure theory (TST) calculations of $\eta_{HCP,\gamma}$

Table 22
RELATIVE TL RESPONSE OF FISSION FRAGMENTS AND ALPHA PARTICLES

TLD population	Annealing	$\bar{\eta}_{ff,\alpha}$	$\bar{\eta}_{\alpha\gamma}$	$\bar{\eta}_{ff,\gamma}$
TLD-600	Air	1.25 ± 0.31	0.21 ± 0.02	0.27 ± 0.07
TLD-700	Air	1.24 ± 0.28	0.30 ± 0.03	0.37 ± 0.09
TLD-100	Air	1.47 ± 0.34	0.17 ± 0.03	0.25 ± 0.07
TLD-600				
TLD-700	N_2	0.58 ± 0.10	0.50 ± 0.045	0.29 ± 0.05

Table 23
RELATIVE TL RESPONSE OF He, Ne, AND Kr IONS IN LiF-TLD

Particle type	Range of energies (MeV μ^{-1})	Range of (1/ρ) LET $_\infty$ (MeV g^{-1} cm^2)	Range of values of $\eta_{HCP,\gamma}$
^{4}He	0.8—1.0	1.11—1.14	0.024—0.045
^{20}Ne	2.7—11	6.3—10.3	0.036—0.17
^{85}Kr	1.4—9.4	37—43	0.006—0.19

via TL dose-response studies establishes that variations in f(D) naturally lead to variations in $\eta_{HCP,\gamma}$.

5. Finally, and especially important from a theoretical standpoint, inspection of the published data assembled in Figures 28 to 31 reveals a very limited correlation of $\eta,\eta'_{HCP,\gamma}$ with $1/\rho\ \overline{LET}_\infty$ for values greater than 10^2 $MeVg^{-1}cm^{+2}$ (i.e., in the region of decreasing relative TL response due to ionization density effects) and implies that other parameters of the radiation (e.g., the HCP velocity) in addition to the specific stopping power are playing an important role in the determination of $\eta_{HCP,\gamma}$.

To emphasize this point consider the work of Horowitz and Kalef-Ezra[237] (Table 22) who irradiated air-annealed LiF-TLDs with degraded ^{252}Cf fission fragments (average energy of the light and heavy fragments 49 and 31 MeV, respectively) and found that the relative TL response of the fission fragments to 5.5-MeV alpha particles was 1.25 ± 0.29 even though the fission fragment stopping power is ~25 times greater than that of the 5.5-MeV alpha particles. Other recent experiments by Montret-Brugerolle[246] report on the irradiation of LiF-TLD (Harshaw) with He, Ne, and Kr ions (Table 23) and also contradict the idea of a single valued unique dependence of $\eta_{HCP,\gamma}$ on $1/\rho\ \overline{LET}_\infty$. Similar results were also obtained by Montret-Brugerolle for CaF_2:Mn and CaF_2:Dy.

In conclusion, the relative TL response of HCPs is dependent on a large number of experimental and material-dependent factors. The variations in relative TL response are especially large in regions of high ionization density where η,η' are significantly less than unity. Although the complexity of the TL mechanism does not allow the theoretical calculations of TL efficiencies, a modified track structure theory is proposed in the following chapter, which, nonetheless, has proven to be remarkably accurate in the calculation of the relative HCP TL response. The essence of the theory is the calculation of the HCP TL properties (e.g., $\eta_{HCP,\gamma}$) via a convolution of the carrier concentration around the HCP track with the TL dose response generated from a carefully selected electron or gamma test radiation.

REFERENCES

1. **Broerse, J. J. and Mijnheer, B. J.,** in Proc. 8th Cong. Int. Soc. Fr. Radioprot. Aspects Fondamentoux Appl. Dosimetrie, Saclay, 1975, 641.
2. **Puite, K. J.,** A thermoluminescence system for the intercomparison of absorbed dose and radiation quality of X-rays with a HVL of 0.1 to 3.0 mm Cu, *Phys. Med. Biol.*, 21, 216, 1976.
3. **Greening, J. R.,** Dosimetry of low energy X-rays, in *Topics in Radiation Dosimetry*, Attix, F. H., Ed., Academic Press, New York, 1972, 262.
4. **Burlin, T. E., Chan, F. K., Zanelli, G. D., and Spiers, F. W.,** General cavity theory, *Nature (London)*, 221, 1047, 1969.
5. **Chan, F. K. and Burlin, T. E.,** The energy size dependence of the response of TLDs to photon irradiation, *Health Phys.*, 18, 325, 1970.
6. **Endres, G. W. R., Kathren, R. L., and Kocher, L. F.,** Thermoluminescence personnel dosimetry at Hanford. II. Energy dependence and application of TLD materials in operational health physics, *Health Phys.*, 18, 665, 1970.
7. **Bassi, P., Busuoli, G., and Rimondi, O.,** Calculated energy dependence of some RTL and RPL detectors, *Int. J. Appl. Radiat. Isot.*, 27, 291, 1976.
8. **Pradhan, A. S., Kher, R. K., Dere, A., and Bhatt, R. C.,** Photon energy dependence of $CaSO_4$:Dy embedded teflon TLD discs, *Int. J. Appl. Radiat. Isot.*, 29, 243, 1978.
9. **Pradhan, A. S. and Bhatt, R. C.,** Effects of phosphor proportion and grain size on photon energy response of $CaSO_4$:Dy teflon discs, *Nucl. Instrum. Methods*, 161, 243, 1979.
10. **Pradhan, A. S. and Bhatt, R. C.,** Metal filters for the compensation of photon energy dependence of the response of $CaSO_4$:Dy-teflon TLD discs, *Nucl. Instrum. Methods*, 166, 497, 1979.
11. **Tochilin, E., Goldstein, N., and Lyman, J. T.,** The quality and LET dependence of three thermoluminescent dosimeters and their potential use as secondary standards, in Proc. 2nd Int. Conf. Luminescence Dosimetry, U.S. A.E.C. CONF-680920, NTIS, Springfield, Va., 1968, 424.
12. **Jayachandran, C. A.,** The response of thermoluminescent dosimetric lithium borates equivalent to air, water and soft tissue and of LiF TLD-100 to low energy X-rays, *Phys. Med. Biol.*, 15, 325, 1970.
13. **Storm, E. and Israel, H. I.,** Photon Cross Sections from 0.001 to 100 MeV for Elements 1 through 100, Los Alamos National Laboratory Internal Rep. No. LA-3753, Los Alamos, N.M., 1967.
14. **Millar, R. H. and Greening, J. R.,** A set of accurate X-ray interaction coefficients for low atomic number elements in the energy range 4-25 keV, *J. Phys. B*, 7, 2345, 1974.
15. **Millar, R. H.,** Atomic number dependence of the photoelectric cross section for photons in the energy range from 4.5 to 25 keV, *J. Phys. B*, 8, 2015, 1975.
16. **Budd, T., Marshall, M., People, L. H. J., and Douglas, J. A.,** The low and high temperature response of LiF dosimeters to X-rays, *Phys. Med. Biol.*, 24, 71, 1979.
17. **Hubbell, J. H.,** Photon mass attenuation and mass energy absorption coefficients for H, C, N, O, Ar and seven mixtures from 0.1 keV to 20 MeV, *Radiat. Res.*, 70, 58, 1977.
18. **Hubbell, J. H.,** Photon mass attenuation and energy absorption coefficients from 1 keV to 20 MeV, *Int. J. Appl. Radiat. Isot.*, 11, 1269, 1982.
19. **Carlsson, C. A.,** A criticism of existing tabulations of mass energy transfer and mass energy absorption coefficients, *Health Phys.*, 20, 653, 1971.
20. **Reddy, A. R. and Mehta, S. C.,** Dosimetry of very low energy photons, *Health Phys.*, 36, 175, 1979.
21. International Commission on Radiation Units and Measurements, Radiation Dosimetry: X-Rays Generated at Potentials of 5 to 150 kV, Rep. No. 17, ICRU Publications, Washington, D.C., 1970.
22. **Jayachandran, C. A.,** Calculated effective atomic number and kerma values for tissue-equivalent and dosimetry materials, *Phys. Med. Biol.*, 16, 617, 1971.
23. **Klick, C. C., Claffy, E. W., Gorbics, S. G., Attix, F. H., Schulman, J. H., and Allard, J. G.,** Thermoluminescence and colour centers in LiF:Mg, *J. Appl. Phys.*, 38, 3867, 1967.
24. **Becker, K., Cheka, J. S., and Oberhofer, M.,** Thermally stimulated exo-electron emission, thermoluminescence and impurities in LiF and BeO, *Health Phys.*, 19, 391, 1970.
25. **Portal, G., Berman, F., Blanchard, Ph., and Prigent, R.,** Improvement of sensitivity and linearity of radiothermoluminescent lithium fluoride, in Proc. 3rd Int. Conf. Luminescence Dosimetry, Risö Rep. No. 249, IAEA/AEC, Risö, Denmark, 1971, 410.
26. **Rossiter, M. J.,** The use of precision thermoluminescence dosimetry for intercomparison of absorbed dose, *Phys. Med. Biol.*, 20, 735, 1975.
27. **Liu, N. H., Gilliam, J. D., and Anderson, D. W.,** Response of LiF thermoluminescent dosimeters to ^{99m}Tc gamma rays, *Health Phys.*, 38, 359, 1980.
28. **Thomas, W. V., Mailii, H. D., and Mermagen, H.,** The whole and partial body dosimetry of the rat exposed to 250 kVp X-irradiation, *Health Phys.*, 14, 365, 1968.

29. **Gorbics, S. G. and Attix, F. H.,** LiF and CaF_2:Mn thermoluminescence dosimeters in tandem, *Int. J. Appl. Radiat. Isot.*, 19, 81, 1968.
30. **Pendurkar, H. K., Boulenger, R., Ghoos, L., Nicosi, W., and Martens, E.,** Energy response of certain TLDs and their application to the dose measurement, in Proc. 3rd Int. Conf. Luminescence Dosimetry, Risö Rep. No. 249, IAEA/AEC, Risö, Denmark, 1971, 1089.
31. **Nollman, C. E. and Thomasz, E.,** Study and application of properties of CaF_2:Dy dosimeters, *Nucl. Instrum. Methods*, 175, 68, 1980.
32. **Law, J.,** The dosimetry of low energy X-rays using LiF, *Phys. Med. Biol.*, 18, 38, 1973.
33. **Hankins, D. E.,** The energy response of TLD badges located on personnel, *Health Phys.*, 28, 80, 1975.
34. **Horowitz, Y. S. and Kalef-Ezra, J.,** Relative TL response of LiF-TLD to 4 keV X-rays, *Nucl. Instrum. Methods*, 188, 603, 1981.
35. **Storm, E., Lier, D. W., and Israel, H. I.,** Photon sources for instrument calibration, *Health Phys.*, 26, 179, 1974.
36. **Lasky, J. B. and Moran, P. R.,** Thermoluminescent response of LiF (TLD-100) to 5-30 keV electrons and the effect of annealing in various atmospheres, *Phys. Med. Biol.*, 22, 852, 1977.
37. **Lasky, J. B. and Moran, P. R.,** Thermoluminescent response of LiF (TLD-100) to 0.1-5 keV electrons: an energy range relationship and comparison of the TL glow curve with TSEE glow curves, *J. Appl. Phys.*, 50, 4951, 1979.
38. **Horowitz, Y. S.,** The theoretical and microdosimetric basis of thermoluminescence and applications to dosimetry, *Phys. Med. Biol.*, 26, 765, 1981.
39. **Mieke, S. and Nink, R.,** LiF:Ti as a material for thermoluminescence dosimetry, *J. Luminescence*, 18/19, 411, 1979.
40. **Mayhugh, M. R. and Fullerton, G. D.,** Altering the energy dependence of LiF-TLDs by pre-irradiation, *Med. Phys.*, 1, 275, 1974.
41. **Shinde, S. S. and Shastry, S. S.,** Energy dependence of sensitized $CaSO_4$:Dy TL phosphor, *Int. J. Appl. Radiat. Isot.*, 30, 501, 1979.
42. **Lakshmanan, A. R. and Bhatt, R. C.,** Photon energy dependence of sensitized TLD phosphors, *Nucl. Instrum. Methods*, 171, 259, 1980.
43. **Tochilin, E., Goldstein, N., and Miller, W. G.,** BeO as a thermoluminescent dosimeter, *Health Phys.*, 16, 1, 1969.
44. **Brunskill, R. T.,** The Preparation and Properties of Thermoluminescent Lithium Borate, UKAEA (PG) Rep. No. 837(W), Her Majesty's Stationery Office, London, 1968.
45. **Binder, W. and Cameron, J. R.,** Dosimetric properties of CaF_2:Dy, *Health Phys.*, 17, 613, 1969.
46. **Thompson, J. J. and Ziemer, P. L.,** Energy response of thermoluminescent dosimeters, *Health Phys.*, 22, 399, 1972.
47. **Thompson, J. J. and Ziemer, P. L.,** The thermoluminescent properties of lithium borate activated by Ag, *Health Phys.*, 25, 435, 1973.
48. **Scarpa, G.,** The dosimetric use of BeO as a TL material: a preliminary study, *Phys. Med. Biol.*, 15, 667, 1970.
49. **Crase, K. W. and Gammage, R. B.,** Improvements in the use of ceramic BeO for thermoluminescent dosimetry, *Health Phys.*, 29, 739, 1975.
50. **Barbina, V., Contento, G., Furetta, C., Malisan, M., and Padovani, R.,** Preliminary results on dosimetric properties of MgB_4O_7:Dy, *Radiat. Eff. Lett.*, 67, 55, 1981.
51. **Cuisimano, J. P., Cipperley, F. V., and Culley, J. C.,** Personnel dosimetry using thermoluminescent dosimeters, in Proc. 2nd Int. Conf. Luminescence Dosimetry, U.S. A.E.C. CONF-680920, NTIS, Springfield, Va., 1968, 733.
52. **Momeni, M. H., Jow, N., Countis, T., Worden, L., and Bradley, E.,** Estimation of effective gamma energy by differential responses of TLD-100 and TLD-200 dosimeters, *Health Phys.*, 28, 809, 1975.
53. **Yamashita, T., Nada, N., Oonishi, H., and Kitahara, S.,** Calcium sulphate activated by thulium or dysprosium for thermoluminescence dosimetry, *Health Phys.*, 21, 295, 1971.
54. **Dixon, R. L. and Watts, F. C.,** The use of BaF_2 thermoluminescence in determining radiation quality, *Phys. Med. Biol.*, 17, 81, 1972.
55. **Dixon, R. L. and Ekstrand, K. E.,** Thermoluminescence of $SrSO_4$:Dy and $BaSO_4$:Dy, *Phys. Med. Biol.*, 19, 196, 1974.
56. **McDougall, R. S. and Rudin, R. S.,** Thermoluminescent dosimetry of aluminium oxide, *Health Phys.*, 19, 281, 1970.
57. **Aypar, A.,** Studies on thermoluminescent $CaSO_4$:Dy for dosimetry, *Int. J. Appl. Radiat. Isot.*, 29, 369, 1978.
58. **McDougall, R. S. and Axt, J. C.,** A preliminary evaluation of $CaSO_4$:Dy thermoluminescent dosimeters, *Health Phys.*, 25, 612, 1973.

59. **Furetta, C. and Gennai, P.,** An extensive study on the dosimetric properties of $CaSO_4$:Dy, TLD-900, in low dose region, *Health Phys.*, 41, 674, 1981.
60. **Cameron, J. R., Suntharalingam, N., and Kenney, G. N.,** *Thermoluminescent Dosimetry*, University of Wisconsin Press, Madison, 1968.
61. **Spurny, Z., Milu, C., and Racoveanu, A.,** Comparison of X-ray beams using thermoluminescent dosimeters, *Phys. Med. Biol.*, 18, 276, 1973.
62. **Puite, K. J. and Crebolder, D. L. J. M.,** Energy dependence of TLDs for X-ray dose and dose distributions in a mouse phantom, *Phys. Med. Biol.*, 19, 341, 1974.
63. **Spurny, Z.,** Some new materials for TLD, *Nucl. Instrum. Methods*, 175, 71, 1980.
64. ICRP, *Recommendations of the International Commission Radiation Protection*, Publ. No. 26, Pergamon Press, Oxford, 1977.
65. **Marshall, M. and Docherty, J.,** Measurement of skin dose from low energy beta and gamma radiation using thermoluminescent discs, *Phys. Med. Biol.*, 16, 503, 1971.
66. **Gibson, J. A. B., Marshall, M., and Docherty, J.,** Comparison of calculated and measured surface dose using LiF discs, *Phys. Med. Biol.*, 16, 283, 1971.
67. **Lowe, D., Lakey, J. R. A., and Tymons, B. J.,** A new development in skin dosimetry, *Nucl. Instrum. Methods*, 169, 609, 1980.
68. **Charles, M. W.,** The development of a practical 5 mg cm^{-2} skin dosimeter, in Proc. 5th Int. Conf. Luminescence Dosimetry, Sao Paulo, Physikalisches Institut, Giessen, 1977, 313.
69. **Charles, M. W. and Khan, Z. U.,** Implementation of the ICRP recommendation on skin dose measurement using TLDs, *Phys. Med. Biol.*, 23, 972, 1978.
70. **Harvey, J. R. and Felstead, S. J.,** Thin layer TLDs based on high temperature self adhesive tape, *Phys. Med. Biol.*, 24, 1250, 1979.
71. **Shaw, K. B. and Wall, B. F.,** Performance Tests on the NRPB Thermoluminescent Dosimeter, Rep. No. R65, NRPB, 1977.
72. **Robertson, M. E. A. and Stewart, J. C.,** Thermoluminescent Dosimeter Patent Specification 1471 893, The Patent Office, London, 1977.
73. **Driscoll, C. M. H.,** Studies of the effect of LET on the TL properties of thin LiF layers, *Phys. Med. Biol.*, 23, 777, 1978.
74. **Uchrin, G.,** A new type of extremity dosimeter, *Nucl. Instrum. Methods*, 175, 173, 1980.
75. **Christensen, P. and Majborn, B.,** Boron diffused thermoluminescent surface layer in LiF-TLDs for skin dose assessment, *Nucl. Instrum. Methods*, 175, 74, 1980.
76. **Lasky, J. B. and Moran, P. R.,** TLD-100 diffused with boron: a "new surface sensitive TL phosphor", in Proc. 5th Int. Conf. Luminescence Dosimetry, Sao Paulo, Physikalisches Institut, Giessen, 1977, 122.
77. **Lowe, D., Lakey, J. R. A., and Yorke, A. V.,** A new thin film dosimeter, *Health Phys.*, 37, 417, 1979.
78. **Kocynski, A., Wolska-Witer, M., Botter-Jensen, L., and Christensen, P.,** Graphite mixed non-transparent LiF and $Li_2B_4O_7$:Mn TL dosimeters combined with a two-side reading system for β-γ dosimetry, in Proc. 4th Int. Conf. Luminescence Dosimetry, Institute of Nuclear Physics, Krakow, 1974, 641.
79. **Pradhan, A. S. and Bhatt, R. C.,** Graphite-mixed $CaSO_4$:Dy teflon TLD discs for beta dosimetry, *Phys. Med. Biol.*, 22, 873, 1977.
80. **Lakshmanan, A. R., Chandra, B., Pradhan, A. S., Kher, R. K., and Bhatt, R. C.,** The development of thin $CaSO_4$:Dy teflon TL dosimeters for beta dosimetry in personnel monitoring, *Int. J. Appl. Radiat. Isot.*, 31, 107, 1980.
81. **Nilsson, B., Schnell, P. D., and Ssengabi, J.,** Build-up studies with thin thermoluminescent dosimeters, in Proc. 4th Int. Conf. Luminescence Dosimetry, Institute of Nuclear Physics, Krakow, 1974, 897.
82. **Ruden, B. I. and Bengtsson, G.,** TLD measurements of dose distribution around a beta ray applicator, *Phys. Med. Biol.*, 19, 186, 1974.
83. **Ehrlich, M.,** Influence of size of CaF_2:Mn TLDs on Co-60 gamma ray dosimetry in extended media, in Proc. 3rd Int. Conf. Luminescence Dosimetry, Risö Rep. No. 249, IAEA/AEC, Risö, Denmark, 1971, 550.
84. **Spencer, L. V. and Attix, F. H.,** A theory of cavity ionization, *Radiat. Res.*, 3, 239, 1955.
85. **National Committee on Radiation Protection and Measurements,** Stopping Powers for Use with Cavity Chambers, Handbook No. 79, National Bureau of Standards, Washington, D.C., 1961.
86. **O'Brien, K.,** Monte Carlo calculations of the energy response of lithium fluoride dosimeters to high energy X-rays (<30 MeV), *Phys. Med. Biol.*, 22, 836, 1977.
87. **Burlin, T. E.,** A general theory of cavity ionization, *Br. J. Radiol.*, 39, 727, 1966.
88. **Bertilsson, G.,** Application of the cavity theory to LiF-teflon dosimeters, in Proc. 4th Int. Conf. Luminescence Dosimetry, Institute of Nuclear Physics, Krakow, 1974, 907.
89. **Horowitz, Y. S. and Dubi, A.,** A proposed modification of Burlin's general cavity theory for photons, *Phys. Med. Biol.*, 27, 867, 1982.

90. **Horowitz, Y. S., Moscovitch, M., and Dubi, A.,** Applications of modified general cavity theory to thermoluminescent dosimetry, in Proc. World Conf. Med. Phys. Biomed. Eng., Hamburg, 1982.
91. **Ogunleye, O. T., Attix, F. H., and Paliwal, B. R.,** Comparison of Burlin cavity theory with LiF-TLD measurements for Co-60 gamma rays, *Phys. Med. Biol.*, 25, 203, 1980.
92. **Eggermont, G., Janssens, A., Jacobs, R., and Thielens, G.,** A discussion on the validity of cavity theories and a comparison with experimental results, in Proc. 4th Symp. Microdosimetry, CEC, Italy, 1973, 733.
93. **Tobata, T., Ito, R., and Okabe, S.,** Generalized semi-empirical equation for the extrapolated range of electrons, *Nucl. Instrum. Methods*, 103, 85, 1972.
94. **Paliwal, B. R. and Almond, P. R.,** Electron attenuation characteristics of LiF, *Health Phys.*, 31, 151, 1976.
95. **Simons, G. G. and Jule, T. J.,** Gamma ray heating measurements in zero power fast reactors with TLDs, *Nucl. Sci. Eng.*, 53, 162, 1974.
96. **Storm, E. and Israel, H. I.,** Photon cross sections from 1 keV to 100 MeV for elements Z = 1 to Z = 100, *Nucl. Data Tables*, A7, 565, 1970.
97. **Berger, M. J. and Seltzer, S. M.,** Tables of Energy Losses and Ranges of Electrons and Positrons, NASA SP-3012, 1964; Additional Stopping Powers and Range Tables for Protons, Mesons and Electrons, NASA SP-3036, NTIS, Springfield, Va., 1966.
98. **Costrell, L.,** Scattered radiation from large Cs-137 sources, *Health Phys.*, 8, 491, 1962.
99. **Evans, R. D.,** *The Atomic Nucleus*, McGraw-Hill, New York, 1955.
100. **Kalef-Ezra, J., Horowitz, Y. S., and Mack, J. M.,** Electron backscattering from low z thick absorbers, *Nucl. Instrum. Methods*, 195, 587, 1982.
101. **Baltakmens, T.,** Energy loss of beta particles on backscattering, *Nucl. Instrum. Methods*, 125, 169, 1975.
102. **Kalef-Ezra, J. and Horowitz, Y. S.,** Backscattering corrections for beta dose-rate measurements using thermoluminescent dosimetry, *PACT J. Eur. Study Group Phys. Chem. Math. Tech. Appl. Archaeol.*, 3, 428, 1979.
103. **Burlin, T. E., Snelling, R. J., and Owen, B.,** The application of general cavity ionization theory to the dosimetry of electron fields, in Proc. 2nd Symp. Microdosimetry, CEC, Italy, 1969, 455.
104. **Almond, P. R. and McCray, K.,** The response of LiF, CaF_2 and $Li_2B_4O_7$:Mn to high energy radiations, *Phys. Med. Biol.*, 15, 335, 1970.
105. **Paliwal, B. R. and Almond, P. R.,** Applications of cavity theories for electrons to LiF dosimeters, *Phys. Med. Biol.*, 20, 547, 1975.
106. **Almond, P. R. and McCray, K.,** The energy response of LiF, CaF_2 and $Li_2B_4O_7$:Mn to high energy radiations, *Phys. Med. Biol.*, 15, 746, 1970.
107. **Burlin, T. E.,** The energy response of LiF, CaF_2 and $Li_2B_4O_7$:Mn to high energy radiations, *Phys. Med. Biol.*, 15, 558, 1970.
108. International Commission on Radiation Units and Measurements, Radiation Dosimetry: Electrons with Initial Energies between 1 and 50 MeV, Publ. No. 21, ICRU Publications, Washington, D.C., 1972.
109. **Ogunleye, O. T. and Paliwal, B. R.,** An investigation of the high energy dependence of LiF dosemeters in low Z media, *Int. J. Appl. Radiat. Isot.*, 31, 63, 1980.
110. **Shiragai, A.,** An approach to an analysis of the energy response of LiF-TLD to high energy electrons, *Phys. Med. Biol.*, 22, 490, 1977.
111. **Lowe, D.,** The application of cavity theory to the dosimetry of electron fields, *Phys. Med. Biol.*, 24, 162, 1979.
112. **Fregene, A. O.,** A comparison of LiF and $FeSO_4$ dosimetry with cavity theory for high energy electrons, *Radiat. Res.*, 65, 20, 1976.
113. **Holt, J. G., Edelstein, G. R., and Clark, T. E.,** Energy dependence of the response of LiF-TLD rods in high energy electron fields, *Phys. Med. Biol.*, 20, 559, 1975.
114. **Rossi, B.,** *High Energy Particles*, Prentice-Hall, Englewood Cliffs, N.J., 1956.
115. **Fregene, A. O.,** unpublished (reported on in Reference 117), 1977.
116. **Suntharalingam, N. and Cameron, J. R.,** The response of LiF to high energy electrons, *Ann. N.Y. Acad. Sci.*, 161, 77, 1969.
117. **Ogunleye, O. T.,** An experimental examination of cavity theory for electrons, *Phys. Med. Biol.*, 27, 573, 1982.
118. **Suntharalingam, N. and Mansfield, C. M.,** The use of TL dosimeters in high energy electron and photon beam clinical dose measurements, in Proc. 4th Int. Conf. Luminescence Dosimetry, Institute of Nuclear Physics, Krakow, 1974, 859.
119. **Suntharalingam, N. and Mansfield, C. M.,** The use of TL dosimeters in high energy electron and photon beam clinical dose measurements, IAEA Symp. Adv. Biomed. Dosimetry, IAEA-SM-193/42, IAEA, Vienna, 1975, 859.

120. **Gantchew, M. G. and Toushlekova, K.,** The influence of the composition of LiF-TLD materials on their sensitivity to high energy electrons, *Phys. Med. Biol.*, 21, 300, 1976.
121. **Binks, C.,** Energy dependence of LiF dosimeters at electron energies from 10 to 35 MeV, *Phys. Med. Biol.*, 14, 327, 1969.
122. **Almond, P. R., Wright, A., and Lontz J. F., II,** The use of LiF-TLDs to measure the dose distribution of a 15 MeV electron beam, *Phys. Med. Biol.*, 12, 389, 1967.
123. **Fregene, A. O.,** The influence of the composition of LiF-TLD materials on their sensitivity to high energy electrons, *Phys. Med. Biol.*, 22, 372, 1977.
124. **Gantchew, M.,** The influence of the composition of LiF TLD materials on their sensitivity to high energy electrons, *Phys. Med. Biol.*, 22, 374, 1977.
125. **Horowitz, Y. S., Fraier, I., Kalef-Ezra, J., Pinto, H., and Goldbart, Z.,** Non-universality of the TL-LET response in thermoluminescent LiF: the effect of batch composition, *Phys. Med. Biol.*, 24, 1268, 1979.
126. **Horowitz, Y. S., Fraier, I., Kalef-Ezra, J., Pinto, H., and Goldbart, Z.,** Non-universality of the TL-LET response in thermoluminescent $Li_2B_4O_7$: the effect of batch composition, *Nucl. Instrum. Methods*, 165, 27, 1979.
127. **Horowitz, Y. S., Kalef-Ezra, J., Moscovitch, M., and Pinto, H.,** Further studies on the non-universality of the TL-LET response in thermoluminescent LiF and $Li_2B_4O_7$: the effect of high temperature TL, *Nucl. Instrum. Methods*, 172, 479, 1980.
128. **Goldstein, N.,** Dose-rate dependence of LiF for exposures above 15,000 R per pulse, *Health Phys.*, 22, 90, 1972.
129. **Gorbics, S. G., Attix, F. H., and Kerris, K.,** TL dosimeters for high dose applications, *Health Phys.*, 25, 499, 1973.
130. **Pinkerton, H. H., Holt, J. G., and Almond, P. R.,** Direct intercomparison of absorbed dose by TL dosimetry, in Proc. 1st Int. Conf. Luminescence Dosimetry, U.S. A.E.C. CONF-650637, Springfield, Va., 1965, 363.
131. **Pinkerton, A. P., Laughlin, J. S., and Holt, J. G.,** Energy dependence of LiF dosemeters at high electron energies, *Phys. Med. Biol.*, 11, 129, 1966.
132. **Eggermont, G., Jacobs, R., Jansens, A., Segart, O., and Thielens, G.,** Dose relationship, energy response and rate dependence of LiF-100, LiF-7 and $CaSO_4$:Mn from 8 keV to 30 MeV, in Proc. 3rd Int. Conf. Luminescence Dosimetry, Risö Rep. No. 249, IAEA/AEC, Risö, Denmark, 1971, 444.
133. **Crosby, E. H., Almond, P. R., and Shalek, R. J.,** Energy dependence of lithium fluoride dosemeters at high energies, *Phys. Med. Biol.*, 11, 129, 1966.
134. **Turner, A. P. and Anderson, D. W.,** Thermoluminescent response of LiF to high energy photons, *Phys. Med. Biol.*, 18, 46, 1973.
135. **Benner, S., Johansson, J. M., Lindskoug, B., and Nyman, P. T.,** A miniature LiF dosemeter for in vivo measurements, in Proc. Symp. Solid State Chem. Radiat. Dosimeters Biol. Med., SM-78/25, 1967, 65.
136. **Bistrovic, M., Maricic, Z., Greenfield, M. A., Breyer, B., Dvornik, I., Slaus, I., and Tomas, P.,** The energy response of LiF, film and a chemical dosimeter to high energy photons and electrons, *Phys. Med. Biol.*, 21, 414, 1976.
137. **Fregene, A. O.,** LiF response to high energy X-rays and electrons relative to Co-60 photons, *Int. J. Appl. Radiat. Isot.*, 28, 965, 1977.
138. **Liu, B. M. and Bagne, F.,** The response of LiF to high energy photons and electrons, in Proc. 4th Int. Conf. Luminescence Dosimetry, Institute of Nuclear Physics, Krakow, 1974, 937.
139. **Antoku, S., Sunayashiki, T., Takeoka, S., and Takeshita, K.,** Energy dependence of LiF for high energy electrons, *Phys. Med. Biol.*, 15, 889, 1973.
140. **Law, J. and Naylor, G. P.,** Comparison of ionization, ferrous sulphate and thermoluminescence dosimetry for megavoltage photons, *Phys. Med. Biol.*, 16, 67, 1971.
141. **Nakajima, T., Hiraoka, T., and Habu, T.,** Energy dependence of LiF and CaF_2 TLDs for high energy electrons, *Health Phys.*, 14, 266, 1968.
142. **Ehrlich, M.,** Response of TLDs to 15 MeV electrons and Co-60 gamma rays, *Health Phys.*, 18, 287, 1970.
143. **Rossow, J. and Streeter, H. D.,** Interaction of LiF-TLD with high energy electrons, in Proc. 4th Annu. Meet. Fachverband fur Strahlenschutz, Berlin, 1969, 471.
144. **Degner, W., Hegewald, H., and Windelband, R.,** Dosimetry of high energy electrons using LiF thermoluminescent dosimetry, *Radiobiol. Radiother.*, 113, 355, 1972.
145. **Pinkerton, A.,** Comparison of calorimetric and other methods for the determination of absorbed dose, in *Proc. Symp. High Energy Radiation Therapy Dosimetry*, N.Y. Academy of Sciences, New York, 1968, 235.
146. **Becker, K.,** *Solid State Dosimetry*, CRC Press, Cleveland, 1973.

147. **Furuta, Y. and Tanaka, S.,** Neutron dosimetry by thermoluminescence dosimeter, in Proc. Int. Symp. Radiat. Phys., Bhabha Atomic Research Center, Bombay, India, 1974, 209.
148. **Griffith, R. V., Hankins, D. E., Gammage, R. B., Tommasino, L., and Wheeler, R. V.,** Recent developments in personnel neutron dosimeters, *Health Phys.*, 36, 235, 1979.
149. **Becker, K.,** Some advances in solid state fast neutron dosimetry, in Proc. 2nd Symp. Neutron Dosimetry Biol. Med., Burger, G. and Ebert, H. G., Eds., EUR 5273-d-e-f, EURATOM, Brussels, 1975, 479.
150. **Ayyangar, K., Reddy, A. R., and Brownell, G. L.,** Some studies on thermoluminescence from LiF and other materials exposed to neutrons and other radiation, in Proc. 2nd Int. Conf. Luminescence Dosimetry, U.S. A.E.C. CONF-680920, Springfield, Va., 1968, 525.
151. **Lowe, D.,** Recommendations for the efficient use of LiF-TLDs in mixed neutron-gamma fields, *Nucl. Instrum. Methods*, 177, 577, 1980.
152. **Horowitz, Y. S., Dubi, A., and Ben Shahar, B.,** Self-shielding factors for TLD-600 and TLD-100 in an isotropic flux of thermal neutrons, *Phys. Med. Biol.*, 21, 976, 1976.
153. **Horowitz, Y. S., Ben Shahar, B., Mordechai, S., Dubi, A., and Pinto, H.,** Thermoluminescence in LiF induced by mono-energetic 13.8 meV and 81.0 meV diffracted neutrons: the intrinsic TL response per absorbed neutron, *Radiat. Res.*, 69, 402, 1977.
154. **Horowitz, Y. S., Freeman, S., and Dubi, A.,** Limitations of the paired LiF TLD-600 and TLD-700 technique for the estimation of γ ray dose in mixed n-γ radiation fields, *Nucl. Instrum. Methods*, 160, 317, 1979.
155. **Horowitz, Y. S., Freeman, S., and Dubi, A.,** Monte Carlo calculated absorption probabilities for cylindrical and rectangular TLD probes in isotropic thermal neutron fluences, *Nucl. Instrum. Methods*, 160, 313, 1979.
156. **Scarpa, G.,** Neutron dosimetry via TLD, in Proc. 2nd Int. Congr. Radiat. Prot. Assoc., Brighton, U.K., 1970, 255.
157. **Reddy, A. R., Ayyangar, K., and Brownell, G.,** TL response of LiF to reactor neutrons, *Radiat. Res.*, 40, 552, 1969.
158. **Distenfeld, C., Bishop, W., and Colvett, D.,** Thermoluminescent neutron dosimetry system, in Proc. 1st Int. Conf. Luminescence Dosimetry, U.S. A.E.C. CONF-650637, NTIS, Springfield, Va., 1965, 457.
159. **Dua, S. K., Boulenger, R., Ghoos, L., and Martens, E.,** Mixed neutron-gamma ray dosimetry, in Proc. 3rd Int. Conf. Luminescence Dosimetry, Risö Rep. No. 249, IAEA/AEC, Risö, Denmark, 1971, 1074.
160. **Simpson, R. E.,** Response of LiF to reactor neutrons, in Proc. 1st Int. Conf. Luminescence Dosimetry, U.S. A.E.C. CONF-650637, NTIS, Springfield, Va., 1965, 444.
161. **Wallace, R. H. and Ziemer, P. L.,** Studies on the thermoluminescence of Mn activated $Li_2B_4O_7$, in Proc. 2nd Int. Conf. Luminescence Dosimetry, U.S. A.E.C. CONF-680920, NTIS, Springfield, Va., 1968, 140.
162. **Scarpa, G., Benincasa, L., and Ceravolo, L.,** BeO Thermoluminescence Applied to Radiation Dosimetry, IAEA-SM-160/39, IAEA, Vienna, 1973, 207.
163. **Ayyangar, K., Lakshmanan, A. R., Chandra, B., and Ramadas, K.,** A comparison of thermal neutron and gamma ray sensitivities of common TLD materials, *Phys. Med. Biol.*, 19, 665, 1974.
164. **Puite, K. J.,** Thermoluminescent sensitivity of CaF_2:Mn in a mixed neutron-gamma ray field, *Health Phys.*, 20, 437, 1971.
165. **Prokic, M.,** Determination of the sensitivity of the CaF_2:Mn TLD to neutrons, in Proc. 3rd Int. Conf. Luminescence Dosimetry, Risö Rep. No. 249, IAEA/AEC, Risö, Denmark, 1971, 1051.
166. **Horowitz, Y. S., Ben Shahar, B., Dubi, A., and Pinto, H.,** Thermoluminescence in CaF_2:Dy and CaF_2:Mn induced by monoenergetic, parallel beam, 81.0 meV diffracted neutrons, *Phys. Med. Biol.*, 22, 500, 1977.
167. **Rossiter, M. J., Lewis, V. E., and Wood, J. W.,** The response of TLDs to fast (14.7 MeV) and thermal neutrons, *Phys. Med. Biol.*, 22, 731, 1977.
168. **Beach, J. L. and Huang, C. Y.,** Mixed field dosimetry with $CaSO_4$:Tm and $CaSO_4$(Tm):Li, *Health Phys.*, 31, 452, 1976.
169. **Mehta, S. K. and Sengupta, S.,** Mixed field dosimetry with Al_2O_3(Si,Ti) TL phosphor, *Nucl. Instrum. Methods*, 187, 515, 1981.
170. **Vlasov, M. F., Dunford, C. L., Schmidt, J. J., and Lemmel, H. D.,** Status of Neutron Cross Section Data for Reactor Radiation Measurements. I. Reactions of High Priority, IAEA Document INDC (NDS)-47/L, IAEA, Vienna, 1972.
171. **Tanaka, S. and Furuta, Y.,** Usage of a thermoluminescent dosimeter as a thermal neutron detector with high sensitivity, *Nucl. Instrum. Methods*, 133, 495, 1976.
172. **Horowitz, Y. S. and Dubi, A.,** Comment on the use of thermoluminescent dosimeters as thermal neutron detectors, *Nucl. Instrum. Methods*, 146, 435, 1977.
173. **Beckurtz, K. H. and Wirtz, K.,** *Neutron Physics*, Springer-Verlag, Basel, 1964.
174. Harshaw Patent Specification, No. 1,059,518, The Patent Office, London, U.S. Patent No. 3,320,180, 1967.

175. **Woodley, R. G. and Johnson, N. M.,** Thermoluminescence induced by low energy alpha particles, in Proc. 1st Int. Conf. Luminescence Dosimetry, U.S. A.E.C. CONF-650637, NTIS, Springfield, Va., 1965, 502.
176. **Piesch, E., Burgkhardt, B., and Sayed, A. M.,** Activation and damage effects in TLD-600 after neutron irradiation, *Nucl. Instrum. Methods,* 157, 179, 1978.
177. **Lakshmanan, A. R. and Ayyangar, K.,** Thermoluminescence response of $CaSO_4$:Dy and $Li_2B_4O_7$:Mn to polonium-210 alpha radiation, *Health Phys.,* 31, 284, 1976.
178. **Goldstein, N., Tochilin, E., and Miller, W. G.,** Millirad and megarad dosimetry with LiF, *Health Phys.,* 14, 159, 1968.
179. **Podgorsak, E. B., Moran, P. R., and Cameron, J. R.,** Interpretation of resolved glow curve shapes in LiF (TLD-100), in Proc. 3rd Int. Conf. Luminescence Dosimetry, Risö Rep. No. 249, IAEA/AEC, Risö, Denmark, 1971, 1.
180. **Attix, F. H.,** Luminescence and exoelectron dosimetry in personnel monitoring, in Proc. Symp. Adv. Phys. Biol. Radiat. Detectors, IAEA-SM-143/25, IAEA, Vienna, 1971, 3.
181. **Horowitz, Y. S.,** The thermal neutron sensitivity of LiF (TLD-700:Harshaw): the effect of sample size and batch origin, *Phys. Med. Biol.,* 23, 340, 1978.
182. **Way, K. Ed.,** Nuclear Data Tables, Section A, Vol. 3, No. 4-6, and Vol. 5, No. 1-2, Academic Press, London, 1976 and 1968.
183. **Rockwell, T., Ed.,** *Reactor Shielding Design Manual,* Van Nostrand Reinhold, London, 1956, 20.
184. **James, M. F., Steel, B. G., and Story, J. S.,** UK Atomic Energy Authority Report, AERE-M-640, Harwell, Buck., 1960.
185. **Lederer, C. M., Ed.,** *Table of Isotopes,* 6th ed., John Wiley & Sons, Chichester, 1967.
186. **Ayyangar, K., Lakshmanan, A. R., and Chandra, B.,** Mixed field dosimetry with $CaSO_4$:Dy, *Phys. Med. Biol.,* 19, 656, 1974.
187. **Lowe, D.,** Tritium production in LiF thermoluminescent dosimeters, *Int. J. Appl. Radiat. Isot.,* 31, 787, 1980.
188. **Busuoli, G., Cavallini, A., Fasso, A., and Rimondi, O.,** Mixed radiation dosimetry with LiF(TLD-100), *Phys. Med. Biol.,* 15, 673, 1970.
189. **Mason, E. W.,** The effect of thermal neutron irradiation on the TL response of Conrad type 7 LiF, *Phys. Med. Biol.,* 15, 79, 1970.
190. **Furuta, Y. and Tanaka, S.,** Response of ^{6}LiF and ^{7}LiF TL dosimeters to fast neutrons, *Nucl. Instrum. Methods,* 104, 365, 1972.
191. **Endres, G. W. R. and Lucas, A. C.,** The use of a high temperature trap in ^{7}LiF for fast neutron dosimetry, in Proc. 4th Int. Conf. Luminescence Dosimetry, Institute of Nuclear Physics, Krakow, 1974, 1141.
192. **Nash, A. E. and Johnson, T. L.,** LiF(TLD-600) thermoluminescence detectors for mixed thermal neutron and gamma ray dosimetry, in Proc. 5th Int. Conf. Luminescence Dosimetry, Sao Paulo, Physikalisches Institut, Giessen, 1977, 393.
193. **Tuyn, J. W. N.,** Response of LiF (TLD-700) thermoluminescence detectors to pions and high energy neutrons, *Nucl. Instrum. Methods,* 175, 40, 1980.
194. **Hoffman, W. and Moller, G.,** Heavy particle dosimetry with high temperature peaks of TL materials, *Nucl. Instrum. Methods,* 175, 205, 1980.
195. **Morata, S. P. and Nambi, K. S. V.,** Development of hydrogen doped TL phosphors for fast neutron dosimetry, in Proc. 5th Int. Conf. Luminescence Dosimetry, Sao Paulo, Physikalisches Institut, Giessen, 1977, 155.
196. **Horowitz, Y. S. and Freeman, S.,** Response of ^{6}LiF and ^{7}LiF thermoluminescent dosimeters to neutrons incorporating the TL-LET dependency, *Nucl. Instrum. Methods,* 157, 393, 1978.
197. **Tanaka, S. and Furuta, Y.,** Estimation of gamma ray exposure in mixed n-γ fields by ^{6}LiF and ^{7}LiF in pair use, *Nucl. Instrum. Methods,* 117, 93, 1974.
198. **Tanaka, S. and Furuta, Y.,** Revised energy response of ^{6}LiF and ^{7}LiF thermoluminescence dosimeters to neutrons, *Nucl. Instrum. Methods,* 140, 395, 1977.
199. **Furuta, Y. and Tanaka, S.,** The relation between light conversion efficiency and stopping power of charged particles in thermoluminescent dosimetry, in Proc. 4th Int. Conf. Luminescence Dosimetry, Institute of Nuclear Physics, Krakow, 1974, 97.
200. **Wingate, C. L., Tochilin, E., and Goldstein, N.,** Response of LiF to neutrons and charged particles, in Proc. 1st Int. Conf. Luminescence Dosimetry, U.S. A.E.C. CONF-650637, NTIS, Springfield, Va., 1965, 421.
201. **Goldstein, N., Miller, W. G., and Rago, P. F.,** Additivity of neutron and gamma ray exposures for TL dosimeters, *Health Phys.,* 18, 157, 1970.
202. **McGinley, P. H.,** Response of LiF to fast neutrons, *Health Phys.,* 23, 105, 1972.
203. **Spurny, F., Medioni, R., and Portal, G.,** Energy transfer to TLD materials by neutrons, *Radioprotection,* 11, 219, 1976.

204. **Handloser, J. S.,** Fast neutron TL dosimetry, in Personnel Dosimetry for Radiation Accidents, STI/PUB/99, IAEA, Vienna, 1965, 115.
205. **Blum, E., Bewley, D. K., and Heather, J. D.,** Thermoluminescent dosimetry for fast neutron beams using CaF_2:Mn, *Phys. Med. Biol.*, 18, 226, 1973.
206. **Attix, F. H., Theus, R. B., Shapiro, P., Surratt, R. E., Nash, A. E., and Gorbics, S. G.,** Neutron beam dosimetry at the NRL cyclotron, *Phys. Med. Biol.*, 16, 497, 1973.
207. **Smathers, J., Otte, V., Smith, A., and Almond, P. R.,** *Med. Phys.*, 2, 153, 1975.
208. **Tochilin, E., Shumway, B. W., and Kohler, G. D.,** Response of photographic emulsions to charged particles and neutrons, *Radiat. Res.*, 4, 467, 1956.
209. **Wagner, E. B. and Hurst, G. S.,** A Geiger-Muller gamma ray dosimeter with low neutron sensitivity, *Health Phys.*, 5, 20, 1961.
210. **Barschall, H. H. and Goldberg, E.,** Response of tissue equivalent ionization chamber to neutrons, *Med. Phys.*, 4, 141, 1977.
211. **Alsmiller, R. G., Jr. and Barish, J.,** The calculated response of ^{6}LiF albedo dosimeters to neutrons with energies less than 400 MeV, *Health Phys.*, 26, 13, 1974.
212. **Spurny, F., Medioni, R., and Portal, G.,** Energy transfer to some TLD materials by neutrons: comparison of theoretical and experimental data, *Nucl. Instrum. Methods*, 138, 165, 1976.
213. **Spurny, F., Medioni, R., Pescoyre, R., and Portal, G.,** Theoretical and experimental studies of energy transfer by fast neutrons to some luminescent materials, in Proc. 3rd Symp. Neutron Dosimetry in Biology and Medicine, Burger, G. and Ebert, H. G., Eds., EUR 5848 DE/EN/FR, EURATOM, Brussels, 1977, 747.
214. **Rinard, P. M. and Simons, G. G.,** Calculated neutron sensitivities of CaF_2 and ^{7}LiF thermoluminescent dosimeters, *Nucl. Instrum. Methods*, 158, 545, 1979.
215. **Patrick, J. W., Stephens, L. D., Thomas, R. H., and Kelly, L. S.,** The efficiency of ^{7}LiF TLDs for measuring 250 MeV/amu $^{+6}$C ions, *Health Phys.*, 28, 615, 1975.
216. **Jahnert, B.,** The response of TLD-700 thermoluminescent dosimeters to protons and alpha particles, *Health Phys.*, 23, 112, 1971.
217. **Bewley, D. K. and Parnell, C. J.,** The fast neutron beam from the MRC cyclotron, *Br. J. Radiol.*, 42, 281, 1969.
218. **Oltman, B. G., Kastner, J., Tedeschi, P., and Beggs, J. N.,** The effects of fast neutron exposure on the ^{7}LiF TL response to gamma rays, *Health Phys.*, 13, 918, 1967.
219. **Mason, E. W., Harrison, N. T., and Linsley, G. S.,** Neutron TL dosimetry, in Proc. 2nd Int. Congr. Radiat. Prot. Assoc., Brighton, U.K., 1970, 112.
220. **Wallace, R. H., Ziemer, P. L., Kastner, J., and Oltman, B. G.,** The relationship between encapsulation and apparent fast neutron induced fading in TLD, *Health Phys.*, 20, 221, 1970.
221. **Bloch, P. H. and Weber, R. J.,** Changes in gamma sensitivity of ^{7}LiF dosimeters, *Health Phys.*, 23, 123, 1972.
222. **Pearson, D. W. and Moran, P. R.,** Fast Neutron Activation Dosimetry with Thermoluminescent Dosimeters, USERDA Rep. No. COO-1105-227, NTIS, Springfield, Va., 1976.
223. **Lakshmanan, A. R., Rajendran, K. V., and Ayyangar, K.,** Thermal neutron and gamma ray mixed field dosimetry with $Li_2B_4O_7$:Mn, *Health Phys.*, 30, 489, 1976.
224. **Schumacher, H. and Krauss, O.,** Thermoluminescence dosimetry (TLD) of a mixed neutron and gamma ray field, in Proc. 3rd Symp. Neutron Dosimetry Biol. Med., Burger, G. and Ebert, H. G., Eds., EUR 5848, DE/EN/FR, EURATOM, Brussels, 1977, 713.
225. **Pradhan, A. S., Bhatt, R. C., Lakshmanan, A. R., Chandra, B., and Shinde, S. S.,** A thermoluminescent fast neutron dosimeter based on pellets of $CaSO_4$:Dy mixed with sulfur, *Phys. Med. Biol.*, 23, 723, 1978.
226. **Mayhugh, M. R. and Watanabe, S.,** Fast neutron detection by phosphor activation, *Health Phys.*, 27, 305, 1974.
227. **Mayhugh, M. R., Watanabe, S., and Muccillo, R.,** Thermal neutron dosimetry by phosphor activation, in Proc. 3rd Int. Conf. Luminescence Dosimetry, Risö Rep. No. 249, IAEA/AEC, Risö, Denmark, 1971, 1040.
228. **Bhatt, R. C., Lakshmanan, A. R., Chandra, B., and Pradhan, A. S.,** Fast neutron dosimetry using sulfur activation in $CaSO_4$:Dy thermoluminescence dosimeters, *Nucl. Instrum. Methods*, 152, 527, 1978.
229. **Santos, E. N. D. and Muccillo, R.,** Intermediate neutron detection by thermoluminescence, *Nucl. Instrum. Methods*, 165, 561, 1979.
230. **Henson, A. M. and Thomas, R. H.,** Measurements of the efficiency of ^{7}LiF TLDs of heavy ions, *Health Phys.*, 34, 389, 1978.
231. **DeCastro, T. M., McCalm, J. B., Smith, A. R., and Thomas, R. H.,** Efficiency of ^{7}LiF TLDs to 798 MeV protons relative to Co-60 photons, Internal Report, HPN No. 52, Lawrence Berkeley Laboratory, Health Physics Department, Berkeley, Calif., 1976.

232. **Pernicka, F. and Spurny, F.,** Efficiency of some luminescent detectors for 600 MeV protons, *Nucl. Instrum. Methods,* 172, 435, 1980.
233. **Parker, C. V., Blake, K. R., and Nelson, J. B.,** The response of TLDs to protons with energy to 137 MeV, in Proc. 2nd Int. Conf. Luminescence Dosimetry, U.S. A.E.C. CONF-680920, NTIS, Springfield, Va., 1968, 438.
234. **Hubner, K., Henniger, J., and Negwer, D.,** LET dependence of the TL response and range measurements of heavy charged particles in thermoluminescence detectors, *Nucl. Instrum. Methods,* 175, 34, 1980.
235. **Patrick, J. W., Stephens, L. D., Thomas, R. H., and Kelly, L. S.,** The efficiency of ^{7}LiF TLDs to high LET particles relative to Co-60 gamma rays, *Health Phys.,* 30, 295, 1976.
236. **Henson, A. and Thomas, R. H.,** Internal Report, HPN No. 58, Lawrence Berkeley Laboratory, Health Physics Department, Berkeley, Calif., 1976.
237. **Horowitz, Y. S. and Kalef-Ezra, J.,** Relative thermoluminescent response of LiF-TLD to fission fragments, *Nucl. Instrum. Methods,* 187, 519, 1981.
238. **Bartlett, D. T. and Edwards, A. S.,** The light conversion efficiency of TLD-700 for alpha particles relative to Co-60 gamma radiation, *Phys. Med. Biol.,* 24, 1276, 1979.
239. **Fain, J., Montret, M., and Sakraoui, L.,** Thermoluminescent response of CaF_2:Dy and LiF:Mg,Ti under heavy ion bombardment, *Nucl. Instrum. Methods,* 175, 37, 1980.
240. **Harvey, J. R. and Townsend, S.,** The measurement of dose from a plane alpha source, in Proc. 3rd Int. Conf. Luminescence Dosimetry, Risö Rep. No. 249, IAEA/AEC, Risö, Denmark, 1971, 1015.
241. **Majborn, B., Botter-Jensen, L., and Christensen, P.,** On the relative efficiency of TL phosphors for high LET radiation, in Proc. 5th Int. Conf. Luminescence Dosimetry, Sao Paulo, Physikalisches Institut, Giessen, 1977, 124.
242. **Lowder, S. B. and De Planque, G.,** Health and Safety Laboratory, USERDA Internal Rep. No. 102, 1977.
243. **Cooke, D. W. and Hogstrom, K. R.,** Thermoluminescent response of LiF and $Li_2B_4O_7$:Mn to pions, *Phys. Med. Biol.,* 25, 657, 1980.
244. **Aitken, M. J., Tite, M. S., and Fleming, J.,** Thermoluminescent response to heavily ionizing radiations, in Proc. 1st Int. Conf. Luminescence Dosimetry, U.S. A.E.C. CONF-650637, NTIS, Springfield, Va., 1965, 490.
245. **Lucas, A. C. and Rainbolt, A.,** Response of CaF_2 and LiF to alpha particles, in Proc. 2nd Int. Conf. Luminescence Dosimetry, U.S. A.E.C. CONF-680920, NTIS, Springfield, Va., 1968, 456.
246. **Montret-Brugerolle, M.,** Ph.D. thesis, University of Clermont-Ferrand II, France, 1980.
247. **Zimmerman, D. W.,** Relative thermoluminescence effects of alpha and beta radiation, *Radiat. Eff.,* 14, 81, 1972.
248. **Jahnert, B.,** Thermoluminescence research of protons and alpha particles with LiF (TLD-700), in Proc. 3rd Int. Conf. Luminescence Dosimetry, Risö Rep. No. 249, IAEA/AEC, Risö, Denmark, 1971, 1031.
249. **Kalef-Ezra, J. and Horowitz, Y. S.,** Heavy charged particle thermoluminescence dosimetry: track structure theory and experiments, *Int. J. Appl. Radiat. Isot.,* 11, 1085, 1982.
250. **Janssens, A.,** Comments on "A proposed modification of Burlin's general cavity theory for photons", *Phys. Med. Biol.,* in press.
251. **Kearsley, E.,** General Cavity Theory for Photon and Neutron Dosimetry, Ph.D. Thesis, University of Wisconsin Medical School, Madison, 1982.
252. ICRU Report 21, Radiation Dosimetry: Electrons with Initial Energies between 1 and 50 MeV, ICRU Publications, Washington, D.C., 1972.
253. NBS Handbook 85, Physical Aspects of Irradiation, National Bureau of Standards, Washington, D.C., 1964.

Chapter 3

TRACK STRUCTURE THEORY AND APPLICATIONS TO THERMOLUMINESCENCE

Yigal S. Horowitz

TABLE OF CONTENTS

I. INTRODUCTION

The response of many physical, chemical, and biological systems to ionizing heavy charged particles (HCP) results from the secondary and higher generation electrons ejected from the atoms and the molecules of the system by the incident primary radiation. Since gamma rays also interact with the medium through secondary electrons, the differences in the observed effects associated with gamma-ray irradiation and with heavy charged particles arises [according to track structure theory (TST)] from the time scale of the dissipation of energy by the irradiations from their spatial distribution, or both. The direct atomic displacements by the primary HCP are assumed to have a negligible effect in comparison with the radiation effect resulting from the ejected electrons. The microscopic distribution of energy deposition in matter is very different for gamma rays and electrons compared with HCPs. For gamma rays and electrons, energy deposition may be characterized as a spatially uniform low density sea of initially free electrons, holes, and excitons. In contrast, electronic energy deposition by HCPs occurs in a very different manner. The immediate vicinity of an ion track is characterized by a very high instantaneous density of electrons and holes combined with the probability of fairly high energy secondary electrons (delta rays) penetrating to much larger radial distances from the HCP track. It is the HCP velocity as well as the linear energy transfer (LET) which determine the spatial extent of electronic energy deposition about the HCP track and the partitioning of the energy among the various modes of lattice energy deposition, e.g., ionization, exciton, and plasmon formation. On a macroscopic level, gamma rays and electrons as compared to HCPs often produce different radiation end effects and one of the purposes of TST is to attempt to describe the variation in the radiation end effect in terms of the differences in the microscopic distribution of energy depositions of the two types of radiation. Thus the radial distribution of absorbed dose imparted by the ejected electrons around the path of the HCP determines the track structure, and for dose rate-independent systems the dose-response function of the system to electrons (or gamma rays) is coupled with the spatial distribution of dose from secondary electrons to yield the response of the system to HCPs.

The major premise of TST is thus that the concentration of liberated charge carriers around the path of the HCP is the only parameter that governs the dependence of the relative TL properties on the type of HCP radiation. In conventional TST[1] the dose response of the system following gamma irradiation (usually ^{60}Co or ^{137}Cs) is folded into the radial distribution of absorbed dose, D(r), around the track of the HCP to determine the relative TL response. This approach has achieved considerable success in various HCP radiation yield calculations, but the choice of "high energy" gamma rays or electrons is particularly unfortunate in TL HCP response calculations because of the well-known dependence of f(D) on electron energy. This latter dependence, therefore, requires that the test electron spectrum be matched as closely as possible to the initial energy spectrum of the electrons ejected during the HCP slowing down ($E_{max} \simeq$ a few keV for $\sim$1MeV amu^{-1} HCP). This approach was successfully applied to the calculation of HCP-relative TL properties by Horowitz and Kalef-Ezra[2] and Kalef-Ezra and Horowitz.[3] Further refinements incorporated into their modified TST (MTST) approach are careful matching of all the relevant experimental parameters in the measurement of $\eta_{HCP,\gamma}$ and f(D) and simulation of the density of "low energy" charge carriers around the HCP track in "tissue-equivalent" dosimeters using published experimental data for "tissue-equivalent" gas rather than the use of approximate theoretical calculations.

Track structure calculations are generally based on a two-step formalism: the emission of secondary electrons by direct projectile-target interactions followed by secondary electron energy dissipation. The factors that govern these mechanisms are discussed briefly in the

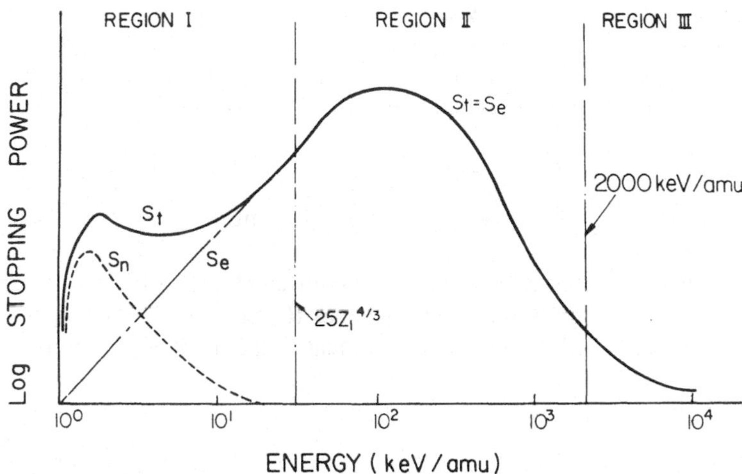

FIGURE 1. Heavy charged particle stopping power characterization into three energy regions. (Adapted from Gouard, P., Chemtob, M., Nguyen, V. D., and Parmentier, N., *Proc. 6th Symp. Microdosimetry*, Booz, J. and Ebert, H. G., Eds., Harwood Academic Publishers, London, 1979, 707.)

following sections. For a more detailed description of HCP stopping power, see, for example, Ziegler.[4]

II. HEAVY CHARGED PARTICLE STOPPING POWER

It is convenient to characterize the energy losses from heavy charged particles by distinguishing three energy regions as shown in Figure 1 following Gouard et al.[5] For protons region II is bounded by approximately 0.2 MeV amu^{-1} and 0.02 MeV amu^{-1} and the energies of these boundaries increases approximately with the atomic number, Z_1, of the incident ion in proportion to $Z_1^{4/3}$. In region I the ion is completely stripped of its orbital electrons and the stripped nucleus interacts with, and loses energy to, the electrons of the surrounding medium. In region III the ion has retained almost all of its orbital electrons and the energy loss occurs predominantly via the interaction between the electronic structures of the ion and the atoms of the medium. Region II is a transition region in which energy losses will include the complex mechanism of capture and loss of electrons from the medium to the partially stripped ion.

A. High Energy Region I

The theory for the stopping power of heavy charged particles in this energy region was first derived by Bethe[6] and a review of various aspects of its applicability has been given by Fano.[7] The Bethe formula for stopping power can be written

$$\frac{dE}{dX} = \frac{4\pi e^4 Z_1^2}{mv^2} \cdot Z_2 \left| \ln\left\{\frac{2mv^2}{I_0(1-\beta^2)} - \beta^2\right\} - \frac{C}{Z_2}(v,Z_2) - \delta(v,Z_2) \right| \tag{1}$$

where e and m are the charge and mass of the electron, Z_1e and v are the charge and velocity of the heavy ion, β is the velocity of the heavy ion relative to the velocity of light, Z_2 is the atomic number of the target atom, $C(v,Z_2)$ is a shell correction, and δ (υ,Z_2) is a density correction term which becomes important only at very high energies.

Equation 1 is derived in the framework of a model of the energy loss mechanism, which considers the heavy charged particle as giving rise to an electromagnetic pulse which acts on the target atom. The Fourier components of the pulse (virtual photons) interact with the orbital electrons so that the mean excitation energy, I_0 is given by

$$\ln I_0 = \int_{E_0}^{\infty} \frac{df}{dE} \ln E \, dE \Big/ \int_{E_0}^{\infty} \frac{df}{dE} dE \tag{2}$$

where E_0 is the threshold energy and df/dE is the differential dipole oscillator strength of the target atom for energy transfers E. The Thomas-Reiche-Kuhn sum rule states that the sum of the dipole oscillation strengths must be equal to the number of electrons belonging to the target atom, i.e.,

$$\int_{E_0}^{\infty} \frac{df}{dE} = Z_2 \tag{3}$$

The shell correction term, important at lower ion energies, arises because of the absence of virtual photons of high energy which damps energy transfers to the inner more tightly bound electrons. The density correction term arises from polarization of the target atoms close to the HCP track which reduces the intensity of the electromagnetic impulse on distant atoms. The polarization effect is important for HCP energies greater than approximately 1000 MeV amu^{-1}. More detailed information on the mean excitation energy, shell corrections, and the density effect have been compiled by Brodsky.[8]

B. Transition Region II

The applicability of Equation 1 in region II has been questioned by Armitage and Hotton.[9] A phenomenological approach is to use the RMS equilibrium charge Z_r instead of Z_1. Values of Z_r (derived from experimental stopping powers) can be compared with measured charge distributions on ions after passing through the appropriate layer of stopping material. Good agreement is obtained for ions passing through gases, but in solids the ions appear to be more heavily stripped.[10] This would imply lower stopping powers for gases than for solids, however, experimental data indicate that the stopping powers are equal within experimental error or that solids have the lower stopping power. Betz and Grodzins[11] have suggested that in solids the excited electrons do not have time to return to their ground states between collisions as they do in gases. The higher charges observed are thus due to autoionization after the ions emerge from the solid. This explanation is partially supported by charge states of ions in channeling studies and the X-ray spectra emitted by both the ions and the target atoms.[12] Other investigators[13-14] have more recently modified the theory of Lindhard and Bohr[15] (proposed for low energy region III) to stopping powers in energy region II with considerable success.

C. Low Energy Region III

In this energy region the nuclear stopping power is not negligible compared to the electronic stopping power. According to the theory of Lindhard and Scharff,[16] the nuclear stopping cross section due to elastic energy losses of the screened nuclei is given by

$$d\sigma = \pi a^2 \frac{f(t^{1/2})}{2t^{3/2}} dt \tag{4}$$

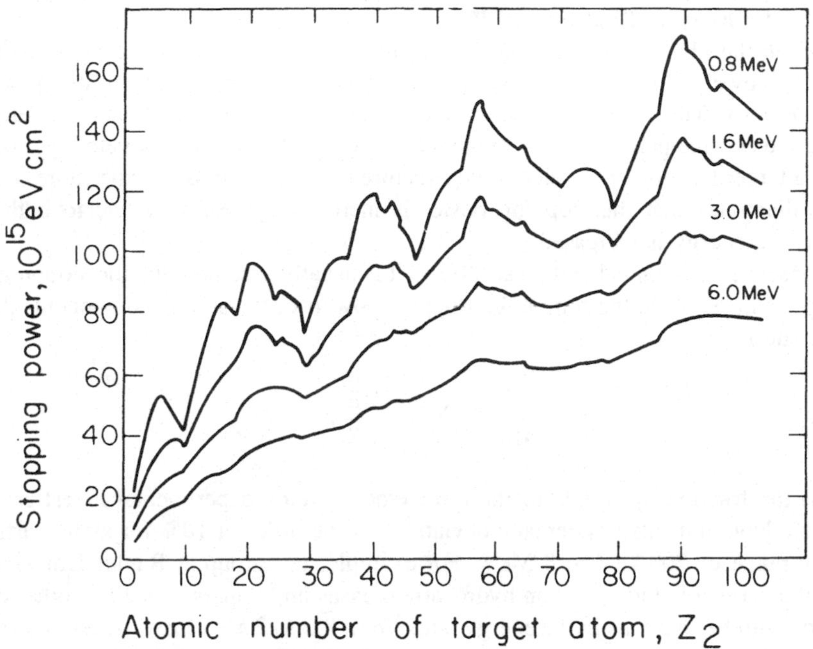

FIGURE 2. Theoretical stopping powers for helium ions as a function of the atomic number of the target atom. (Adapted from Chu, W. K. and Powers, D., *Phys. Rev.*, 187, 478, 1969.)

where

$$a = 0.8853a_0(Z_1^{2/3} + Z_2^{2/3})^{-1/2},\ t = \epsilon^2 \frac{T}{T_m} \tag{5}$$

and

$$\epsilon = E \frac{aM_2}{Z_1Z_2e^2(M_1 + M_2)} \tag{6}$$

where a_0 is the Bohr radius of the hydrogen atom, T and T_m are the energy transfer and the maximum energy transfer, respectively, and $f(t^{1/2})$ is a numerically computed function from Fermi's function.

The electronic stopping power in this energy region is predicted to be proportional to the ion velocity,[15-18] i.e.,

$$\frac{dE}{dX} = -K\ E^{1/2} \tag{7}$$

The constant, K, has been experimentally shown to be an oscillating function of both Z_1 and Z_2[19] and some experiments have also indicated that the stopping power may increase with energy faster than $E^{1/2}$.[20,21] Chu and Powers[22] were able to explain the oscillatory dependence of stopping power on Z_2 over a wide energy region by using Hartree-Foch-Slater wave functions in the formulation of Lindhard and Winther.[18] Typical results are shown in Figure 2.

D. Effects of Chemical Binding and Phase

Changes in the physical and chemical states of the target atoms can be expected to alter the stopping power most significantly at lower ion velocities and for lower atomic numbers of the target ions. This arises because these changes usually affect the outermost electrons, which are those principally involved in the interactions with ions of low velocity, also, they represent a larger fraction of the electronic structure of low Z_2 atoms. Furthermore, Equation 1 clearly illustrates that the stopping power is increasingly less sensitive to both I_0 and $C(v,Z_2)$ as the energy is increased.

In the absence of chemical and phase effects it is usually assumed that the stopping power of a compound is equal to the sum of the stopping powers of the atomic components (Bragg's additivity rule)

$$\frac{dE}{dX} = \sum_j f_j \left(\frac{dE}{dX}\right)_j \tag{8}$$

where f_j is the fraction by weight of the j-th element. Many experimental investigations of Bragg's rule have indicated systematic deviations of the order of 10% for alpha particles in the energy range of 200 keV to 8 MeV. For example, the group at Baylor University has carried out a series of studies[23-30] on hydrocarbon gases and vapors as well as other organic compounds with helium ions in the energy range from 0.3 to 2 MeV. Estimated experimental uncertainty in the stopping powers was less than 1%. For the hydrocarbons, deviations from Bragg's rule depended on the nature of the chemical bond (single, double, triple bonds, ring structure) and similar results were obtained using organic molecules in which oxygen was incorporated in different bond states. Other investigations using 0.080 to 0.80 MeV ^7Li ions[31] have found deviations of as much as 50% between stopping powers derived from compounds and those measured in elemental substances. The deviations from Bragg's rule increase as the disparity between the atomic numbers of the atomic constituents of the molecule is increased. Recently some theoretical attempts[14,32,33] have met with some success in explaining the above effects using the wave functions of the electron orbitals arising from the molecular binding.

Aside from the well-known channeling effect,[34] an allotropic effect on stopping power has been convincingly illustrated for graphite, vapor-deposited carbon, and diamond.[35,36] The greatest difference is about 20% for 0.4 MeV alpha particles decreasing to 6% at 2 MeV. Many investigators have observed differences between stopping powers of vapor and gaseous phases and the liquid and solid phases. For alpha particles the differences range from about 9 to 12% at 0.5 MeV to 3 to 6% at 8 MeV for organic compounds with possibly even larger differences for water vapor and water.[37] Theoretical stopping power values based on the Thomas-Fermi statistical model of the atom have been calculated by Thwaites and Watt,[38] whereas the more realistic Hartree-Foch-Slater potentials have been employed by Ziegler and Chu.[39] The results of the two groups for low atomic number targets are shown in Figure 3.

In conclusion, both theoretical and experimental investigations have established the influence of chemical binding effects on stopping power that increase as the ion energy is decreased below 0.2 MeV amu^{-1} $(Z_1)^{-4/3}$. Significant progress has already been achieved in relating these effects quantitatively to the nature of the chemical binding. Similar evidence, both theoretical and experimental, also clearly indicates the existence of phase effects on stopping power, however, a systematic quantitative explanation of these effects has yet to evolve.

E. Secondary Electron Emission Spectrum

Track structure theory is usually applied to condensed phase systems, however, due to

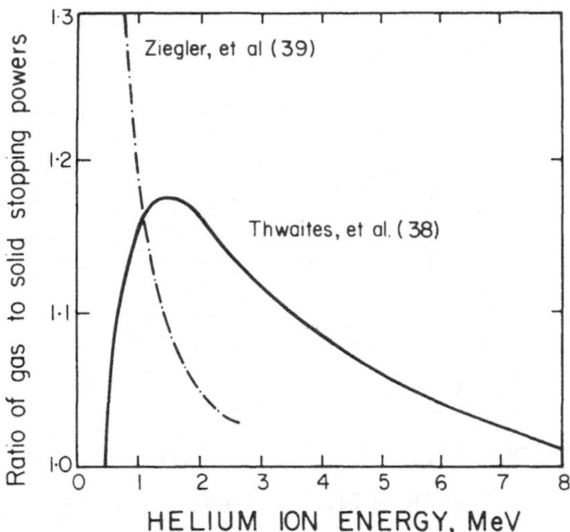

FIGURE 3. Ratios of gas-to-solid stopping powers of low atomic number elements (Z < 15). (Adapted from Thwaites, D. I. and Watt, D. E., *Proc. 6th Symp. Microdosimetry*, Booz, J. and Ebert, H. G., Eds., Harwood Academic Publishers, London, 1978, 777 and Ziegler, J. F. and Chu, W. K., *At. Data Nucl. Data Tables*, 13, 463, 1974.

the extremely swift time scale and the very short range of the secondary electrons, measurements of the radial distribution of dose around the HCP track are not possible. Thus, evaluation of track structure effects in the condensed phase must rely on measurements in the gaseous phase. The majority of the existing data on energy dissipation cross sections for low energy electrons refer to the gaseous phase. Only limited data for solids are available; e.g., Cole[40] measured low energy electron dissipation characteristics in collodion. In the gaseous phase, many investigations have been carried out on the cross section, differential in ejection angle and energy, for electron emission in materials with interaction properties similar to those of tissue, by protons with energy between 0.2 and 2 MeV (see, e.g., Figure 4 following Toburen[41]). The general features of this and other investigations[42-46] include a maximum in the cross sections (multiplied by the energy of the ejected electrons) at very low electron energies with a gradual decrease with increasing energy, a K Auger peak, and a broad peak for electrons ejected at forward angles with energies corresponding to the kinematic maximum energy transfer. The energies at which maxima occur in the high energy region of the forward angle electron spectra are approximately equal to the maximum energies that can be transferred in a binary collision between the incident proton and a free electron.

Bragg's additivity rule has been extended by various groups in the following manner. The ionization cross section for a molecule is given as the sum of the corresponding ionization cross section of the constituent atoms. Lynch et al.[45] proposed a simple scaling technique for the continuum emission (excluding Auger emission) for low Z materials, i.e., the cross section is scaled by the number of outer weakly bound electrons per target molecule. Both additivity and scaling have been reported[45,46] to work adequately within experimental error for single differential cross sections for proton-ejected electron energies greater than approximately 100 eV. Below 100 eV, however, molecular structure seems to play an important role in the determination of the electron energy spectrum. These observations are intended to illustrate that important differences in the total ionization cross sections between gaseous and condensed phase systems probably do not exist due to the weak intermolecular binding. On the other hand, extrapolation of gaseous data to condensed phase data concerning the

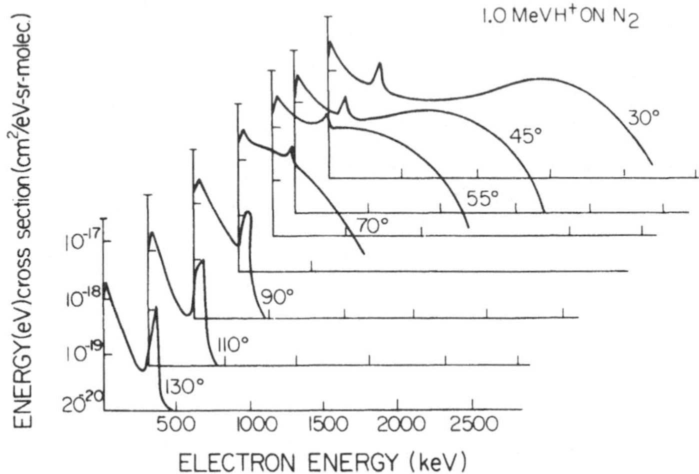

FIGURE 4. Cross sections, differential in electron energy and emission angle for ejection of electrons by 1.0-MeV protons in molecular nitrogen. (Adapted from Toburen, L. H., *Phys. Rev.*, A3, 216, 1971.)

secondary electron spectrum below approximately 100 eV may be compromised due to the important role played by molecular binding at very low electron energies. Since charge capture in thermoluminescence may be, or probably is, significant below 100 eV, the extrapolation of energy dissipation characteristics from the gaseous to the condensed phase in TST applied to thermoluminescence is a critical approximation.

III. TRACK STRUCTURE THEORY MODELS

A. The "Biological" TST Model of HCP Radiation Action

In conventional TST the radial distribution of dose (defined as a point quantity) deposition by HCP-induced secondary electrons is converted into a radial distribution of the final radiation effect via application of the dose-response function generated by ^{60}Co or ^{137}Cs gamma rays.[1,47-49] The δ-ray spectrum is calculated relatively crudely assuming electron ejection perpendicular to the HCP track.[50] Other angular distributions have also been used, e.g., a $\cos^2\theta$ distribution or an angular distribution based on classical kinematics between the HCP and a free electron. The energy deposition around the HCP path is then calculated using empirical relations such as transmission of electrons or range energy relations of the type R = kT. In its simplest, most widely used form, the δ-ray spectrum, dn/dT, ejected by the HCPs is given by:

$$\frac{dn}{dT} = \frac{2\pi Ne^4Z^2}{m_ec^2\beta^2}\frac{1}{T^2} \quad \text{for } T < T_{max} = \frac{2m_ec^2\beta^2}{1 - \beta^2} \tag{9}$$

where Z is the effective charge, β is the relative velocity, and N is the number of electrons per unit volume. The dose at a point distance t from the HCP path is given by

$$D_\delta(Z,\beta,t) = \frac{1}{\rho}\frac{Ne^4}{m_ec^2}\frac{Z^2}{\beta^2}\frac{1}{t}\left(\frac{1}{t} - \frac{1}{kT_{max}}\right) \tag{10}$$

This formulation has been applied to the HCP relative response of NaI(Tl) scintillators,[47] thermoluminescent dosimeters,[51] the width of heavy ion tracks in emulsions,[49] and the formation of etchable tracks in dielectrics.[48]

In a later version,[1] Katz assumed the existence of sensitive elements called "sites" or "targets" in which the radiation effect is initiated. An interaction or "hit" with this sensitive element initiates the changes which lead to the observable effect. The hits in the sites (treated as black boxes) are independent of each other and follow the cumulative Poisson distribution. The various systems (biological, chemical, or physical) are classified according to their "hittedness", the number of hits in a target required for the site to respond (multihit systems) or according to the number of targets that have to be hit simultaneously for the observable radiation effect to take place (multitarget systems). It is further assumed that sensitive elements between adjacent isodose contours respond to energy dissipation by the δ-rays as if those elements were a part of a larger system uniformly irradiated with gamma rays of the same dose. The response is averaged over sufficiently long ion track segments to ensure that the noise due to fluctuation in the δ-ray production is negligible or the response is averaged over a sufficiently large number of particles in short track segments to achieve the same results. The probability that the site of a one-hit system (of radius a_0 whose center is at a distance t from the path of an ion in a medium of characteristic dose D_0) is activated by the passing ion is given by

$$1 - \exp(-\overline{D}(Z,\beta,t,a_0)/D_0) \tag{11}$$

where D_0 is the gamma-ray dose at which 37% of the elements is not influenced by the irradiation. In the framework of this model, it can be shown that when the HCP passes at distance $t > 3a_0$ the dose is independent of the site radius and the point-target approximation (as in the earlier version) can be used. When the HCP passes through the sensitive site, i.e., $t < a_0$, the dose in the site itself varies inversely with a_0^2. In all cases the dose varies as Z^2/β^2. In this manner the response of a physical system to HCPs can be parametrized by the "hittedness" (c), the characteristic dose, D_0, and the size of the sensitive element, a_0.

The "biological" model has been applied to radical production, lyoluminescence,[52] inactivation of enzymes and viruses, optical rotation of arabinose solutions, Fricke dosimeters, thermoluminescence[53,54] and various other systems.

B. The "Chemical" Model of HCP Radiation Action

The "chemical" model of track structure developed by Chatterjee et al.[55-57] distinguishes between two regions: the core and penumbra (the sharp division into two regions is an artificial concept introduced mainly to facilitate the analytical calculations). The core is described as a narrow central zone of radius R_c (in tissue less than $1.2\ 10^{-6}$ g cm^{-2}) where energy dissipation occurs mainly via excitation and collective oscillations of the electrons (approximately 50% of the total energy dissipation is assumed to occur in this manner). The penumbra is a peripheral zone, of maximum radius R_p, enveloping the core where energy dissipation occurs mainly via ionization by secondary electrons released by the primary particle. The physical dimensions of the penumbra for a particle with velocity β is determined by the range R_{max} and the diffusion distances of the knock-on electrons ($E < 1.6$ keV). For water the following relations have been proposed:

$$R_{max}(\mu m) = 0.768\ E - 1.925\ E^{1/2} + 1.251 \tag{12}$$

$$R_c\ (\mu m) = 0.0116\ \beta,\ E > 2\ \text{MeV nucleon}^{-1} \tag{13}$$

Rutherford's cross section (modified for relativistic particles), classical angle of electron ejection ($E > 1.6$ keV), and Fermi's theory were used for the interaction of the HCP with the stopping material. The dissipation of electron energy is calculated by assuming that the electrons with energies greater than 1.6 keV undergo diffusive motion, whereas the lower energy electrons travel in straight paths until their energy is reduced below 1.6 keV. The fraction, F, of the total energy dissipation within a cylinder of radius, r, is then given by

$$F = 0.5 + \frac{1 + 2\ln(r/R_c)}{4\ln(R_p/R_c)^{1/2}} \tag{14}$$

This model has been applied to the calculation of chemical yields,[57] the radiation inactivation of dilute ribonuclease solutions[58] and the dosimetry of HCPs with nuclear emulsions.[56]

A model, similar in some aspects to the chemical model, has been described by Fain et al.[58,59] In this model "primary energy" refers to the energy released within the molecules directly by the incident HCP (this particular form of energy transfer takes place in a core of radius the same order of magnitude as the atomic radius and the energy deposition is divided into "excitation primary energy" and the deposited energy required to overcome the binding energy of the ejected electrons). "Secondary energy" deposition refers to the production of the secondary electrons and subsequent energy dissipation. The classical binary encounter theory is used for the calculation of the secondary electron energy spectrum, with angular distributions following classical kinematics. The secondary electron energy dissipation was calculated using Berger's scaling[60] for a line source (even though this scaling procedure does not apply for electrons with energy less than 5 keV[61]) and values for the R_{CSDA} by Cole[40] and ICRU (1970).[62] Aside from radial dose an attempt was also made to calculate the partition of energy between ionizations, excitations, and intramolecular vibrations. This model has been applied to the calculation of average W values of HCPs in various gases and water, the percentage of spurs, blobs, and short tracks around a heavy ion trajectory[63] and more recently to thermoluminescent HCP efficiencies.[64]

IV. EXPERIMENTAL MEASUREMENTS OF RADIAL DOSE DISTRIBUTIONS

Various authors have used various approximations in calculating radial dose distributions around HCP tracks but the results differ significantly depending on the assumptions invoked to enable the calculations. In the past decade, various experiments[65-69] have been carried out to measure the radial energy deposited around the path of HCP beams in tissue-like materials. These experiments employed a variable pressure ionization chamber filled with tissue-equivalent gas (64.4% methane, 32.4% CO_2 and 3.2% N_2 by volume: occasionally H_2 gas has also been employed) into which a beam of monoenergetic heavy ions was directed. Inside this chamber, a moveable, mesh ionization chamber (probe) was used to measure ionization at various radial distances from the ion path. The simulated radius in tissue was changed by varying the probe position and/or gas pressure in the chamber. A potential of the order of 10 V was applied between the mesh wall and the copper central wire of the probe. Thus, what is actually measured in this technique is the density of low energy electrons at various distances from the HCP path. This quantity is converted to absorbed dose by multiplying it with the energy, W_{HCP}, needed to create an ion pair in the gas. To the best of our knowledge, no attempt was made to take into account the variation of W with radial distance which arises from variations in the electron slowing-down spectrum as a function of radial distance (this may explain why the experimental results for D(r) at small distances are smaller than those predicted by most of the theoretical models). The W_{HCP} values used for the various HCPs studied by the group at Brookhaven National Laboratory are listed in

Table 1
MEASUREMENTS OF THE RADIAL DISTRIBUTION OF DOSE

Ion	Energy (MeV)	W_{HCP} (eV)	Material	Ref.
^{1}H	1	31.3	Tissue-Equivalent gas	65
^{1}H	3	31.3	Tissue-equivalent gas	65
^{4}He	1	31.0	Tissue-equivalent gas	65
^{4}He	2	31.0	Tissue-equivalent gas	65
^{4}He	3	31.0	Tissue-equivalent gas	65
^{4}He	930	33.73	Air	66
^{16}O	38.4	36.7	Tissue-equivalent gas	67, 68
^{20}Ne	7540	33.73	Air	69
^{127}I	33.25	36.7	Tissue-equivalent gas	67, 68
^{127}I	61.9	36.7	Tissue-equivalent gas	67, 68

Table 1. From the experimental data and most of the theoretical calculations the average dose in tissue at intermediate distances is given approximately by

$$D(r) = \frac{D_0}{r^2} \tag{15}$$

where D_0 is directly proportional to the square of the effective charge and inversely proportional to the energy per atomic mass unit. In the case of low energy protons and alpha particles the overall experimental error in values of D(r) for radii in tissue greater than 50 Å is estimated by Wingate and Baum[65] as $\pm$ 20% and for smaller radii, $\pm$40%. Figure 5 shows a comparison of calculated and measured radial dose distributions for 1- and 3-MeV alpha particles in tissue-equivalent gas following Paretzke.[70] The calculations of Baum et al.[71] and Butts and Katz[50] are also shown together with the results of Monte Carlo calculations using the program MOCA 3.[70] The agreement between the Monte Carlo calculations and the experimental data is generally excellent. This good agreement extends even to much heavier ions (e.g., ^{127}I), although at very small radii the experimental values tend to be somewhat smaller than the theoretical calculations. Figure 6 shows D(r) vs. r for 33.25- and 61.9-MeV ^{127}I ions in tissue-equivalent gas. In conclusion, there is an astonishing degree of correspondence between the experimental measurements of D(r) in tissue-equivalent gas and Monte Carlo calculations. In the following section we develop a modified track structure approach[2,3] to the calculation of HCP-induced thermoluminescence which employs the D(r) values measured in tissue-equivalent gas with appropriate scaling procedures.

V. TRACK STRUCTURE THEORY APPLICATIONS TO THERMOLUMINESCENCE

A. Introduction

The TL properties of HCPs are significantly different from those of gamma-ray and electron-induced TL:

1. The TL dose-response curve, $f_\alpha(D)$, is almost always linear-sublinear rather than linear-supralinear-sublinear as observed with gamma rays and electrons. See, e.g., Zimmerman[51] who measured $f_\alpha(D)$ for natural fluorite:MBLE type 6, CaF_2:Tb, CaF_2:Mn, $CaSO_4$:Mn, and Norwegian quartz. In all cases $f_\alpha(D)$ = 1 up to approximately 5 $\times$

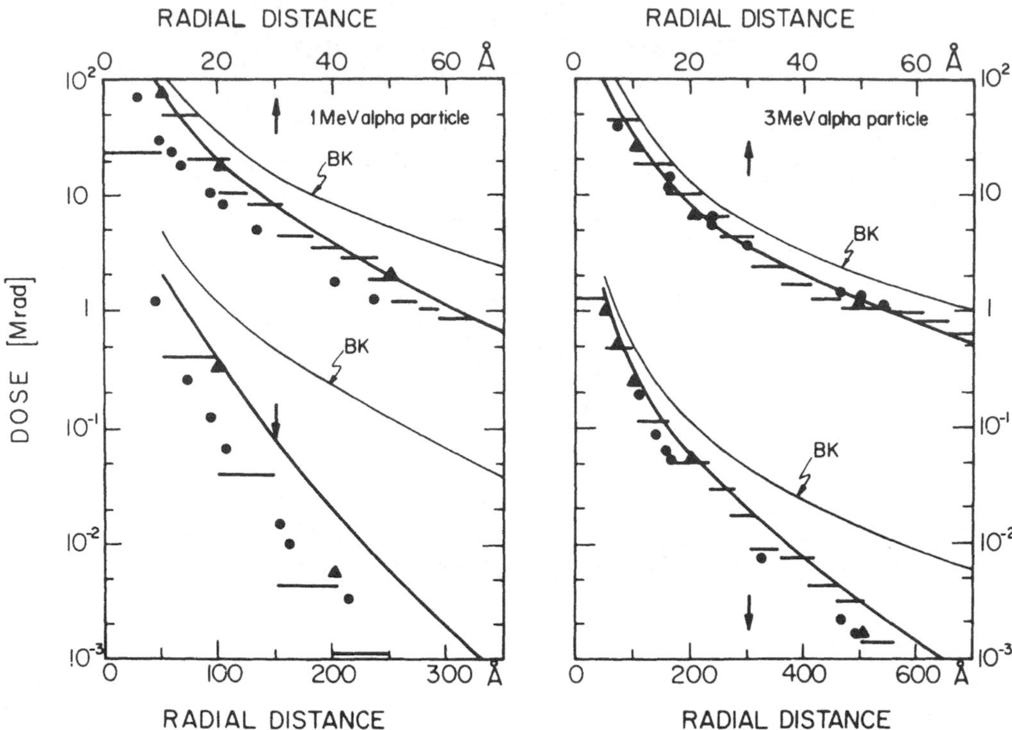

FIGURE 5. Comparison of calculated and measured radial dose distributions for 1- and 3-MeV alpha particles and protons in tissue-equivalent gas: BK, Reference 50; dark solid line, result of CONT; triangles, results of MOCAMIX, histogram, results of MOCA 3; and solid circles, experimental data of Wingate and Baum.[65] (Adapted from Paretzke, H. G., *Proc. 4th Symp. Microdosimetry*, Booz, J., Ebert, H. G., Eickel, R., and Waker, A., Eds., EURATOM, Brussels, 1973, 141.)

10^3 Gy followed by a sublinear region. There are, however, exceptions to this rule. For example, although peak (4 + 5) in LiF:Mg,Ti is linear-sublinear for alpha particles (~1 MeV amu^{-1}) the dose response for thermal neutrons is definitely linear-supralinear-sublinear.[72] Also, peak 8 in LiF-TLD is slightly supralinear for alpha particles over the dose range 50 to 1000 Gy.[73,74] Similarly Lakshmanan et al.[75] have observed slight supralinearity in the TL alpha particle dose-response curves of the 145°C and 200°C glow peaks in CaF_2:Dy. In any event, the supralinearity induced by low energy HCPs is always very small compared to gamma- and electron-induced supralinearity and there are no known exceptions to this latter rule.

2. The relative TL response of HCPs is usually less than unity. For example, $\eta_{\alpha\gamma}$ lies between 0.05 and 0.5 for peaks (4 + 5) in LiF-TLD.[72] Here too, however, there are exceptions. For example, $\eta_{n_{th},\gamma} \simeq 1$ in $Li_2B_4O_7$.[76,77] The relative TL response of HCPs has recently been treated by Horowitz and Kalef-Ezra[2] and Kalef-Ezra and Horowitz[3] in a point-target track structure theory approach. In this approach the relative HCP TL response is given by

$$\eta_{HCP,\gamma} = \eta_{\delta\gamma} \frac{W_\gamma}{W_{HCP}} \frac{\int_0^{R_{max}} \int_0^{r_{max}} f_\delta(D)\, n(r,\ell,E)\, 2\pi r\, dr\, d\ell}{\int_0^{R_{max}} \int_0^{r_{max}} n(r,\ell,E)\, 2\pi r\, dr\, d\ell} \tag{16}$$

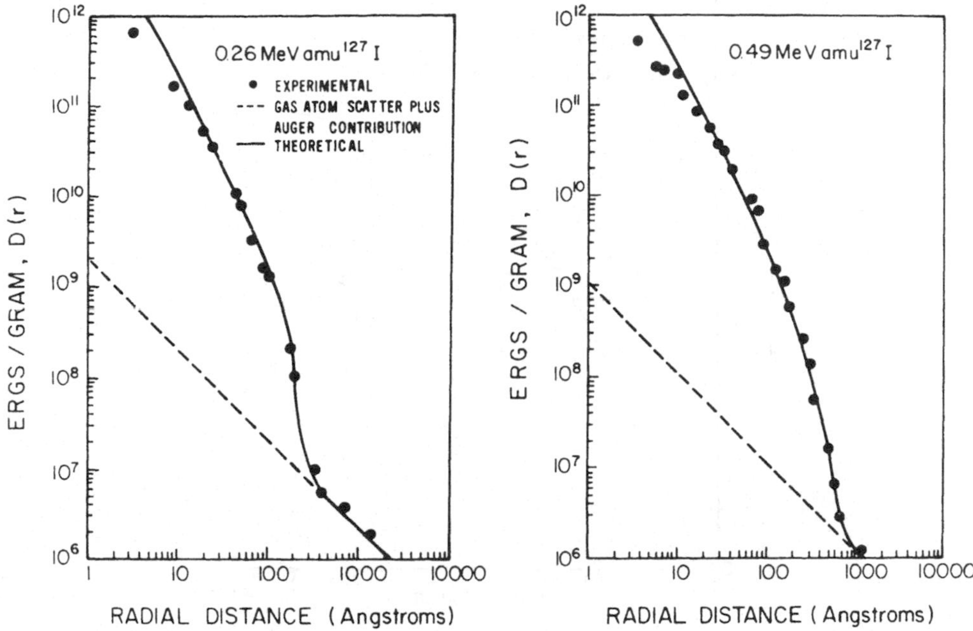

FIGURE 6. D(r) vs. r for 33.25-and 61.9-MeV ^{127}I ions in tissue-equivalent gas. Lower dashed curve is the sum of calculated gas atom scattering and heavy ion Auger electron contributions. Solid curve is the sum of Paretzke's delta-ray dose prediction plus the gas atom and Auger contribution. Points are experimental data. (Adapted from Baum, J. W., Varma, M. N., Wingate, C. L., Paretzke, H. G., and Kuehner, A., Proc. 4th Symp. Microdosimetry, Booz, J., Ebert, H. G., Eickel, R., and Waker, A., Eds., EURATOM, Brussels, 1973, 93.)

where $\eta_{\delta\gamma}$ is the relative TL response of the HCP-ejected secondary electron spectrum to the gamma reference radiation, W_γ and W_{HCP} are the average energies required to produce an electron-hole pair by the gamma reference radiation and the HCP radiation respectively, R_{max} and r_{max} are the maximum axial and radial distances of penetration of the charge carriers from the HCP path (in the radiation absorption stage); $f_\delta(D)$ is the TL dose-response measured with a test electron radiation chosen to mimic the radiation action of the HCPs as closely as possible, and finally, $n(r,\ell,E)$ is the density of "low energy" charge carriers around the HCP path taken from direct experimental measurements[65-69] of the density of "low energy" charge carriers in tissue-equivalent gas. The agreement between theory (Equation 16) and experimental measurements in both LiF and BeO was very good for a large variety of experimental measurements (alpha particles, thermal neutrons, and fission fragments). Figure 7 shows the integral TL radial distribution for 3-MeV alpha particles stopping in LiF-TLD. For $r < 50$ Å, $f(D) < 1$ so that approximately 50% of the TL arises from the "saturation" region close to the track. The supralinear region $[f(D) > 1]$ is defined by the region 50 Å $<$ $r <$ 230 Å in which an additional approximately 45% of the TL arises. Only 5% of the total TL arises from the linear $[f(D) = 1]$ region at the radial extremities of the track for which $r <$230 Å.

3. The shape of the glow curve (i.e., the relative population of the various glow peaks, ϵ_i, is also radiation-type dependent). In LiF-TLD, e.g., the high temperature peaks (peaks 6 to 8) are more strongly populated via alpha particle irradiation than low dose gamma radiation. For example, Horowitz et al.[77] report $\epsilon^\alpha = 0.35$ and $\epsilon^\gamma = 0.035$ for peaks 6 + 7 in TLD-700 at low dose. This behavior is also easily explained via

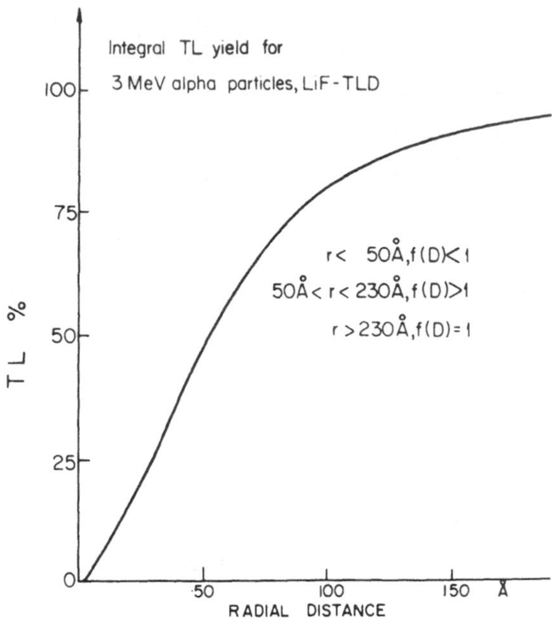

FIGURE 7. Integral TL yield for 3-MeV alpha particles stopping in LiF. Approximately 50% of the TL is produced at radial distances less than 50 Å where f(D) < 1; the other 50% is produced at radial distances between 50 and 230 Å where f(D) > 1. Only a few percent of the total TL yield arises from radial distances greater than 230 Å where f(D) = 1.

track structure theory since the alpha particle-induced glow curve can be calculated via a convolution of the radial dose distribution around the HCP track with the glow curve shape induced by low energy electrons matched as closely as possible in energy to the secondary electron spectrum ejected by the HCP. It is well known[72] that $f_e(D)$ for the high temperature glow peaks in LiF-TLD is more strongly supralinear than for peak (4 + 5) so that the relative population of the high temperature peaks increases with increasing electron dose (Figure 8). Since the alpha particle track is heavily weighted towards high dose (80% of the energy is deposited within 50 Å of the track axis at an average dose of approximately 0.25 MGy), TST naturally predicts preferential population of the high temperature peaks via HCPs. Calculations of this nature have been carried out by Kalef-Ezra and Horowitz[3] with fairly satisfactory results (Figure 9). For technical reasons the low energy electron-induced glow curves were not measured beyond 0.16 MGy which, we believe, is the main contributing factor to the discrepancy between theory and experiment. More detailed experiments of this nature are currently underway in the Radiation Physics Laboratory of the Ben Gurion University of the Negev. As an additional example, in $Li_2B_4O_7$:Mn the glow curve shape is nearly constant as a function of gamma ray or electron dose and, as predicted by TST, the alpha– and gamma ray-induced glow curves are also essentially identical.[77]

We have seen that the relative HCP TL response and the HCP-induced glow curve can be adequately explained via point-target TST. The following sections report on our investigations into HCP-induced TL-dose response curves. It turns out that the linear-sublinear behavior yields interesting information on the TL "effective volume" influenced by the

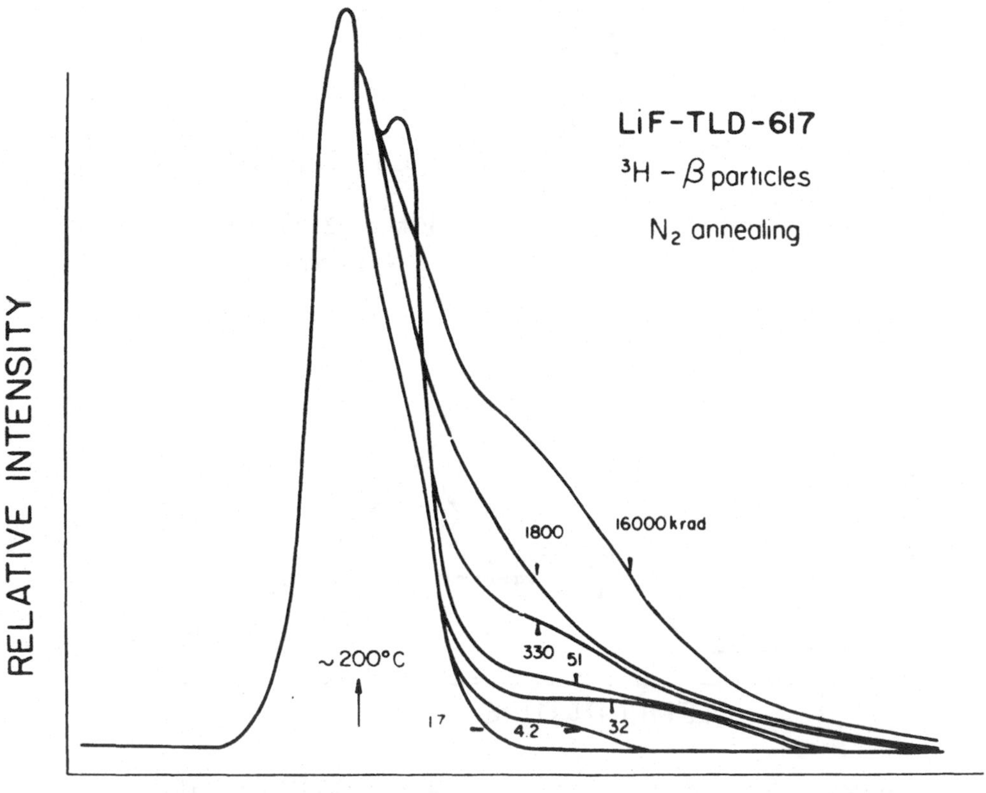

FIGURE 8. Normalized TL glow curves of N_2-annealed LiF TLD-600 [(TLD-617) is the laboratory number given to a specific TLD] following ^3H beta particle irradiation at various dose levels. The glow curves were normalized by setting the height of glow peak 5 equal to unity.

HCP track. In the final section we discuss in detail the modified TST calculations[2,3] for the estimation of the HCP relative TL response.

B. HCP-Induced TL Dose Response

Since the production of TL by ionizing events certainly involves several intermediate processes, it is obligatory to ask whether the radial dose distribution (or distribution of ejected charge carriers) around the HCP track is significantly different from the final radiation effect distribution (i.e., emission of TL photons). During glow curve heating, e.g., diffusion of charge carriers away from their region of initial capture or diffusion of TL-related defects could result in a final radiation effect distribution which extended to significantly greater radial distances from the HCP track than the radial extension of the radial dose distribution. Kalef-Ezra and Horowitz[3] have used conventional scaling techniques to calculate the radial dose distribution for low energy alpha particles in LiF based on the radial dose distribution measurements in tissue-equivalent gas. The results are presented in Figure 10 for 3-MeV alpha particles stopping in LiF ($\rho = 2.64$ g cm^{-3}). It can be seen that 80% of the dose is deposited within 50 Å of the track, 92% within 100 Å and more than 95% within 150 Å of the track. We shall see that this latter figure of 150 Å corresponds exactly to the value of r_{eff} for peak (4 + 5) deduced from the sublinear behavior of the HCP TL dose response curve.

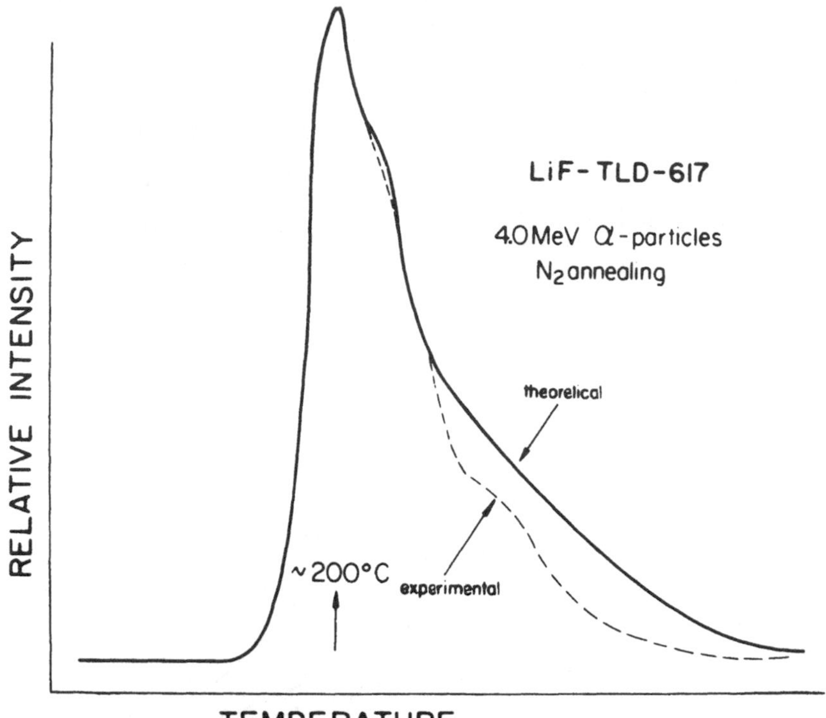

FIGURE 9. Normalized TL glow curve of N_2-annealed LiF TLD-600 after irradiation at "low dose" by 4.0-MeV alpha particles. The irradiation dose was approximately 1 Gy. The theoretical curve is obtained by a convolution of the alpha particle radial dose distribution with the dose-dependent glow curve shapes induced by H^3 beta particles shown in Figure 8.

It is fairly straightforward to show that the nearly universal result of $f_\alpha(D) = 1$ up to approximately 5×10^3 Gy supports the concept that the TL "radiation effects" of a single alpha particle are limited to a small volume along the alpha particle track. The growth curve must then be linear until the tracks begin to overlap. Consider the situation illustrated in Figure 11 where O is an arbitrary point in the crystal volume and R is the radius of an imaginary circle centered at the point O. We wish to determine the probability that the dose at point O will be influenced by a single alpha particle, i.e., will not fall in the overlap region between intersecting tracks. For an alpha particle fluence of n particles cm^{-2} the total probability that the region πR^2 will be intercepted by either no tracks or one track (of effective radius R) is given by

$$P_{0+1} = \left\{\frac{\pi R^2}{A_0}\right\}^0 \left\{1 - \frac{\pi R^2}{A_0}\right\}^n + {}^nC_1 \left\{\frac{\pi R^2}{A_0}\right\}^1 \left\{1 - \frac{\pi R^2}{A_0}\right\}^{n-1} \tag{17}$$

where $A_0 = 1$ cm^2. Using the approximation $(1\text{-}p)^n \simeq \exp(\text{-}pn)$ it is easily shown that Equation 17 reduces to

$$P_{0+1} = \left\{1 + \left(\pi R^2 \frac{n}{A_0}\right)\right\} \exp - \left(\pi R^2 \frac{n}{A_0}\right) \tag{18}$$

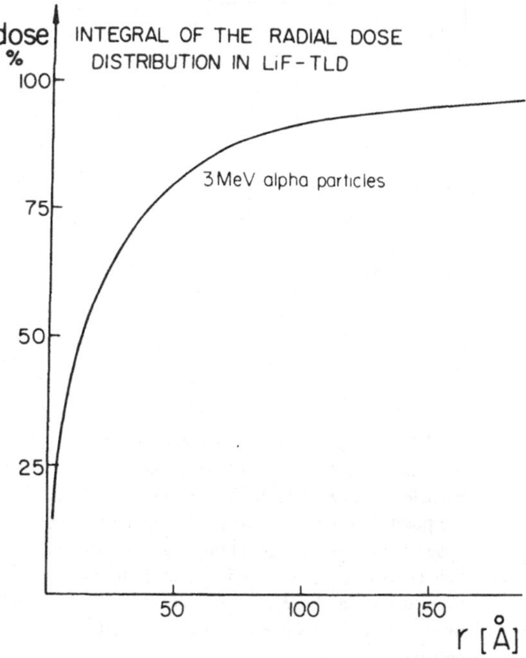

FIGURE 10. The integral of the radial dose distribution in LiF-TLD due to 3-MeV alpha particle penetration. The value of $100 \int_0^{r} D(r)\, 2\pi r\, dr / \int_0^{r_{max}} D(r)\, 2\pi r\, dr$ is plotted on the vertical axis.

For $D = 5 \times 10^3$ Gy, $n = 3 \times 10^{10}$ alpha particles cm^{-2} for 3 MeV alpha particles in LiF, and choosing $R = 150$ Å we obtain $P_{0+1} = 0.979$. For $D = 10$ Gy (the approximate dose for onset of gamma-induced supralinearity) P_{0+1} is essentially unity. Thus in the dose region where gamma-induced supralinearity is normally observed (10 to 5000 Gy) the probability of overlapping alpha particle tracks rises from essentially 0% to a maximum value of only 2%. In the framework of TST, therefore, there can be no possibility of significant alpha particle-induced supralinearity in this dose region. As the dose is further increased the probability of overlapping tracks $(1 - P_{0+1})$ increases, however, the highly localized nature of the radial dose distribution of the HCP track ensures that the loss in TL due to saturation in the overlapping regions is always equal to or greater than the gain in TL due to supralinearity arising from grazing track intersections. This statement can be proven in the following manner in the framework of a more advanced variable ionization density model currently under development at the Radiation Physics Laboratory of the Ben Gurion University of the Negev.[116] The model calculates the integral

$$\int_0^{\infty} \int_0^{2\pi} D(r,\theta) f_e\{D(r,\theta)\}\, r\, dr\, d\theta \tag{19}$$

using a Riemann integration technique with interval adjustment. The integral is calculated as a function of the distance d between two alpha particle tracks where

$$D(r,\theta) = \frac{k}{r^2} + \frac{k}{r^2 + d^2 - 2rd\cos\theta} \tag{20}$$

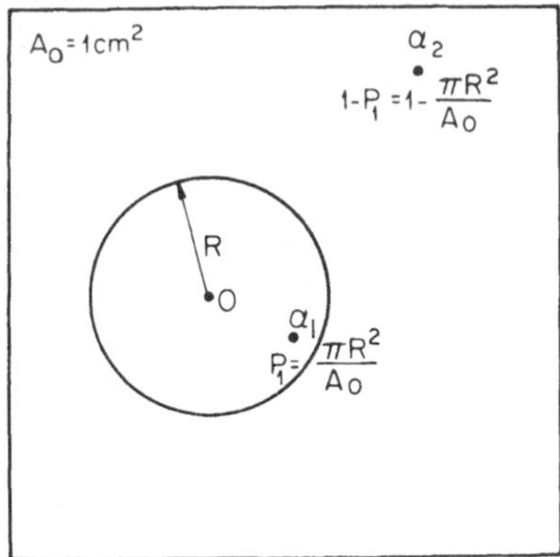

FIGURE 11. Schematic geometry for the calculation of Equation 17. α_1 is an alpha particle track axis which will deliver dose to the point 0, whereas α_2 will not. The radial dose distribution is a function of alpha particle energy as can be seen in Figure 5, however, the cone shape of the alpha particle-irradiated volume is not incorporated into the present model.

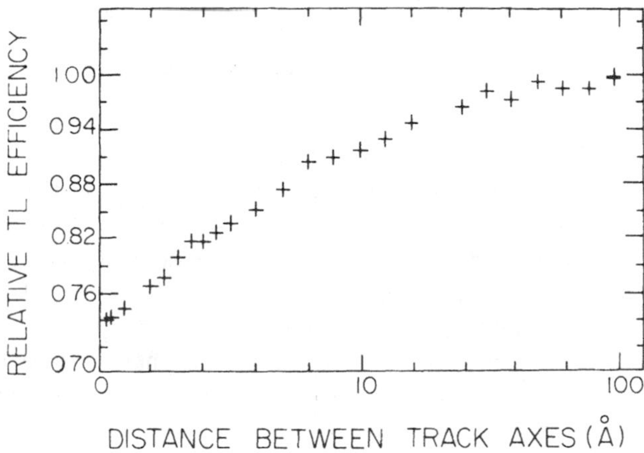

FIGURE 11A. The relative TL efficiency of intersecting alpha particle tracks as a function of the distance between the track axes.

is the summed dose from the two alpha particle tracks and $f_e(D)$ is the TL dose response generated by low energy electrons matched as closely as possible in energy to the primary electron spectrum liberated by the alpha particles. Calculations using $f_e(D)$ generated by tritium electrons and 20 kV_p X-rays show (Figure 11A) that there is no distance d for which the total TL yield for two intersecting alpha particle tracks is greater than twice the TL yield of a single alpha particle track, i.e., the loss in TL yield owing to saturation (at points close to the track axis) is always greater than the gain in TL owing to supralinearity (at points relatively far from the track axis). The immediate conclusion is that alpha particle supralinearity cannot exist for glow peaks possessing $f_e(D)$ curves normally encountered in most TL materials (i.e., $f_e(D)$ = 1 to ~ 5 Gy and $f_e(D)_{max} \simeq 3$ to 5 at 10^2 to 10^3 Gy). Thus the track intersection model based on realistic values of D(r) in condensed phase LiF correctly describes the lack of supralinearity observed in most alpha particle TL dose-response curves. Further studies are in progress at RPL-BGU to determine whether the slight alpha-induced supralinearity observed for peak 8 in LiF-TLD can be understood in the framework of this model.

The onset of saturation at approximately 10^3 Gy can yield information about the effective radius, r_{eff}, which parametrizes the maximum distance from the alpha particle track from which TL photons originate. For the purposes of illustration, it is convenient to define a fluence TL response curve, f(n), given by

$$f(n) = \frac{S(n)/n}{S(n_0)/n_0} \tag{21}$$

where n_0 corresponds to an alpha particle fluence for which the growth curve is linear and S is the intensity of the TL signal at alpha particle fluence n. The correspondence between alpha particle fluence and dose is given by

$$D(n) = \frac{n\,E_\alpha\,10^{-10}}{R_\alpha\,0.6\rho} \tag{22}$$

where D is in Gy, E_α in MeV and R_α is the range of the alpha particle in cm. In these initial stages of our work, f(n) is calculated under the assumption that the HCP track can be described by a cylindrical volume whose height corresponds to the HCP range and that overlapping track regions are completely saturated. Under these assumptions

$$f(n) = V(n)/nV_\alpha \tag{23}$$

where V(n) is the "relative irradiated volume", i.e., the ratio of the irradiated volume to the total volume of the TLD for an alpha particle fluence, n, and V_α is the "relative irradiated volume" for a single alpha particle ($V_\alpha = \pi r^2_{eff} R_\alpha/V$). We calculate V(n) by analytic as well as Monte Carlo methods. The calculation is based on homogeneous sampling of points in the irradiated TLD so that the ratio of the number of points included in the irradiated volume to the total number of sampled points is an estimator of V(n). For the i-th track, the probability, P_i, that a randomly sampled point will be included exclusively in the i-th track is

$$P_i = p\,(1 - p)^{i-1} \tag{24}$$

where p is the probability that a randomly sampled point will be included in a single track. P(n), which is an estimator of V(n), is the probability that a randomly sampled point will be included in at least one track and is given by

$$P(n) = \sum_{i=1}^{n} P_i = 1 - (1 - p)^n \tag{25}$$

so that

$$V(n) = 1 - (1 - V_\alpha)^n \tag{26}$$

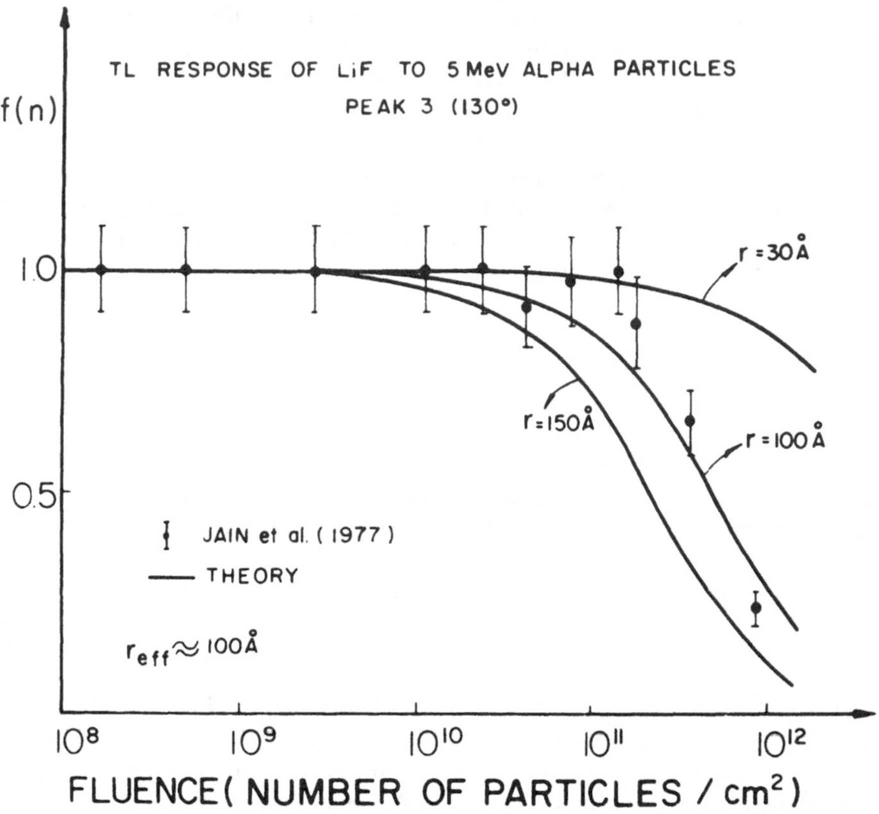

FIGURE 12. The TL fluence (dose) response of LiF-TLD to 5-MeV alpha particles for peak 3. The experimental points are from Jain and Ganguly.[73] A value of r_{eff} = 100 Å is in good agreement with the experimental data (X^2/n = 1.3).

thus

$$f(n) = \frac{1 - (1 - V_\alpha)^n}{nV_\alpha} \tag{27}$$

Figures 12 to 15 illustrate the experimental alpha particle fluence TL response curves for peaks 3, 5, 7, and 10 in LiF-TLD[73] fitted by Equation 27 with r_{eff} as a single free parameter. The values for r_{eff} are 100, 150, 85, and 75 Å, respectively. A similar analysis for the data of Aitken et al.[78] yields 200, 75, and 70 Å for peaks 5, 7, and 10, respectively.

It is interesting to note that r_{eff} = 150 to 200 Å corresponds very closely to the radial extension of the radial dose distribution as illustrated in Figure 10. This would indicate that charge carrier diffusion during the glow peak heating stage for peak 5 is negligible or not more than the order of tens of Angstroms, i.e., the thermally liberated charge carriers recombine to form TL photons at sites close to the initial trapping centers. This could be interpreted as additional evidence for spatially coordinated trapping and luminescent centers as suggested by Jain and Ganguly[73] and Horowitz.[72]

The values of r_{eff} = 70 to 85 Å for the higher temperature glow peaks are more difficult to understand but could indicate that these glow peaks are associated with trapping centers populated at high ionization density (perhaps multiple charge carrier capture) closer to the

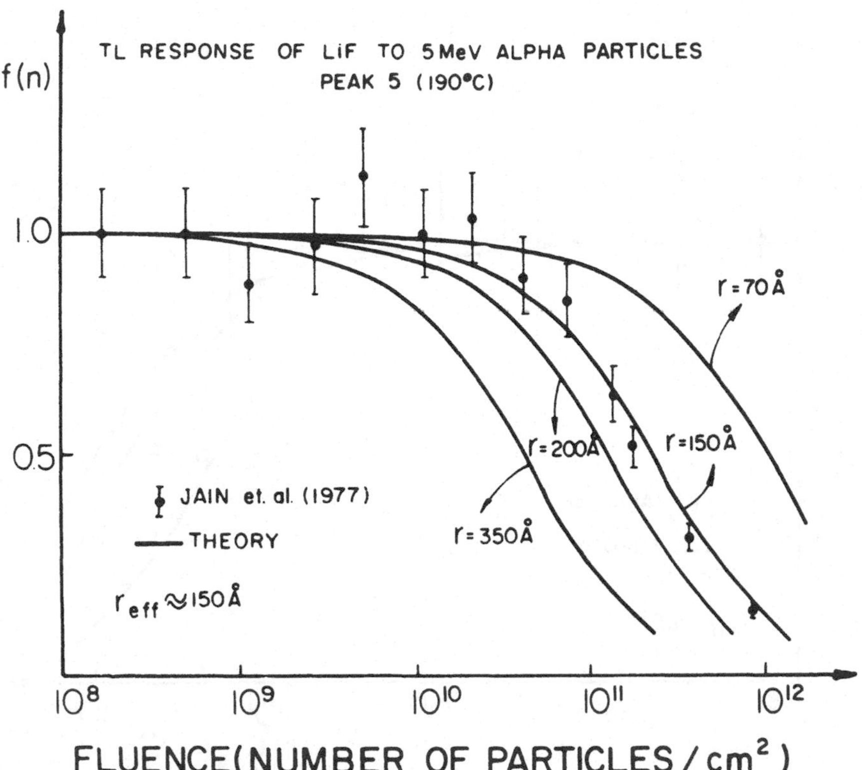

FIGURE 13. TL fluence (dose) response of LiF-TLD to 5-MeV alpha particles for peak 5. The experimental points are from Jain and Ganguly.[73] A value of r_{eff} = 150 Å is in good agreement with the experimental data (χ^2/n = 0.81).

HCP axis. This explanation is supported by the strong gamma-induced supralinearity of these high temperature peaks which also suggests "multiple-hit" processes.[54] Another possibility is that the trapping centers corresponding to these high temperature peaks capture relatively high energy secondary electrons. The secondary electron spectrum is expected to shift upwards in energy for small radial distances from the HCP track, although detailed calculations on this subject are not available.

In conclusion we have illustrated that the linear behavior of alpha particle TL dose-response curves arises naturally from the nature of the particle track, i.e., a very high density of electron-hole pairs created very close to the track axis with over 95% of the energy deposited within 150 Å of the track axis. The sublinear behavior at high dose is shown to yield values of r_{eff} for peak 5 in very good agreement with the maximum radial extension of the radial dose distribution. The lower values of r_{eff} for the higher temperature peaks are suggested to arise from preferential population of high temperature trapping centers at high ionization density or via higher energy secondary electrons. Both processes could lead to lower values of r_{eff}. The assumption that every overlapping region is totally saturated is a simplification that does not allow the possibility of slight supralinear behavior. In order to explain the presence of thermal neutron supralinearity (coupled with the lack of alpha particle-induced supralinearity for peak 5) as well as slight alpha particle-induced supralinearity for peak 8 in LiF-TLD and for the 145 and 200°C peaks in CaF_2:Dy, we are developing a more realistic model that takes into account the radial dose distribution around the axis of the HCP track.[116]

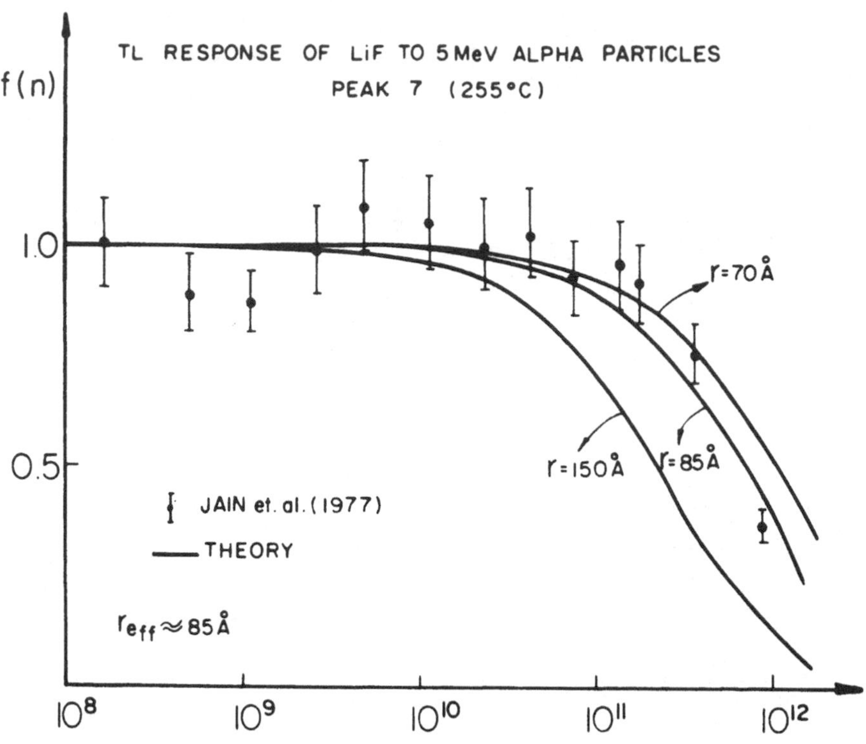

FIGURE 14. TL fluence (dose) response of LiF-TLD to 5-MeV alpha particles for peak 7. The experimental points are from Jain and Ganguly.[73] A value of r_{eff} = 85 Å is in good agreement with the experimental data (χ^2/n = 1.07).

C. HCP Relative TL Response

1. General Considerations

In HCP interaction with matter, the two mechanisms of ionization and excitation are the main channels of energy loss for HCP specific energies greater than approximately 0.1 MeV amu^{-1}. The direct energy transfer from the HCP (first generation of energy transport agents) takes place within only a few Å of the HCP path following which the ejected electrons also mainly dissipate their kinetic energy via ionizations and excitations (second- and higher-order generation of energy transport agents). The physical track of the HCP is thus determined by the spatial distribution of the primary localized events* (e.g., capture of charge carriers at particular defect centers) following the physical stage of the radiation action and the subsequent localization of the energy transport agents. The time required for the localization is believed to be approximately 10^{-10} sec. The final effect produced by the radiation action (e.g., emission of TL light, radical production, mutation, etc.) is believed to be strongly influenced by the spatial distribution of the localized primary events around the HCP path.

Assuming that collective effects do not play a role in the HCP efficiency and neglecting the continuity of matter, the response of a system to an HCP which imparts energy, ϵ, is given by

$$\frac{1}{\epsilon}\int_V P(n)\, dV \qquad (28)$$

* The spatial distribution of the primary localized events can be expected to be closely related to the radial distribution of dose around the HCP path.

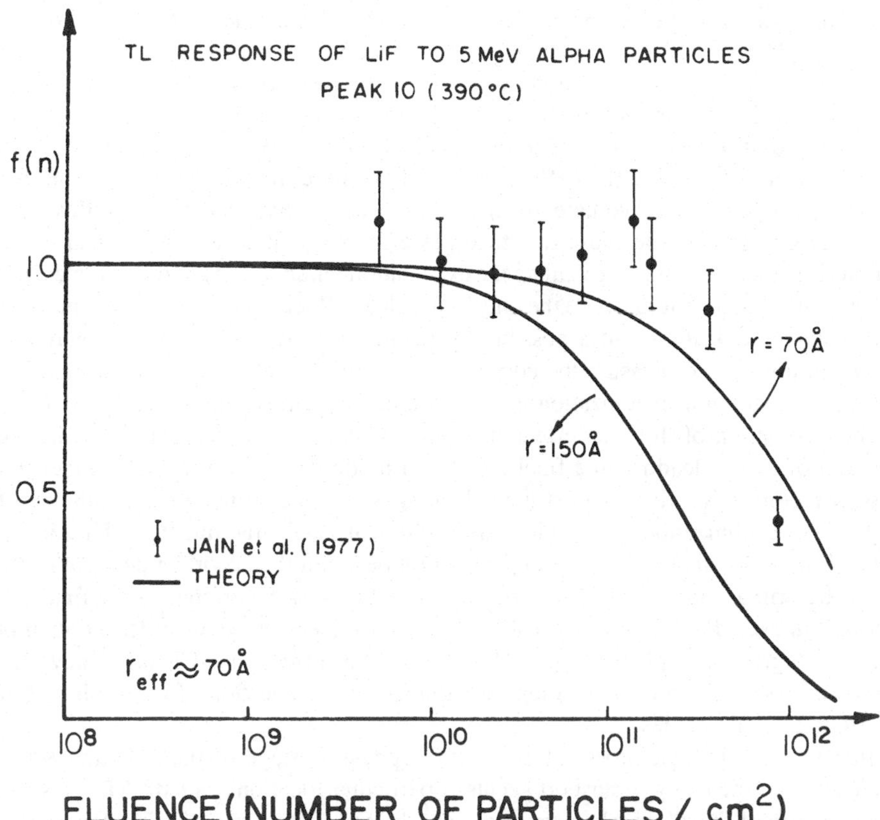

FIGURE 15. TL fluence (dose) response of LiF-TLD to 5-MeV alpha particles for peak 10. The experimental points are from Jain and Ganguly.[73] A value of r_{eff} = 70 Å is in good agreement with the experimental data (χ^2/n = 2.0).

where ndV is the probability of creation of a localized primary event by a single HCP in the volume element dV at a radial distance r from the HCP path and P(n)dV is the probability of creation of the final effect in the same volume. If we further assume that an "average HCP"* of initial energy E interacts with the system without any significant change in its direction of motion and imparts energy $\bar{\epsilon}_{HCP}$ to the system in such a way that there is radial symmetry $P\{n(r,\ell,E)\}$ we obtain

$$\frac{1}{\bar{\epsilon}_{HCP}} \int_0^\infty \int_0^\infty P(n(r,\ell,E))\, 2\pi r \, dr \, d\ell \qquad (29)$$

for the system response. For systems that exhibit a smooth dependence of P on $n(r,\ell,E)$ and for sufficiently "low" HCP fluence to ensure negligible intersection of the HCP tracks, the efficiency of the system to a particular type** of HCP can be approximated with that of the corresponding "average HCP".

* Certain systems require the reference to an "average HCP" due to the very low probability of creation for a particular final effect. In TL efficiency studies even at the lower limit of detection the distinction is usually superfluous because of the large number of generated TL photons.

** The HCP type refers to all the relevant particle parameters (e.g., atomic number, mass, initial energy, and total energy loss by the HCP to the medium).

The main problem in the determination of the HCP efficiency is in the estimation of the final effect probability distribution function. Our theoretical knowledge does not usually permit direct calculation of P(n), moreover, direct measurement of P(n) is practically impossible due to the very rapid dependence on r requiring an experimental microscopic probe with spatial resolution of a few Å. The fundamental aim of TST is to bypass this problem via direct measurement of the final effect probability using uniform electron irradiation (test radiation) of a macroscopic volume so that the processes that take place in this volume mimic as closely as possible those that take place in the volume element dV under HCP irradiation. It is then possible to calculate the expectation value of P(n) in the volume element dV. For example, in a system where the only factor that influences the final effect probability is the density of ionizations, it is possible by gamma ray irradiation under conditions of electronic equilibrium to measure the correlation between these two quantities. Knowledge of the radial distribution of ionization events around the path of the "average" HCP then allows the calculation of the final effect efficiency of the system under HCP irradiation.

The chain of events leading to a final effect in a particular system may be influenced by a variety of factors (e.g., dose rate, structural changes due to atomic collisions, the partition of energy between ionizations and excitations, biological repair, fading, etc.). Thus it cannot be assumed, *a priori,* that there is a unique relation between the absorbed dose (or even the density of the primary localized events) and the probability of occurrence of the final effect. The various intermediate factors for each system must be studied in order to attempt to evaluate the degree of applicability of TST to the final effect of HCP efficiency. In the following discussion these considerations are applied to the question of the applicability of TST to HCP TL efficiencies.

The first stage in the TL process following the physical stage of the radiation action is the localization of the energy transport agents at particular locations called "TL traps". The multistage nature of the TL mechanism leading to the final effect (i.e., TL photon emission) does not allow the possibility of measuring or calculating P(n)dV directly, in fact, even the calculation of the first stage probability, i.e., the number of charge carriers trapped in TL traps as a function of radial distance from the HCP path, is prohibitively difficult since we lack knowledge of most of the details of the trapping mechanism. As previously discussed, the application of TST to HCP relative TL efficiencies requires, therefore, direct measurement of the final effect probability using *uniform electron irradiation so that the processes that take place mimic as closely as possible those that take place under HCP irradiation.* We therefore require some knowledge of the energy and mode of action of the energy transport agents associated with the TL mechanism in order to apply TST. Furthermore, since application of TST will involve the use of data on low energy charge carrier densities arising from *HCP ionization* alone, we must attempt to justify the exclusion of nonionizing processes from the intermediate stages leading to TL emission.

In inorganic crystalline insulators the TL traps arise from crystal imperfections which give rise to energy levels that lie within the energy gap, E_g, between the valence and conduction bands. For most TL materials, the presence of ionization* processes during the physical stage of the radiation action is a necessary condition for "normal" TL photon emission. This statement is substantiated by the experimental observation that irradiation with UV photons in a number of TL materials (e.g., LiF, BeO) at energies less than E_g leads to negligible relative TL efficiencies with considerable evidence to the contrary for incident energies greater than or approximately equal to E_g. For example, $E_{hf} > (E_g = 4.9$ eV) leads to considerable TL emission in Zirconia,[79] and similar results have been obtained in LiF, BeO, NaCl, CsBr, and CsI for 40 eV $< E_{hf} <$ 240 eV.[80,81] In LiF and BeO (the

* The term "ionization" in crystalline insulators refers to any energy transfer process in which a valence or a more tightly bound electron is raised in energy above the top of the energy gap and into the conduction band.

two materials to which we have applied TST), the band-gap energies are approximately 13 and 11 eV, respectively.[82] Muller-Sievers,[83] irradiating undoped LiF with continuous ultrasoft X-rays (tube voltage between 10 and 200 V), observed significant TL production at 15 eV and higher. No TL production was observed at 10 eV implying a threshold between 10 and 15 eV. Irradiation of TLD-100 with 4.9-eV photons[84-86] and BeO with 3.2- to 5-eV photons[87-89] results in very slight or negligible TL emission. Furthermore, in TLD-100, 4.9-eV irradiation results in a single glow peak at 130°C drastically different from the complex higher temperature glow curve structure observed following ionizing radiation ($E_{hf} > E_g$). Finally the intrinsic TL response to UV photons of many TL materials has been reported to be strongly exposure-rate dependent.[88] These observations thus confirm that:

1. In LiF and BeO, E_{hf} must exceed E_g for significant TL emission to occur, i.e., the ionization process is a necessary condition for TL photon emission.
2. The drastically different glow curves imply that even the very slight TL emission observed in LiF and BeO arising from nonionizing processes proceeds via pathways very different from ionization-induced TL (for our point of view, "normal TL").
3. Even in the other TL materials where $E_{hf} < E_g$ leads to somewhat greater relative TL efficiencies, the strong exposure rate dependence also implies pathways leading to TL emission very different from those pathways leading to TL emission following irradiation by ionizing radiation.

Current theoretical models of pair production usually postulate that electrons and holes are generated primarily as a result of ionization involving secondaries whose energy is a few times larger than the band-gap energy.[90] Having established that ionization is a necessary condition for TL photon emission in LiF and BeO, general quantum mechanical considerations indicate that "low energy" electrons, close in energy to the bottom of the conduction band, initiate the localization stage, i.e., are most likely to be trapped in TL traps following the physical stage of the radiation action. The application of TST requires, therefore, the determination of: (1) $n(r,\ell,E)$ where n is the "low energy" electron radial distribution density and (2) P(n), representing the probability of TL photon emission.

As previously discussed the TL production efficiency of the HCP δ-rays* can be measured using an electron test radiation chosen to mimic as closely as possible the radiation action of the HCP. This requirement poses formidable obstacles since it demands matching the electron spectrum generated by the HCP with the electron test spectrum as well as matching of the HCP and electron test spectrum-irradiated volumes. The former is necessary since the TL efficiencies of LiF, BeO, and $Li_2B_4O_7$ are known to be strongly energy dependent.[72] In LiF-TLD, f(D) is strongly energy dependent, but reported measurements of $\eta_{e\gamma}$ are inconsistent. The maximum reported deviation from unity, e.g., over the energy range of 10 keV to 1 MeV is approximately 30% although many authors have reported data consistent with unity over this energy range. The latter requirement is necessary since the TL properties are not necessarily homogeneous over the TLD volume, indeed, surface effects are a well-known problem in TL dosimetry.[72,89] Let us examine these dual requirements in greater detail. The classical maximum energy transfer to free electrons is given by

$$Q_{m,f} = 2\, M_e\, c^2\, \beta^2\, \gamma^2 \tag{30}$$

where m_e is the mass of the electron, v is the velocity of the incident ion, $\beta = v/c$ and γ

* HCP δ-rays refer to all the first generation electrons ejected due to passage of the HCP with no lower energy limit restriction.

$= (1 - \beta^2)^{1/2}$. The maximum energy transfer $Q_{m,b}$ to initially bound electrons, however, depends on their binding energy U and, therefore, effects of different fractions of electrons in inner and outer shells and their binding energies may come into play. The nonrelativistic maximum energy transfer $Q_{m,b}$ can approximately be written as

$$Q_{m,b} = Q_{m,f} + 2 (Q_{m,f} \cdot U)^{1/2} \tag{31}$$

so that it can be easily seen that the binding effect is very small and therefore of no practical importance even for relatively low HCP specific energies of approximately 1 MeV amu^{-1}. Thus, the maximum energy that a 5 MeV amu^{-1} HCP can transfer to a free electron is approximately 10 keV. Wilson and Toburen[91] reviewed existing data on delta ray production by protons up to a few MeV in low Z gases and concluded that the mean energy of the delta rays with energy greater than 30 eV is approximately 120 eV and does not depend to a significant extent on the chemical form of the medium. For 0.5-MeV protons in water vapor the mean energy of all the ejected delta rays is 56 eV.[92] Since the Gruens range[93] of even a relatively high energy 1-keV electron is only approximately 450 Å in LiF and is more than 2 orders of magnitude less than the range of an approximately 1 MeV amu^{-1} HCP, the volume matching is exceedingly poor and direct ultralow electron irradiation (approximately 100 eV) aside from many other technical difficulties (e.g., electron back-scattering) must therefore be rejected as a viable test radiation.

On the other hand, ultrasoft X-rays are a better although not ideal candidate for the determination of $\eta_{\delta\gamma}$ for the following reasons:

1. The volume matching is greatly improved; the inverse of the energy absorption coefficient for 2.7-keV photons in LiF-TLD is approximately equal to the range of 4-MeV alpha particles.
2. As opposed to direct electron irradiation, electric charging of the TLD during the irradiation does not alter the impinging fluence and accurate calibration of the X-ray fluence is relatively easily accomplished.
3. The backscattering losses are not significant (the backscattering coefficient of even 2-keV electrons, on the other hand, is 0.14 (Kalef-Ezra et al.[94])), and a certain percentage of the low energy electrons is lost from the TLD as true secondary electrons.

A possible disadvantage from the use of ultrasoft X-rays arises from the simultaneous creation of two electrons following ultrasoft X-ray irradiation. In LiF, e.g., irradiated with 2.7-keV photons, 90% of the interacting photons produces fluorine K photoelectrons followed by 0.7-keV Auger electron emission. On the contrary, the majority of HCP interactions liberate weakly bound valence electrons with smaller average energy transfer per collision. An additional important consideration in the choice of the test radiation is the question of the energy dependence of $\eta_{\delta t}$ (t = test) and $f_\delta(D)$. As previously mentioned the results quoted in the literature are somewhat controversial for LiF-TLD; indeed the energy behavior of both η and f(D) is batch dependent. We have therefore carried out experimental investigations of both $\eta_{\delta t}$ and $f_\delta(D)$ in the identical TLDs for which η_{HCP} was measured.

2. Determination of $\eta_{\delta t}$

Two batches of LiF-TLD (TLD-600, 700: Harshaw) annealed in dry N_2 were irradiated in vacuum with ultrasoft X-rays generated from a ^3H beta source (tritium adsorbed in a thin layer of titanium) covered with a thin mylar foil to stop the beta particles. The photon energy spectrum consisted of mainly Ti K_α X-rays (4.5 keV) superimposed on the bremsstrahlung continuum. The energy fluence was calibrated using a Si(li) detector of known efficiency in geometry identical to that of the TLD irradiation. Lower energy X-rays (e.g., 2.7 keV) were not employed because of increasing contribution to the overall error arising from the high absorption probability of the low energy photons in the Si dead layer (the energy of the Si K absorption edge is 1.839 keV). For a more detailed description of the experimental technique, see Horowitz and Kalef-Ezra.[95] The dose averaged mean energy of the interacting photons with the 0.8 cm thick TLDs was determined to be 4.1 keV. Since only approximately 5% of the impinging photons do not interact with the TLD, possible error in the attenuation coefficients do not contribute significantly to the overall error. To increase confidence in the experimental technique additional measurements were carried out with a slightly harder photon spectrum (4.7 keV) obtained by covering the ^3H source with a 1.52 mg cm^{-2} thick Ni foil.

For both photon spectra, the relative TL response compared to ^{60}Co gamma rays were consistent with unity

$$\eta_{x\gamma} = 0.97 \pm 0.08 \text{ and } \eta_{x\gamma} = 1.09 \pm 0.07$$

respectively. The quoted error is 1 S.D. of the overall estimated error. Similar results consistent with unity were obtained with three batches (TLD-100, 600, 700) of air-annealed TLDs irradiated with ultra-soft X rays and ^{85}Kr beta particles ($E_{av.} \simeq 250$ keV, range in LiF approximately twice the range of 4-MeV alpha particles). To summarize up to this point, we have shown that ultra-soft X rays approximately fulfill the requirements of the test radiation required by TST. That is, (1) The irradiated HCP and test radiation TLD volumes are approximately matched and (2) The significantly greater average energy of the electrons liberated via ultrasoft X-ray irradiation compared to approximately 1 MeV amu^{-1} HCP irradiation is acceptable in LiF-TLD since $\eta_{\delta\gamma}$ (at least for the Harshaw TLD materials in our possession) appears to be independent of electron energy and equal to unity over the entire range of energies from approximately 4 keV to 1 MeV. It seems reasonable to assume that this constant behavior of η with energy can be extrapolated to even the lower electron energies liberated by HCPs, however, this point requires proof and is currently under investigation in the Radiation Physics Laboratory of the Ben Gurion University of the Negev. It is worthwhile pointing out that in $Li_2B_4O_7$-TLD, e.g., this latter requirement is not fulfilled and our measurements of $\eta_{x\gamma}$ for ultrasoft X-rays yield results significantly greater than unity.

3. Determination of $f_\delta(D)$

The experimental determination of $f_\delta(D)$ is achieved with a test radiation where the conditions of electron spectrum and volume matching with the HCP radiation action are required as previously discussed. In the case that the TLD thickness, d, is greater than the mean HCP range, R, we require that the depth dose distribution of the test radiation be of the form

$$\begin{aligned} D(x) &= D_0 \qquad 0 < x < R \\ &= 0 \qquad R < x < d \end{aligned} \tag{32}$$

in order to approximately simulate the HCP depth dose distribution. A method that appears attractive at first glance is to use TLDs for which d is slightly greater than R but much smaller than the inverse of the attenuation coefficient of the test radiation. Unfortunately, this solution will usually require the use of relatively high energy photons as the test radiation and must therefore be rejected because of the dependence of $f_e(D)$ on the initial electron

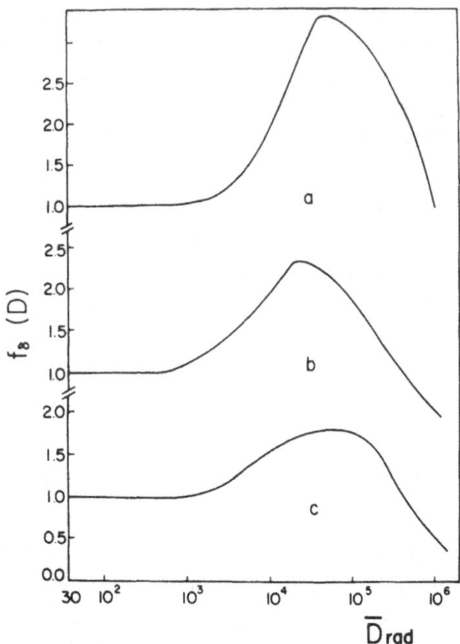

FIGURE 16. The TL dose response of air-annealed LiF-TLDs. (a) ^{60}Co gamma rays; (b) 50 kV_p X-rays; (c) 20 kV_p X-rays.

energy spectrum. This dependence is illustrated in Figure 16; $f(D)_{max}$ for the air-annealed LiF-TLDs used in this study equals approximately 3.35 for ^{60}Co gamma rays, but is only 1.8 for 20 kV_p X-rays(the 20 kV_p X-ray spectrum consisted mainly of an 8.0 keV X-ray from the Cu anode of the X-ray tube). A similar dependence of f(D) on energy was observed for N_2-annealed LiF-TLDs and for both air and N_2-annealed $Li_2B_4O_7$ TLDs. In this context it is important to point out that the choice of ^{60}Co gamma rays[1,53,54,64,96,97] or ^{90}Sr/^{90}Y electrons[51] as the test radiation in earlier studies of the applicability of TST to TL is unfortunate and without physical justification! *The dependence of f(D) on electron energy in TL demands a test radiation sufficiently low in energy to reasonably approximate the HCP-liberated electron spectrum.* The importance of the correct choice of $f_e(D)$ is illustrated from Figure 7 which shows that approximately 50% of the TL comes from the supralinear dose region of the radial dose distribution around the HCP track.

An obvious second choice might therefore be a test radiation identical to that used for the determination of $\eta_{\delta t}$, i.e., 4 keV X-rays. Unfortunately the resulting depth dose distribution in 0.08-cm thick LiF-TLDs used in these studies is very nonuniform (entrance dose/exit dose = 50), contrary to the requirement expressed in Equation 32, and furthermore results in unacceptable distortion of f(D) at high dose due to very significant TL light self-absorption. To compromise between the conflicting requirements of energy matching and depth dose uniformity, 20 kV_p X-rays (average photon interaction energy approximately twice that of the 4-keV reference radiation) and ^{3}H beta particles were finally selected as the ''best compromise'' test radiation. In the latter case the depth dose distribution was calculated using a scaling technique first introduced by Everhart and Hoff[98]

$$D(y) = \psi(1 - \eta f) \frac{1}{\rho R_G} \lambda(y) \qquad (33)$$

where $y = x/R_G$ and D(y) is almost insensitive to T and Z for 1 to 100 keV electrons impinging normally on low Z material. In addition, ψ is the impinging energy fluence, ηf is the backscattered fraction of the impinging energy fluence, R_G is the specific range of the electrons and $\lambda(y)$ is an empirical function fitted to experimental data. The resulting depth dose distribution is illustrated in Figure 17 for ^3H beta particles irradiating LiF-TLDs. The average macroscopic dose was calculated somewhat arbitrarily using the approximation

$$D(x) = \begin{cases} D_0 & 0 < x < 0.142 \text{ mg cm}^{-2} \\ 0 & 0.142 < x < d \end{cases} \tag{34}$$

where 80% of the dose is absorbed in a surface layer of thickness 0.142 mg cm^{-2}. The estimated deviation from the actual ^3H depth dose distribution is approximately 25%. Our "best compromise" test radiation is obviously less than ideal because of the dose nonuniformity and the fact that 0.142 mg cm^{-2} is roughly 25 times smaller than the range of 4.0 MeV alpha particles used as the HCP radiation. The "not-extreme" dose nonuniformity does not introduce significant error in the calculation of $\eta_{HCP,e}$ because of the fairly slow dependence of $f_e(D)$ on D. As for the nonideal matching of the test radiation and HCP irradiated volumes we can only comment that our glow curve and spectral emission studies using ^3H beta particles, 4 keV and 20 kV_p X-rays indicate good uniformity of the nature of the TL traps and luminescent centers over the entire TLD volume. It should be mentioned, however, that these radiations would not be sensitive to nonuniformities in these characteristics in the first few dozen surface atomic layers.

4. *Dose Rate Effects*

A potentially serious problem in the determination of P(n) is the question of the production rate of the "low energy" electrons via the HCP and test radiations. HCP-induced ionization events occur within less than 10^{-13} sec. On the other hand, the duration of the test irradiation with ^3H beta particles at 10^5 Gy, e.g., was 10^6 sec, i.e., 19 orders of magnitude greater than the slowing down time of a single HCP* and its corresponding secondary electrons. Since pulsed test radiations of the appropriate characteristics do not exist, our only recourse is to examine published data on the TL response dose rate dependence. Karzmark et al.[99] reported that the TL efficiency of LiF following 15-MeV electron or 15 MV_p X-rays was dose rate independent from 5 Gy to 2×10^6 Gy sec^{-1} in agreement with Tochilin and Goldstein[100] who reported dose rate independence following approximately 1 MeV X-ray irradiation up to 2×10^9 Gy sec^{-1} in TLD-700. On the other hand, Gorbics et al.[101] and Goldstein[102] observed a decrease in response for TLD-700 of 11 and 36%, respectively, at approximately the same dose rate. Unfortunately these experiments are not described in sufficient detail to allow critical evaluation so that the discrepancy between the two sets of experiments cannot be presently resolved. Until further, more precise results are available our operational assumption has been to assume dose rate independence.

5. *Atomic Displacements*

An additional characteristic of the radiation action of HCPs and the test radiation which may affect P(n) lies in their different respective abilities to induce changes in the crystalline structure of the TLD via the creation of atomic displacements. The reader is referred to Wardle and Murray[103] for a discussion of the defects induced by HCP bombardment of alkali

* The electron slowing down time (test radiation) is also of the order of 10^{-13} sec so that on the microscopic level the disparity in dose rates is similar to the disparity in the dose level of the HCP and test radiation microscopic dose distributions.

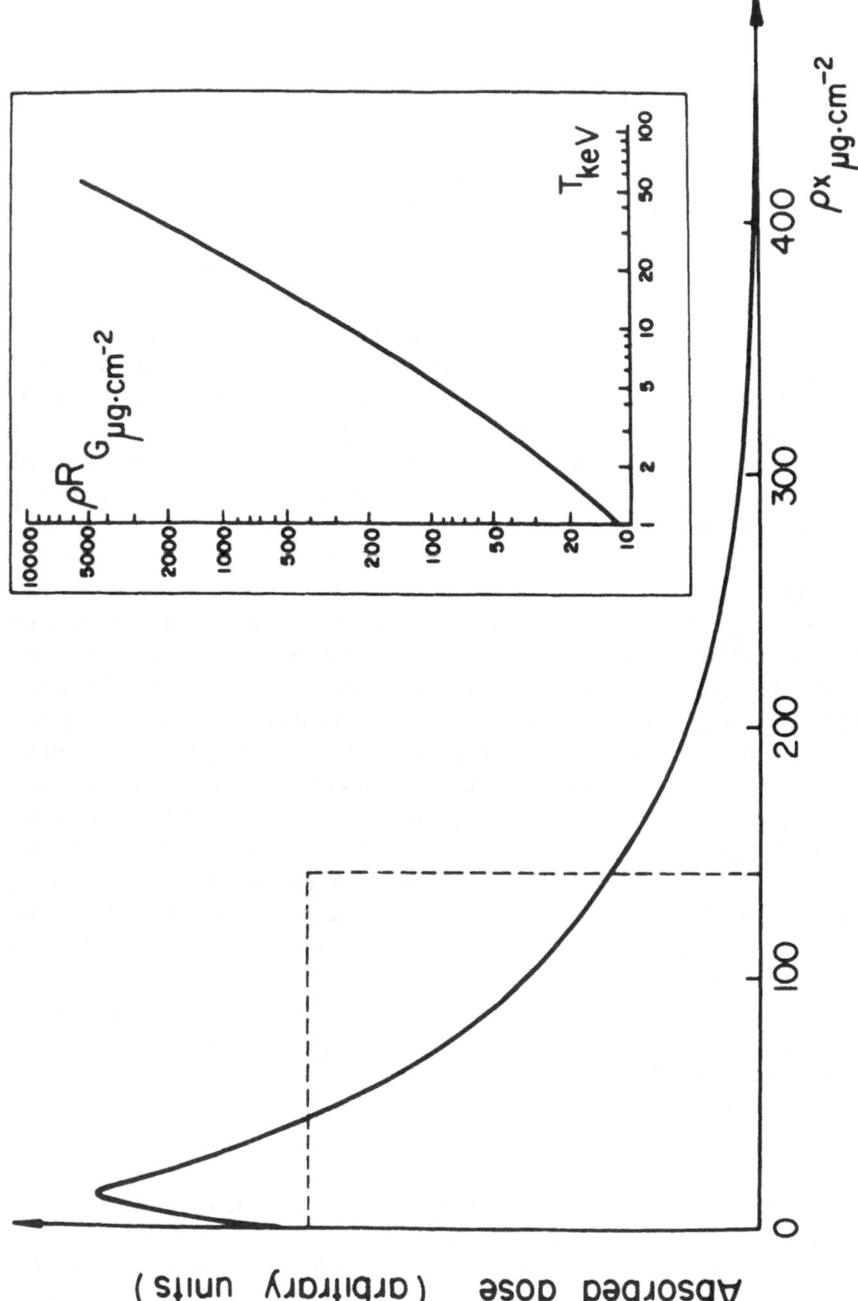

FIGURE 17. The depth dose distribution of ^{3}H beta particles in LiF-TLD. Insert: The Gruen range energy dependence of electrons in LiF. (Adapted from Kalef-Ezra, J. and Horowitz, Y. S., *Int. J. Appl. Radiat. Isot.*, 11, 1085, 1982.)

halides. Although we cannot rule out the possibility that HCP-induced atomic displacements affect the TL properties, we have carried out fairly extensive glow curve and TL emission spectra measurements which indicate the contrary. In the latter studies the emission spectra of air-annealed LiF-TLDs following irradiation by ^{3}H beta particles, 20 kV_p X-rays (dose region of maximum supralinearity* as well as low dose linear region), ^{60}Co gamma rays, 4-MeV alpha particles, 81-meV neutrons, and fission fragments are essentially identical. In the former studies, all the TL glow peaks observed following HCP irradiation (even in the case of fission fragments** where a considerable fraction of the energy is lost via nuclear collisions) have also been observed after low energy electron irradiation at sufficiently high dose. Finally, Van den Bosch[104] has reported that neutron and gamma-ray irradiation of undoped LiF create the same absorption bands in the region of 200 to 600 nm. On the basis of these observations we assume that the different ability of the HCP and electron test radiation to induce atomic displacements or other structural changes does not significantly influence the TL efficiency.

6. Determination of $n(r,\ell,E)$

As previously discussed, the HCP track can, in first approximation, be described as a cone-shaped region of excitations and ionizations induced directly by the HCP as well as by the ejected electrons around the HCP path and enveloping a cylindrical zone of a few Å radius called the core which does not play a significant role in the TL process. Inspection of Figure 7 indicates that not more than 5% of the TL originates from the first 10 Å surrounding the HCP track. For HCPs with energy greater than 0.1 MeV amu^{-1}, the lateral extension of the track (less than 0.05 μm for 2-MeV alpha particles in LiF) relative to the mean free path for significant angular deflection justifies the neglect of the HCP angular deflection in these calculations. The direct calculation of the "low energy" electron density, $n(r,\ell,E)$ requires knowledge of the following quantities: (1) the absolute cross sections, differential in electron energy and ejection angle for direct ejection of electrons from the constituent atoms of the TLD or from the HCP itself (autoionization) and (2) the cross section for the processes related to the energy transport agents. Since the cross-section data are unavailable in sufficient detail in condensed matter, we are obliged to use similar data obtained in gases with interaction properties similar to the TL material and to apply appropriate scaling techniques. We justify this approach with the following observations: (1) Wilson and Toburen[91] showed that with appropriate scaling the molecular structure in low Z gases does not influence the cross sections for electron ejection with energies greater than 30 eV; (2) with appropriate scaling the transport properties of electrons in low Z materials can be described by nearly universal curves;[60,98,105,106] and (3) P(n) is a smooth function of n, i.e., $f_e(D)$ changes slowly with D so that very high accuracy is not required in the calculation of $n(r,\ell,E)$. On the contrary, accurate calculation of the total number of "low energy" electrons is very important.

There are two main sources of data for $n(r,\ell,E)$ in gases: (1) theoretical analytic calculations of the radial distribution of the absorbed dose[1,55,57-59] and Monte Carlo calculations[61,70] and (2) direct measurement of the density of low energy charge carriers using a small movable ionization chamber of transparent mesh around the HCP beam.[65-69] Since the electron spectrum around the HCP path changes drastically with r, use of the theoretical data requires additional detailed knowledge of the variation of the value of W with electron energy. We

* The study of the TL emission spectra in the sublinear dose-response region is problematic due to significant self-absorption of the TL light.

** When the HCP mass is greater than the mass of the TLD constituent atoms (as is the case with fission fragments), a considerable amount of energy is transferred to scattered atoms (e.g., Li, F). These scattered atoms have been treated as individual HCPs interacting with the TL host lattice.

have therefore preferred to use the direct experimental measurements which do not require knowledge of W(r).

The charge carrier density in the solid state can be calculated from the gas phase using appropriate scaling which corrects for the different densities as well as the differences in the HCP energy loss and the differences in the transport properties of the energy transport agents. The HCP energy loss differences have to be calculated using data which take into account phase effects. For the pattern of energy dissipation of the ejected electrons the scaling technique of Everhart and Hoff[98] was applied with small geometrical corrections to account for the angular distribution of the ejected electrons. To a good approximation the difference in the two materials is given by the ratio of the radial Gruen range in the two media.*

The partition of energy between the ionizing and nonionizing processes** may lead to differences in W values in the two media. Unfortunately no direct information on W values for LiF or BeO is available. A phenomenological model[107] relating W values in solids with Eg yields 37 and 31 eV in LiF and BeO, not significantly different from W = 33 ± 1 eV suggested[108] as the W value for 1-MeV alpha particles in "tissue-equivalent" gas. To a first approximation it seems reasonable, therefore, to assume that differences in the partition of energy in the tissue-equivalent gas and in LiF and BeO can be neglected. On the basis of these considerations we have applied the following scaling

$$r_2 = r_1 \, \overline{(R_{G,2}/R_{G,1})} \tag{35}$$

$$n_2(r,\ell,E) = \frac{(dE/dX)_2}{(dE/dX)_1} \, \overline{(R_{G,1}/R_{G,2})}^2 \, n_1 \, (r,\ell,E) \tag{36}$$

where $R_{G,2}/R_{G,1}$ is the ratio of the radial Gruens range in the TL material (2) to the "tissue equivalent" gas (1) averaged over the energy spectrum of the ejected electrons and dE/dX is the electronic stopping power.

7. *The Calculation of* $\eta_{HCP,\gamma}$

The expectation value of the energy emitted as TL light after uniform gamma irradiation of a volume, V, of TL material is given by

$$\overline{\epsilon_\Phi} = K \, f_\gamma(D) \, n_\gamma \, V \tag{37}$$

where n_γ is the density of the "low energy" charge carriers and K is a constant. For an "average" HCP, the expectation value of the energy emitted as TL light, dE_Φ, in cylindrical

* R_G for electrons in the methane-based "tissue-equivalent" gas was calculated using data from Smith and Booz[109] and Dayashankar.[110]

** The nonionizing processes are thought to contain an "optimal phonon" contribution in addition to thermalization losses. The phonon process reflects the well-substantiated observation that charge carriers capable of pair-producing impacts are coupled to optical modes and may emit Raman phonons. Thermalization losses arise from the fact that the deposition of radiation energy culminates in the production of electrons and holes unable to produce further ionization and, consequently, bound to convert their energy residual into lattice vibrations.

coordinates is given by*

$$\overline{dE_\Phi} = \eta_{\delta\gamma} K f_\delta(D) n(r,\ell,E) 2\pi r\, dr\, d\ell \quad (38)$$

so that the total energy emitted as TL light for an "average" HCP stopping in the TL material is given by

$$\overline{E_\Phi} = \int_0^{R_{max}} \int_0^{r_{max}} \eta_{\delta\gamma} K f_\delta(D)\, n\, (r,\ell,E) 2\pi r\, dr\, d\ell \quad (39)$$

where R_{max} and r_{max} are the maximum lateral and radial distances of penetration of the "low energy" charge carriers from the HCP radiation action. Since the total energy liberated by the HCP as "low energy" charge carriers is given by

$$E = W_{HCP} \int_0^{R_{max}} \int_0^{r_{max}} n(r,\ell,E) 2\pi r\, dr\, d\ell \quad (40)$$

it follows that the TL efficiency of the HCP at low dose is given by

$$\alpha_{HCP}(D_0) = \frac{\int_0^{R_{max}} \int_0^{r_{max}} \eta_{\delta\gamma} K f_\delta(D)\, n(r,\ell,E) 2\pi r\, dr\, d\ell}{W_{HCP} \int_0^{R_{max}} \int_0^{r_{max}} n(r,\ell,E) 2\pi r\, dr\, d\ell} \quad (41)$$

Similarly for the reference radiation the TL efficiency at "low" dose is given by

$$\alpha_\gamma(D_0) = k\, n_\gamma V / W_\gamma n_\gamma\, V \quad (42)$$

* Point target or site approximation: an alternative microdosimetric approach to the microdosimetric "point target" approximation used in this study (Equation 39) involves the possibility of the existence of sensitive sites characteristic of the TL process and possibly underlying certain aspects of the TL mechanism. The site size might somehow be associated with the mean free path of the energy transport agents in either or both stages of irradiation and readout or alternately with the dimensions of a defect complex responsible for TL trapping. Battye et al.[111] calculated that the mean free path of electrons in pure LiF decreases from approximately 13 Å at 400 eV to approximately 7 Å at 100 eV increasing to approximately 22 Å at 20 eV. In heavily doped LiF-TLD we would expect these values to be significantly decreased. The mean escape depth of electrons from LiF:Ag during photostimulated exoelectron emission has been reported[80] to be 30 Å. On the other hand, Bronshteyn and Protsenko[112] found the escape zone of "true" secondary electrons in undoped LiF to be 200-Å thick after bombardment by electrons of initial energy greater than 1 keV. Jamal et al.[113] found that the presence of impurities in alkali halides drastically reduced the exciton (and electron) diffusion in electron-induced sputtering. Obviously these limited observations do not allow any quantitative conclusion on the diffusion length of electrons (and alternately on a site size of possible meaning in LiF-TLD). In the TST model proposed by Katz et al.[1] the site radius is introduced as a phenomenological free parameter to be evaluated from experimental dose-response data. For example, Larsson and Katz[53] and Waligorski and Katz[54] obtained site radii of approximately 100 and 400 Å for peaks 5 and "6" in LiF-TLD, respectively, following a multihit analysis invoked to explain supralinearity. These site radii seem rather large. Indeed, the schism between TL supralinearity and microdosimetry is even more dramatically emphasized when one attempts to explain the dependence of the supralinearity on electron energy, which then requires enormous sensitive sites possibly of the order of magnitude of 10^6 Å! Obviously one must look to other mechanisms to help explain supralinearity. In the absence of any direct experimental or theoretical evidence indicating a relevance of "site size" to TL phenomena we have chosen to avoid the incorporation of "site size" phenomenology in this work.

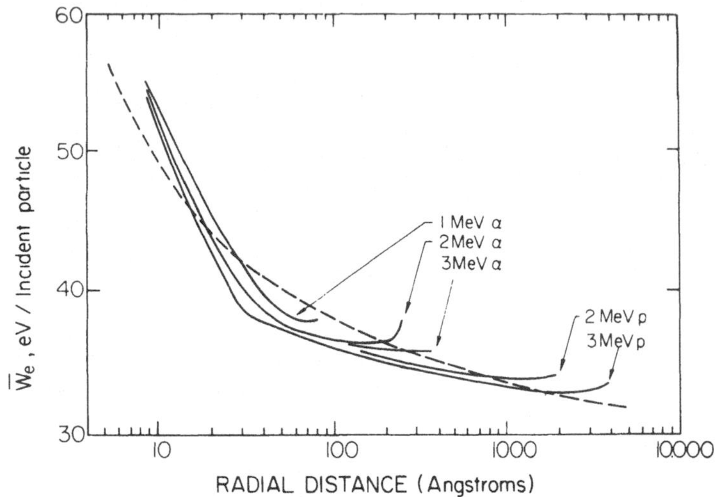

FIGURE 18. Variations of W values with radial distance from the trajectory of 1-, 2-, and 3-MeV alpha particles and 2- and 3-MeV protons. According to the model of Fain et al.[59] there is no significant change in the W values either with respect to the particle or its energy. (Adapted from Fain, J., Monin, M., and Montret, M., *Radiat. Res.*, 57, 579, 1974.)

Dividing Equation 41 by Equation 42 we obtain

$$\eta_{HCP,\gamma} = \eta_{\delta\gamma}\,\frac{\overline{W_\gamma}}{\overline{W}_{HCP}}\;\frac{\int_0^{R_{max}}\int_0^{r_{max}} f_\delta(D)\; n(r,\ell,E)\; 2\pi r\; dr\; d\ell}{\int_0^{R_{max}}\int_0^{r_{max}} n(r,\ell,E)\; 2\pi r\; dr\; d\ell} \qquad (43)$$

where $\eta_{\delta\gamma}$ has been assumed to be constant over the entire electron spectrum generated by the HCP. Figure 18 shows variation of W_α values with radial distance as estimated by Fain et al.[59] theoretically and from the data of Cole.[40] As can be seen W_α is approximately constant over the radial distances from approximately 30 to 300 Å so that the use of average values for W_γ and W_α is not expected to introduce serious error. Only approximately 25% of the TL is emitted for radii less than 30 Å (Figure 7).

8. Results and Conclusions

Table 2 summarizes the experimental and theoretical data used in our TST calculations of $\eta_{HCP,\gamma}$ and, as can be seen, the agreement between theory and experiment in both LiF and BeO is surprisingly good. The experiments with fission fragments with initial energies below the energy corresponding to the Bragg peak are an extreme test of TST due to the very high ionization density and the relatively large fraction of the energy leading to atomic displacements. Even in this case the agreement is more than satisfactory. For a more detailed discussion of the experimental technique with fission fragments, see Horowitz and Kalef-Ezra.[114] Air and N_2 annealing (as a single example of one of the many possible experimental variables which may affect the relative TL response) change the values of $\eta_{HCP,\gamma}$ and $f_\delta(D)$, but again the agreement between theory and experiment is maintained in all cases. In BeO (the only other TL material in which sufficient experimental data have been published) the calculations were carried out using $n(r,\ell,E)$ data for 10.4 MeV amu^{-1} ^{16}O ions at the same

Table 2
COMPARISON OF THEORETICAL AND EXPERIMENTAL MEASUREMENTS OF $\eta_{HCP,\gamma}$

		HCP Radiation (k)		Reference radiation (ℓ)		Test radiation for determination of:		$\eta^{f}_{k\ell}$		
Material	Annealing procedure	Type	Energy (MeV)	Type	Energy (keV_{eff})	$\eta_{\delta\ell}$	$f_{\delta}(D)$	Theoretical	Batch	Experimental
LiF	N_2	Alpha	4.0	X-rays	4	ℓ	H-3 betas	0.56	2, 3	0.50 ± 0.045
	Air	Alpha	3.8	X-rays	4	ℓ	20 kV_p X-rays	0.20[a]	1	0.16 ± 0.03
									2	0.20 ± 0.03
									3	0.28 ± 0.04
	Air	n[b]	13.5×10^{-9}	X-rays	4	ℓ	20 kV_p X-rays	0.31	1	0.32 ± 0.03
									2	0.39 ± 0.06
	N_2	ff[c]	52/32	X-rays	4	ℓ	H-3 betas	0.21	2, 3	0.29 ± 0.05
	Air	ff	49/31	X-rays	4	ℓ	H-3 betas	0.25	1	0.20 ± 0.07
BeO	—	C	125	X-rays	6	ℓ	15 kV_p X-rays	1.42[d]		1.30 ± 0.15[e]
	—	Ne	208	X-rays	6	ℓ	15 kV_p X-rays	1.42		1.41 ± 0.25
	—	Ar	416	X-rays	6	ℓ	15 kV_p X-rays	1.42		1.42 ± 0.18

[a] Possible batch dependence in the theoretical calculations are not indicated because of the difficulties in measuring f(D) in the Mrad region.
[b] HCP creation via the $^{6}Li\ (n,\alpha)\tau$ reaction.
[c] "Average" light and heavy degraded fission fragments from Cf-252.
[d] Calculated values for 0-16 ions of the same velocity.
[e] Experimental data from Tochilin et al.
[f] Preliminary results were reported by Horowitz and Kalef-Ezra.[2,114]

initial velocity as the C, Ne, and Ar ions employed by Tochilin et al.[115] It will be interesting to extend these experiments to TL materials in which the intrinsic UV response is not negligible.

In conclusion, this work has attempted to test the applicability of TST to TL phenomena under more stringent experimental and theoretical restrictions than those that have been attempted in previous studies. Specifically, an effort has been made to significantly improve the matching of the electron spectra used to generate $f_\delta(D)$ with the initial electron spectra ejected by the HCP. It deserves mention that the critical requirement of the matching of the electron spectra is not a peculiarity of the TL phenomenon but exists in all systems where there is significant variation of the dose response with electron energy. The electron spectra used in this study to generate f(D), i.e., ^3H beta particles and 20 kV_p X-rays, may still be significantly mismatched. In fact, an ideal matching would require an electron spectrum identical to the HCP-ejected electron spectra and generated in identical volumes — obviously this dual condition is very difficult to achieve. This difficulty has been, however, compensated by our investigations concerning the relative TL response as a function of electron energy. In this regard we have established that the relative TL response is independent of energy for LiF-TLD (at least for the materials in our possession) over the range of energies from approximately 4 keV to 1 MeV.

A general complication of TL investigations which we have attempted to emphasize is the great variability of even the relative TL properties of what is commonly referred to as LiF:Mg,Ti or instead even LiF-TLD (Harshaw). Both of these materials are, in fact, a family of materials of considerably different relative TL properties. Thus any serious attempt to apply TST must involve the generation of $f_\delta(D)$ and the measurement of $\eta_{HCP,\gamma}$ in identical materials (ideally the same TLD!) and under identical experimental conditions. Again, great effort has been expended in this study to fulfill the latter restriction [at least to the extent that these studies were carried out on TLDs belonging to the same preselected batch of LiF-TLD (Harshaw)]. This restriction has, unfortunately, not been fulfilled to any serious extent in most previous applications of TST to thermoluminescence.

Finally, aside from its basic implications to our understanding of radiation effects, TST can be applied to the calculation of HCP-induced TL properties (the TL efficiency of neutrons and pions are only two examples) with considerable advantages since the TST calculation is many orders of magnitude less time consuming or expensive than direct experimental measurement. Naturally, as we have illustrated, considerable care must be exercised in the theoretical and experimental manipulations. Considering the wide range of assumptions that were necessary to enable these TST calculations, the overall agreement is excellent. Aside, of course, from the possibility of a fortuitous cancellation of errors, this good agreement supports the assumptions used in these investigations (i.e., negligible dose rate dependence, negligible influence of atomic displacements, applicability of "low energy" charge carrier densities measured in the gas phase to the condensed phase). At this stage of our knowledge of electron-induced TL, however, it would be premature to claim that TST is the entire truth underlying HCP-induced TL; it can certainly be said, however, to describe the dominant mechanism.

REFERENCES

1. **Katz, R., Sharma, S. C., and Homayoonfar, M.,** The structure of particle tracks, in *Topics in Radiation Dosimetry,* Suppl. 1, Attix, F. H., Ed., Academic Press, New York, 1972, 317.
2. **Horowitz, Y. S. and Kalef-Ezra, J.,** Relative thermoluminescent yields of heavy charged particles: theory and experiment, *Nucl. Instrum. Methods,* 175, 29, 1980.
3. **Kalef-Ezra, J. and Horowitz, Y. S.,** Heavy charged particle thermoluminescent dosimetry: track structure theory and experiments, *Int. J. Appl. Radiat. Isot.,* 11, 1085, 1982.
4. **Ziegler, J. F.,** Helium: stopping powers and ranges in all elements, in *The Stopping and Ranges of Ions in Matter,* Vol. 4, Pergamon Press, Elmsford, N.Y., 1977.
5. **Gouard, P., Chemtob, M., Nguyen, V. D., and Parmentier, N.,** Une approche possible de l'etude du ralentissement des particules chargees lourdes, in *Proc. 6th Symp. Microdosimetry,* Booz, J. and Ebert, H. G., Eds., Harwood Academic Publishers, London, 1979, 707.
6. **Bethe, H. A.,** Quantenmechanik der Ein- und Zwei-Elektronenprobleme, in *Handbuch der Physik,* Vol. 24, No. 1, Geiger, H. and Scheel, K., Eds., Springer-Verlag, Berlin, 1933, 491.
7. **Fano, U.,** Penetration of protons, alpha particles and mesons, *Annu. Rev. Nucl. Sci.,* 13, 1, 1963.
8. **Brodsky, A.,** *CRC Handbook of Radiation Measurement and Protection,* CRC Press, Boca Raton, Fla., 1978, 214.
9. **Armitage, B. H. and Hotton, B. W.,** Energy loss of oxygen and sulfur ions in matter, *Nucl. Instrum. Methods,* 58, 29, 1968.
10. **Pierce, T. E. and Blann, M.,** Stopping powers and ranges of 5-90 MeV ^{32}S, ^{35}Cl, and ^{79}Br and ^{127}I ions in H_2, H_e, N_2, Ar and Kr: a semi-empirical stopping power theory for heavy ions in gases, *Phys. Rev.,* 173, 390, 1968.
11. **Betz, H. D. and Grodzins, L.,** Charge states and excitation of fast heavy ions passing through solids: a new model for the density effect, *Phys. Rev. Lett.,* 25, 211, 1970.
12. **Datz, S.,** Excitation and ionization states of ions penetrating solids, *Nucl. Instrum. Methods,* 132, 7, 1976.
13. **Sugiyama, H.,** Energy loss and range of electrons below 10 keV, *J. Phys. Soc. Jpn.,* 41, 4, 1976.
14. **Ziegler, J. F.,** The stopping of energetic ions in solids, *Nucl. Instrum. Methods,* 168, 17, 1980.
15. **Lindhard, J. and Bohr, N.,** Electron capture and loss by heavy ions penetrating through matter, *Mat. Fys. Medd. Dan. Vid. Selsk.,* 28, 8, 1954.
16. **Lindhard, J. and Scharff, M.,** Energy loss in matter by fast particles of low charge, *Mat. Fys. Medd. Dan. Vid. Selsk.,* 27, 15, 1953.
17. **Firsov, O. B.,** A qualitative interpretation of the mean electron excitation energy in atomic collisions, *Sov. Phys. JETP.,* 36, 1076, 1959.
18. **Lindhard, J. and Winther, A.,** Stopping power of electron gas and equipartition rule, *Mat. Fys. Medd. Dan. Vid. Selsk.,* 34, 4, 1964.
19. **Dennis, J. A. and Powers, D.,** The dependence of stopping power on physical and chemical states, in *Proc. 6th Symp. Microdosimetry,* Booz, J. and Ebert, H. G., Eds., Harwood Academic Publishers, London, 1979, 661.
20. **Hvelplund, P. and Fastrup, B.,** Stopping cross sections in carbon of 0.2—1.5 MeV atoms with $21 < Z_1 < 39$, *Phys. Rev.,* 165, 408, 1968.
21. **Anderson, H. H., Besenbacher, F., and Knudsen, H.,** Stopping power and straggling of 65-500 keV lithium ions in H_2, He, CO_2, N_2, O_2, Ne, Ar, Kr and Xe, *Nucl. Instrum. Methods,* 149, 121, 1978.
22. **Chu, W. K. and Powers, D.,** On the Z_2 dependence of stopping cross sections for low energy alpha particles, *Phys. Lett.,* 38A, 267, 1972.
23. **Chu, W. K. and Powers, D.,** Alpha particle stopping cross section in solids from 4 keV to 2 MeV, *Phys. Rev.,* 187, 478, 1969.
24. **Bourland, P. D. and Powers, D.,** Bragg-rule applicability of stopping cross sections of gases for alpha particles, *Phys. Rev.,* B3, 3635, 1971.
25. **Powers, D., Chu, W. K., Robinson, R. J., and Lodhi, A. S.,** Measurement of molecular stopping cross sections of halogen-carbon compounds and calculation of atomic stopping cross sections of Halogens, *Phys. Rev.,* A6, 1425, 1972.
26. **Lin, W. K., Olson, H. G., and Powers, D.,** Alpha particle stopping cross sections of silicon and germanium, *J. Appl. Phys.,* 44, 3631, 1973.
27. **Lodhi, A. S. and Powers, D.,** Energy Loss of alpha particles in gaseous C-H and C-H-F compounds, *Phys. Rev.,* A10, 2131, 1974.
28. **Chau, E. K. L., Brown, R. B., Lodhi, A. S., Powers, D., Matteson, S., and Eisenbarth, S. R.,** Stopping cross sections of oxygen for 0.3 — 2.0 MeV alpha particles in saturated alcohols and ethers, *Phys. Rev.,* A16, 1407, 1977.

29. **Chau, E. K. L., Powers, D., Lodhi, A. S., and Brown, R. B.,** Stopping cross sections of three membered ring structure molecules and their double-bonded isomers for 0.3 — 2.0 MeV He^+ ions, *J. Appl. Phys.*, 16, 1407, 1978.
30. **Chau, E. K. L. and Powers, D.,** Measurements of molecular stopping cross sections of aldehydes and ketones and calculation of the atomic stopping cross section of oxygen in double-bonded and three membered ring structure C-H-O compounds, *J. Appl. Phys.*, 49, 2611, 1978.
31. **Neuwirth, W., Pietsch, W., Richter, K., and Hauser, U.,** On the invalidity of Bragg's rule in stopping cross sections of molecules for swift Li ions, *Z. Phys.*, A275, 209, 1977.
32. **Cruz, S. A., Vargas, C., and Brice, D. K.,** Critical analysis of the modified Firsov model sensitivity to the choice of atomic wave functions, *Radiat. Eff. Lett.*, 43, 143, 1979.
33. **Neuwirth, W., Pietsch, W., and Kreutz, R.,** Chemical influences on the stopping power, *Nucl. Instrum. Methods*, 149, 105, 1978.
34. **Jarvis, O. N. and Sherwood, A. C.,** Stopping power for fast channeled alpha particles in silicon, *Phys. Rev.*, B16, 3880, 1977.
35. **Softky, S. D.,** Ratio of atomic stopping power of graphite and diamond for 1.1 MeV protons, *Phys. Rev.*, 123, 1685, 1961.
36. **Matteson, S., Chau, E. K. L., and Powers, D.,** Stopping cross sections of bulk graphite for alpha particles, *Phys. Rev.*, A14, 169, 1976.
37. **Palmer, R. B. J. and Akhaven-Rezayat, A.,** The stopping power of water, water vapour and aqueous tissue equivalent solution for alpha particles over the energy range 0.5 — 8 MeV, *J. Phys. D*, 11, 605, 1978.
38. **Thwaites, D. I. and Watt, D. E.,** Phase effects in stopping power for low energy heavy charged particles, in *Proc. 6th Symp. Microdosimetry*, Booz, J. and Ebert, H. G., Eds., Harwood Academic Publishers, London, 1978, 777.
39. **Ziegler, J. F. and Chu, W. K.,** Stopping cross sections and backscattering factors for ^4He ions in matter, Z = 1 — 92, E(^4He) = 400 — 4000 keV, *At. Data Nucl. Data Tables*, 13, 463, 1974.
40. **Cole, A.,** Absorption of 20 eV to 50,000 eV electron beams in air and plastic, *Radiat. Res.*, 38, 7, 1969.
41. **Toburen, L. H.,** Distribution in energy and angle of electrons ejected from molecular nitrogen by 0.3 to 1.7 MeV protons, *Phys. Rev.*, A3, 216, 1971.
42. **Jacobi, W. and Stolterfoht, N.,** Spatial distribution of ionization and deposited energy along proton tracks in gases, in Proc. 3rd Symp. Microdosimetry, Ebert, H. G., Ed., EURATOM, Brussels, 1971, 109.
43. **Glass, W. A., Toburen, L. H., and Wilson, W.,** Energy ejected in fast proton collisions, in Proc. 3rd Symp. Microdosimetry, Ebert, H. G., Ed., EURATOM, Brussels, 1971, 71.
44. **Wilson, W. and Paretzke, H. G.,** Electron ejection cross sections for hydrocarbon molecules and their implications for phase effects, in Proc. 4th Symp. Microdosimetry, Booz, J., Ebert, H. G., Eickel, R., and Waker, A., Eds., EURATOM, Brussels, 1973, 113.
45. **Lynch, D. J., Toburen, L. H., and Wilson, W.,** Electron emission from methane, ammonia, monomethyamine and dimethylamine by 0.25 and 2.0 MeV protons, *J. Chem. Phys.*, 64, 2616, 1976.
46. **Toburen, L. H. and Wilson, W. E.,** Energy and angular distribution of electrons ejected from water vapour by 0.3 — 1.5 MeV protons, *J. Chem. Phys.*, 66, 5202, 1977.
47. **Katz, R. and Kobetich, E.,** Response of NaI (T1) to energetic heavy ions, *Phys. Rev.*, 170, 397, 1968.
48. **Katz, R. and Kobetich, E.,** Formation of etchable tracks in dielectrics, *Phys. Rev.*, 170, 401, 1968.
49. **Katz, R. and Kobetich, E.,** Particle tracks in emulsions, *Phys. Rev.*, 186, 344, 1969.
50. **Butts, J. J. and Katz, R.,** Theory of RBE for heavy ion bombardment of dry enzymes and viruses, *Radiat. Res.*, 30, 855, 1967.
51. **Zimmerman, D. W.,** Relative thermoluminescent effects of alpha and beta radiation, *Radiat. Eff.*, 14, 81, 1972.
52. **Bartlett, D. T. and Edwards, A. A.,** Investigation of the dependence of the lyoluminescence response on radiation quality and its relationship with gamma dose response function, *Nucl. Instrum. Methods*, 175, 126, 1980.
53. **Larsson, L. and Katz, R.,** Supralinearity of thermoluminescent dosimeters, *Nucl. Instrum. Methods*, 138, 631, 1976.
54. **Waligorski, M. P. R. and Katz, R.,** Supralinearity of peak 5 and peak 6 in TLD-700, *Nucl. Instrum. Methods*, 175, 48, 1980.
55. **Chatterjee, A., Maccabee, H. D., and Tobias, C. A.,** Radial cutoff LET and radial cutoff dose calculations for HCPs in water, *Radiat. Res.*, 54, 479, 1973.
56. **Chatterjee, A. and Schaefer, H. J.,** Microdosimetric structure of heavy ion tracks in tissue, *Radiat. Environ. Biophys.*, 13, 215, 1976.
57. **Chatterjee, A. and Magee, J.,** Relationship of the track structure of heavy particles to the physical distribution and chemical effects of radicals, in *Proc. 6th Symp. Microdosimetry*, Booz, J. and Ebert, H. G., Eds., Harwood Academic Publishers, London, 1978, 283.

58. **Fain, J., Monin, M., and Montret, M.,** Energy density deposited by a heavy ion around its path, in Proc. 4th Symp. Microdosimetry, Booz, J., Ebert, H. G., Eickel, R., and Waker, A., Eds., EURATOM, Brussels, 1973, 169.
59. **Fain, J., Monin, M., and Montret, M.,** Spatial energy distribution around heavy ion path, *Radiat. Res.*, 57, 579, 1974.
60. **Berger, M. J.,** Distribution of dose around point sources of e^- and β particles in water, *J. Nucl. Med.*, 12, 5, 1971.
61. **Berger, M. J.,** Some new transport calculations of the deposition of energy in biological materials by low energy electrons, in Proc. 4th Symp. Microdosimetry, Booz, J., Ebert, H. G., Eickel, R., and Waker, A., Eds., EURATOM, Brussels, 1973, 695.
62. International Commission on Radiation Units and Measurements, Radiation Dosimetry: X-rays Generated at Potentials of 5 to 150 kV, Rep. No. 17, ICRU Publications, Washington, D.C., 1970.
63. **Mozumder, A.,** Charged particle tracks and their structure, *Adv. Radiat. Res.*, 76, 1, 1969.
64. **Fain, J., Montret, M., and Sakraoui, L.,** TL response of CaF_2:Dy and LiF:Mg,Ti under heavy ion bombardment, *Nucl. Instrum. Methods,* 175, 37, 1980.
65. **Wingate, C. L. and Baum, J. W.,** Measured radial distribution of dose and LET for alpha and proton beams in hydrogen and tissue equivalent gas, *Radiat. Res.*, 65, 1, 1976.
66. **Varma, M. N., Paretzke, H. G., Baum, J., Lyman, J. T., and Howard, J.,** Dose as a function of radial distance from a 930 MeV ^{4}He ion beam, in Proc. 5th Symp. Microdosimetry, Booz, J., Ebert, H. G., and Smith, B. G. R., Eds., EURATOM, Brussels, 1975, 75.
67. **Baum, J. W., Varma, M. N., Wingate, C. L., Paretzke, H. G., and Kuehner, A.,** Nanometer dosimetry of heavy ion tracks, in Proc. 4th Symp. Microdosimetry, Booz, J., Ebert, H. G., Eickel, R., and Waker, A., Eds., EURATOM, Brussels, 1973, 93.
68. **Varma, M. N., Baum, J. W., and Kuehner, A. V.,** Energy deposition of heavy ions in a tissue equivalent gas, *Radiat. Res.*, 62, 1, 1975.
69. **Varma, M. N., Baum, J. W., and Klianga, P.,** Microdosimetric results obtained by proportional counter and ionization chamber methods: a comparison, in *Proc. 6th Symp. Microdosimetry,* Booz, J. and Ebert, H. G., Eds., Harwood Academic Publishers, London, 1978, 227.
70. **Paretzke, H. G.,** Comparison of track structure calculations with experimental results, in Proc. 4th Symp. Microdosimetry, Booz, J., Ebert, H. G., Eickel, R., and Waker, A., Eds., EURATOM, Brussels, 1973, 141.
71. **Baum, J. W., Stone, S. L., and Kuehner, A. V.,** Radial Distribution of Dose Along Heavy Ion Tracks, Ebert, H. G., Ed., EURATOM, Brussels, 1969, 653.
72. **Horowitz, Y. S.,** The microdosimetric and theoretical basis of thermoluminescence and applications to dosimetry, *Phys. Med. Biol.*, 26, 765, 1981.
73. **Jain, V. K. and Ganguly, A. K.,** Some Aspects of Thermal, Radiation and LET Effects in the TL of LiF, B.A.R.C./I—466, Bhabha Atomic Research Center, Bombay, 1977.
74. **Jain, V. K.,** High temperature peaks in LiF-TLD: dependence on LET, *Nucl. Instrum. Methods,* 180, 195, 1981.
75. **Lakshmanan, A. R., Bhatt, R. C., and Supe, S. J.,** Mechanisms of non-linearity in the response characteristics of TLDs, *J. Phys. D,* 14, 1683, 1981.
76. **Horowitz, Y. S., Fraier, I., Kalef-Ezra, J., Pinto, H., and Goldbart, Z.,** Non-universality of the TL-LET response in thermoluminescent $Li_2B_4O_7$: the effect of batch composition, *Nucl. Instrum. Methods,* 165, 27, 1979.
77. **Horowitz, Y. S., Kalef-Ezra, J., Moscovitch, M., and Pinto, H.,** Further studies on the non-universality of the TL-LET response in thermoluminescent LiF and $Li_2B_4O_7$: the effect of high temperature TL, *Nucl. Instrum. Methods,* 172, 479, 1980.
78. **Aitken, M .J., Tite, M. S., and Fleming, J.,** Thermoluminescent response to heavily ionizing radiations, in Proc. 1st Int. Conf. Luminescence Dosimetry, U.S. A.E.C. CONF-650637, NTIS, Springfield, Va., 1965, 490.
79. **Jain, V. K.,** Some aspects of the TL of Zircon (sand) and Zirconia, *Indian J. Pure Appl. Phys.*, 15, 601, 1977.
80. **Elango, M. A., Zhurakovskii, A. P., Kadchenko, V. N., and Sorkin, B. A.,** Luminescence and electron emission by ionic crystals exposed to ultra-soft X-rays, *Izv. Akad. Nauk SSSR Ser. Fiz.*, 41, 1314, 1977.
81. **Saar, M. A. E., Maiste, A. A., and Elango, M. A.,** Manifestation of the electron-hole interaction in the K absorption spectrum of Li in LiF, *Sov. Phys. Solid State,* 15, 1663, 1973.
82. **Strehlow, W. H. and Cook, E. L.,** Compilation of energy band-gaps in elemental and binary compound semiconductors and insulators, *J. Phys. Chem. Ref. Data,* 2, 163, 1973.
83. **Muller-Sievers, K.,** Intensitatsmessungen an Rontgen Bremsstrahlung im Energiebereich von 200 bis 10 eV, *Biophysik,* 10, 163, 1973.

84. **Mason, E. W.,** Thermoluminescent response of ^{7}LiF to UV light, *Phys. Med. Biol.*, 16, 303, 1971.
85. **Bassi, L., Busuoli, G., Lembo, L., and Rimondi, O.,** *G. Fis. San. Radioprot.*, 18, 137, 1974.
86. **Henaish, B. W., Sayed, A. M., and Morsy, S. M.,** Light influence on thermoluminescent BeO and other TL phosphors, *Nucl. Instrum. Methods*, 163, 1979, 511.
87. **Hobzova, L.,** UV induced thermoluminescence in BeO, in Proc. 4th Int. Conf. Luminescence Dosimetry, Niewiadomski, T., Ed., Institute of Nuclear Physics, Krakow, 1974, 1081.
88. **Nagpal, J. S.,** UV induced processes in TL phosphors, *Phys. Status Solidi A*, 57, K63, 1980.
89. **Lasky, J. B. and Moran, P. R.,** Thermoluminescent response of LiF (TLD-100) to 0.1 — 5 keV electrons: an energy range relationship and comparison of the TL glow with TSEE glow curves, *J. Appl. Phys.*, 50, 4951, 1979.
90. **Klein, C. A.,** Radiation ionization energies in semiconductors: speculations about the role of plasmons, *J. Phys. Soc. Jpn.*, 21, 307, 1966.
91. **Wilson, W. E. and Toburen, L. H.,** Delta-ray production in ion-atom collisions, in *Proc. 7th Symp. Microdosimetry*, Booz, J., Ebert, H. G., and Hartfiel, H. D., Eds., Harwood Academic Publishers, London, 1980, 435.
92. **Toburen, L. H., Wilson, W. E., and Popowick, R. J.,** Secondary electron emission from ionization of water vapour by 0.3 — 2.0 MeV He^+ and He^{2+} ions, *Radiat. Res.*, 82, 27, 1980.
93. **Gruen, V. A. E.,** Lumineszenz-Photometrische Messungen der Energieabsorption im Strahlungsfeld von Eleckronenquellen, *Z. Naturforsch.*, 12, 89, 1957.
94. **Kalef-Ezra, J., Horowitz, Y. S., and Mack, J. M.,** Electron back-scattering from low Z thick absorbers, *Nucl. Instrum. Methods*, 195, 587, 1982.
95. **Horowitz, Y. S. and Kalef-Ezra, J.,** Relative thermoluminescent response of LiF-TLD to 4 keV X-rays, *Nucl. Instrum. Methods*, 188, 603, 1981.
96. **Montret-Brugerolle, M.,** Ph.D. thesis, University of Clermont-Ferrand II, France,1980.
97. **Fain, J., Montret, M., and Sanzelle, S.,** Thermoluminescence and heavy ion microdosimetry, in *Proc. 7th Symp. Microdosimetry*, Booz, J., Ebert, H. G., and Hartfiel, H. D., Eds., Harwood Academic Publishers, London, 1980, 807.
98. **Everhart, T. E. and Hoff, P. H.,** Determination of kilovolt electron energy dissipation versus penetration distance in solid materials, *J. Appl. Phys.*, 42, 5837, 1971.
99. **Karzmark, J., White, J., and Fowler, F.,** Lithium fluoride TL dosimetry, *Phys. Med. Biol.*, 9, 273, 1964.
100. **Tochilin, E. and Goldstein, N.,** Dose rate and spectral measurements from pulsed X-ray generators, *Health Phys.*, 12, 1705, 1966.
101. **Gorbics, S. G., Attix, F. H., and Kerris, K.,** Thermoluminescence dosimeters for high dose applications, *Health Phys.*, 25, 499, 1973.
102. **Goldstein, N.,** Dose rate dependence of LiF for exposures above 15,000 R per pulse, *Health Phys.*, 22, 90, 1972.
103. **Wardle, M. W. and Murray, R. B.,** Ion bombardment of alkali halides. II. Effects observed at high fluence, *Radiat. Eff.*, 25, 133, 1975.
104. **Van den Bosch, G.,** Gamma radiolysis of lithium fluoride, *Radiat. Eff.*, 19, 129, 1973.
105. **Harder, D.,** in Proc. 2nd Symp. Microdosimetry, Ebert, H. G., Ed., EURATOM, Brussels, 1965, 567.
106. **Cross, W.,** The distribution of absorbed energy from a point beta source, *Can. J. Phys.*, 45, 2021, 1967.
107. **Klein, C.,** Bandgap dependence and related features of radiation ionization energies in semiconductors, *J. Appl. Phys.*, 39, 2029, 1968.
108. International Commission on Radiation Units and Measurements, Average energy required to produce an ion pair, Rep. No. 31, ICRU Publications, Washington, D.C., 1979, 23.
109. **Smith, B. G. R. and Booz, J.,** Experimental results on W-values and transmission of low energy electrons in gases, in *Proc. 6th Symp. Microdosimetry*, Booz, J. and Ebert, H. G., Eds., Harwood Academic Publishers, London, 1978, 759.
110. **Dayashankar, N.,** Calculation of W for low energy electrons in tissue equivalent gas, *Health Phys.*, 33, 465, 1977.
111. **Battye, F. L., Liesegang, J. L., Leckey, R., and Jenkin, J.,** Electron attenuation lengths in alkali halides, *Phys. Rev.*, B13, 2646, 1976.
112. **Bronshteyn, I. M. and Protsenko, A. N.,** *Radio Eng. Electron Phys. (USSR)*, 15, 667, 1969.
113. **Jamal, Y., Pooley, D., and Townsend, P.,** The role of exciton diffusion in the electron induced sputtering of alkali halides, *J. Phys. C*, 6, 247, 1973.
114. **Horowitz, Y. S. and Kalef-Ezra, J.,** Relative thermoluminescent response of LiF-TLD to Cf-252 fission fragments, *Nucl. Instrum. Methods*, 187, 519, 1981.
115. **Tochilin, E., Goldstein, N., and Lyman, J. T.,** The quality and LET dependence of three thermoluminescent dosimeters and their potential use as secondary standards, in Proc. 2nd Int. Conf. Luminescence Dosimetry, U.S. A.E.C. CONF-680920, NTIS, Springfield, Va., 1968, 424.

116. **Horowitz, Y. S. and Moscovitch, M.,** Track Structure Theory - track intersection model for heavy charged particle TL dose response curves, in *Proc. 7th Int. Conf. Solid State Dosimetry,* Nuclear Technology Publishing, England, in press.

Chapter 4

PHOTOSTIMULATED THERMOLUMINESCENCE

Vinod K. Jain

TABLE OF CONTENTS

I. INTRODUCTION

The phenomenon of thermoluminescence when excited by light is known as photostimulated thermoluminescence (PSTL). PSTL has two aspects, i.e., phototransferred thermoluminescence (PTTL)[1] and photo-induced intrinsic thermoluminescence (PITL).[2] PTTL refers to thermoluminescence (TL) induced by light as a result of (photo) transfer of charge carriers from one or more traps to other traps. Usually this transfer takes place from high temperature deep traps which have been filled earlier by irradiation with ionizing radiation, to lower temperature shallow traps which have been emptied thermally, leaving the deep traps filled. PITL is thermoluminescence directly induced by light in a fully annealed material. Most of the materials which exhibit TL on exposure to ionizing radiations also show PSTL. Thus PSTL has been observed in dosimetry phosphors, alkali halides, alkaline earth compounds, archaeological and pottery samples, minerals, green photosynthetic plants, etc. Ultraviolet light is usually most efficient, but sometimes even visible light is effective in causing PSTL.

PSTL has been found useful in the study of electronic processes, the identification of the nature of traps whether electron or hole type, delineation of the mechanism of trap filling, etc. Both PTTL and PITL have been tried for the measurement of ultraviolet radiation. PTTL is instrumental in the use of "memory effects" in TL materials for the reestimation of dose. Light-induced TL is of importance in the study of organic macromolecules and has been successfully employed in understanding the photosynthetic photochemical reactions in green plants.

II. TYPES OF PSTL

A. Photo-Induced Thermoluminescence (PITL)

PITL is thermoluminescence induced by direct exposure to light. If the fundamental absorption band lies in the UV region, as is the case for ZrO_2 (4.9 eV), very high PITL is induced. Usually the band-gap energy is very large and band-to-band transitions involving freeing of charge carriers are not possible, still very high UV sensitivity is observed, e.g., in MgO (7.3 eV). Valence change in impurities or defect structures or charge release from lattice defects or impurities present in the material are some of the factors suggested.[3-5] In fact, two distinct groups of glow peaks have been identified: one group appears after irradiation into the excitation band (EB) and the other after irradiation with UV light of wavelength corresponding to band-to-band transition. The latter depends strongly on impurities and environmental effects such as thermal and pressure pretreatment, etc., is restricted to a thin surface layer, and is attributed to the presence of oxygen impurities and the generation of V_k centers.[6,7] Presence of oxygen as an "impurity" or otherwise appears to play an important part. It is observed that many oxides, irrespective of their band-gap energies, have high PITL. In rare earth (RE)-doped phosphors a center composed of RE ion and a neighboring F center (RE-F) have been proposed from which it is possible to lift an electron into the conduction band (CB) by light of appropriate wavelength (λ). The electron is trapped, released thermally or optically, and recombines over the CB with the ionized RE-F center.[8,9]

Very often the glow curve due to PITL is not very different from the one induced by X-rays. The maximum TL intensity attained, however, is far less. Quite expectedly there is grain size dependence and the PITL response increases with decrease in grain size. There is also, in general, wavelength dependence in the response of most phosphors which depends on the nature of defects and impurities. Dose rate dependence has not been studied by many authors, but a recent report gives dependence on flux of 254 nm UV in the range of 10^{-2} to 10 J m^{-2} sec^{-1}. Except for CaO:Dy and to a lesser extent MgO, the response of Al_2O_3, CaF_2:Dy, $CaSO_4$:Tm, $CaSO_4$:Dy, and Mg_2SiO_4:Tb increases with increase in flux and that of $CaSO_4$:Dy is maximum.[10] Increasing the temperature during UV irradiation also has the effect of increasing the PITL peak height as well as integrated response at least in some phosphors.[10]

B. Phototransferred Thermoluminescence (PTTL)

PTTL is thermoluminescence due to light exposure in a phosphor which had been earlier exposed to ionizing radiation and then partially annealed, leaving some TL as residual (RTL). For inducing PTTL, the phosphor is either irradiated at an elevated temperature so that the lower temperature traps are not filled or after irradiation at a low temperature, it is heated to a higher temperature to empty the lower temperature traps. Illuminating the phosphor now has two effects: (1) peaks which are stable at the temperature of irradiation or annealing are diminished in intensity and (2) the annealed out peaks are regenerated. A schematic representation of these steps is shown in Figure 1. Many times these effects are greatest for illumination in the F band, however, this need not always be so. Absorption characteristics of the residual filled traps are important and PTTL effects in most materials show definite wavelength dependence.

III. APPLICATIONS

A. Study of Electronic Processes

Stoddard was perhaps the first person who studied PTTL systematically.[1] Shining NaCl crystal with F light he found that the PTTL results were not consistent with the existing

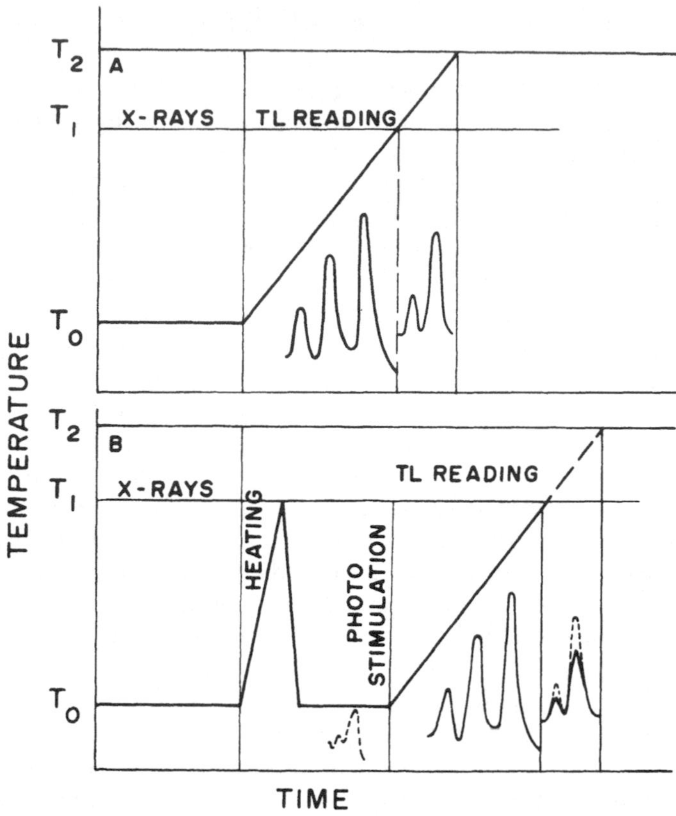

FIGURE 1. Schematics of PTTL. (A) Irradiation at T_0 and TL reading. (B) Irradiation at T_0, heating to T_1 or irradiation at T_1, cooling to T_0, and TL reading.

models for the TL process. It was then taken up by a number of workers, particularly for pursuing the investigation of alkali halides.[11-19] Braner and Israeli[12] X-rayed pure KBr, KI, NaCl, and KCl single crystals at room temperature (RT), cooled them to liquid nitrogen temperature (LNT), and then illuminated them with monochromatic F or V light. In this way they separated electron and hole peaks which together added up to the X-ray-induced glow curve. However, differences in the glow curve for F and V light were not observed by other workers.[13-19] Instead it was noted that:

1. Only one carrier type was involved in the spectral range studied (200 to 800 nm) since the same TL peaks appeared irrespective of the wavelength of the stimulating light.
2. Electrons were photoionized in the conduction band and captured by the traps; in fact the stimulation spectra of different traps and the photoconductivity spectrum of additively colored crystals coincided.
3. In KCl:$SrCl_2$, containing large concentration of V_k centers, the stimulation spectrum followed the absorption spectrum of the V_k band.
4. By PSTL experiments in NaCl, L bands, the excited state of which falls within the conduction band, were detected.
5. In KCl and KBr, a peak was found to be due to F′ centers; for these peaks the PSTL spectrum deviated from the photoconductivity spectrum in the F-band region indicating the tunneling of electrons from the excited F centers to neighboring traps.

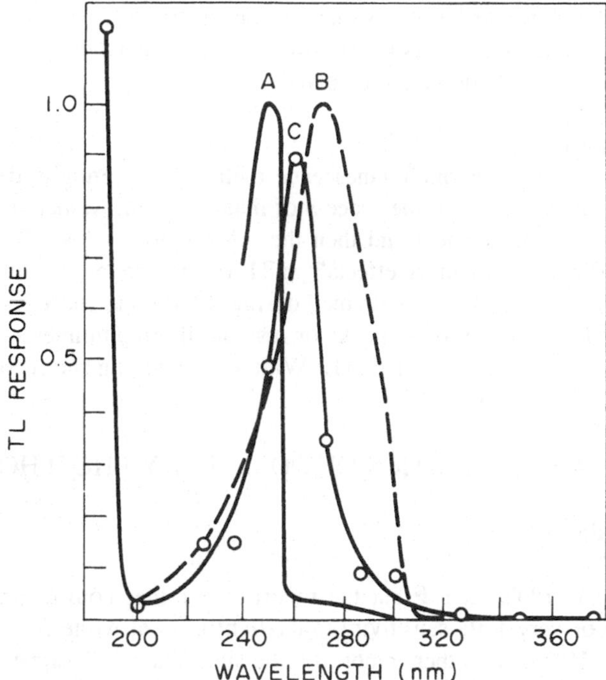

FIGURE 2. Response vs. wavelength of sensitized CaF_2:Dy (curve A) and composite action spectrum for keratitis and erythema as proposed by ACGIH (curve B).[73] Curve C is for CaO. (From Bapat, V. N., Proc. Natl. Symp. Radiat. Phys., June 10 to 12, 1976. With permission.)

In CaF_2 glow peaks have been grouped into families, and in $Cd_2B_4O_7$ coupling of traps has been observed on the basis of PSTL studies.

B. Ultraviolet Dosimetry

Ultraviolet radiation is employed in a wide variety of human activity, in particular biomedical and industrial. It is being increasingly used as a disinfectant and a sterilizing agent. At the same time UV has deleterious effects on the skin[20] and the eye.[21] Overexposure of the skin results in pigmentation, erythema, and even cancer. Excessive exposure of the eye causes an inflammatory condition in the cornea called photokeratitis. These effects are wavelength and dose dependent and accumulate over a period of time. Erythemal effectiveness of UV radiation is expressed by a "standard erythemal curve" extending from 240 to 320 nm. Similarly for ocular effects the effectiveness curve extends from 200 to 320 nm with a peak around 270 to 280 nm. Sliney suggested that an action spectrum for threshold skin and eye effects could be synthesized, as is shown in Figure 2.[22] This action spectrum was adopted by the American Conference of Industrial Hygienists[23] and also forms the basis of the "Standard for Occupational Exposure to Ultraviolet Radiation" recommended by the National Institute for Occupational Safety and Health.[24] The minimum erythemal dose (MED) required to produce barely visible reddening of the skin is about 1 to 3 $\times$ 10^4 μW·sec·cm^{-2} of weighted ultraviolet for the erythemal or keratitic action spectrum. For any dosimeter with spectral sensitivity approximately equal to the respective action spectrum, an absolute calibration per MED is determined uniquely by recording the response for 1 MED for any selected reference wavelength. TL dosimeters based on both PITL and PTTL have been described for the measurement of UV doses. However, due to the lack of coincidence of action spectrum and λ dependence of dosimeter response, these dosimeters rarely measure true MED. For PTTL type dosimeters the sensitivity also changes due to the loss of stored signal particularly if large UV doses are involved.

C. Reestimation of Dose

Dose measurement by the thermoluminescence method is essentially destructive in the sense that the signal is destroyed in the process of measurement. Sometimes the signal can be destroyed by mistake or accident and then the information is lost. To overcome such shortcomings, the so-called "memory effects" in TL dosimeters are useful. The high temperature peaks which are not thermally drained during the normal readout keep a record of the dose received. The record in these peaks or residual thermoluminescence (RTL), as it is called, can be scanned by means of PTTL. Without destroying the record, PTTL gives an estimate of the dose accumulated.[25-27]

IV. CHARACTERISTICS OF DOSIMETRY PHOSPHORS

A. Lithium Fluoride

1. PITL

There may be small PITL in LiF but the reports are rather conflicting. While several authors found none or very little sensitivity to daylight or cool white fluorescent light,[28-30] or even to 366 nm UV,[30] many others report it to be significant.[31-35] Lippert and Mejdahl,[31] using powder bound on a kanthal strip with silicone resin, found that 20 min of exposure to daylight caused 0.13 mGy gamma-equivalent signal initially and then increasing at the rate of about 0.05 mGy $hr.^{-1}$ Bjärngard et al.[32] and Bjärngard and Jones[33] observed in LiF-teflon® discs 0.1 mGy gamma-equivalent signal in 10 min of exposure to normal laboratory fluorescent lighting which increased further on prolonged exposure. Freeswick and Shambon,[34] while measuring doses $\leq$0.03 mGy, detected a signal 0.01 mGy equivalent after 6 hr of exposure 7 ft below the laboratory fluorescent lighting fixtures. The sensitivity of LiF to light could possibly be the result of many factors such as the presence of trace amounts of particular impurities, inadequacy of annealing, and the UV content of the light. Marshall et al.[35] observed that in LiF-teflon® the light sensitivity was due purely to teflon®, i.e., the teflon® itself exhibits light-induced phosphorescence. Mason exposed virgin ^{7}LiF (single crystal and powder) to 254 nm UV light and recorded a signal at 90°C which saturated at 2.2 mGy gamma equivalent.[36]

2. PTTL

a. Response Characteristics

PTTL in LiF has been investigated in detail using different postirradiation annealing treatments.[36-46] The temperatures employed range from 280 to 350°C and duration from 15 min to 1 hr depending on temperature. A residual TL peak near 390°C remains after 15 min of annealing at 350°C, and PTTL is generated by subjecting the phosphor to 254 nm UV light (Figure 3). The graphs shown are for TLD-100, but similar results are observed in other dosimetry-grade LiF samples. UV repopulation, which fills all the lower temperature peaks except no. IV, is λ dependent (Figure 4).[47,48] In this figure the step at 225 nm is to be noted. The PTTL response of peak V follows the gamma response in that it is linear initially then becomes supralinear and saturates at the same original irradiation dose as does the gamma response, unlike the response of residual peak which is supralinear from the lowest dose that is detectable.[48,49] However, PTTL is observable for dose levels very much below the detection threshold of the RTL. At these very low doses PTTL turns out to be proportional to RTL because peak-V response is linear, nonlinearity in RTL would be very small, and PTTL itself is only a few percent of the former. Buckman and Payne found

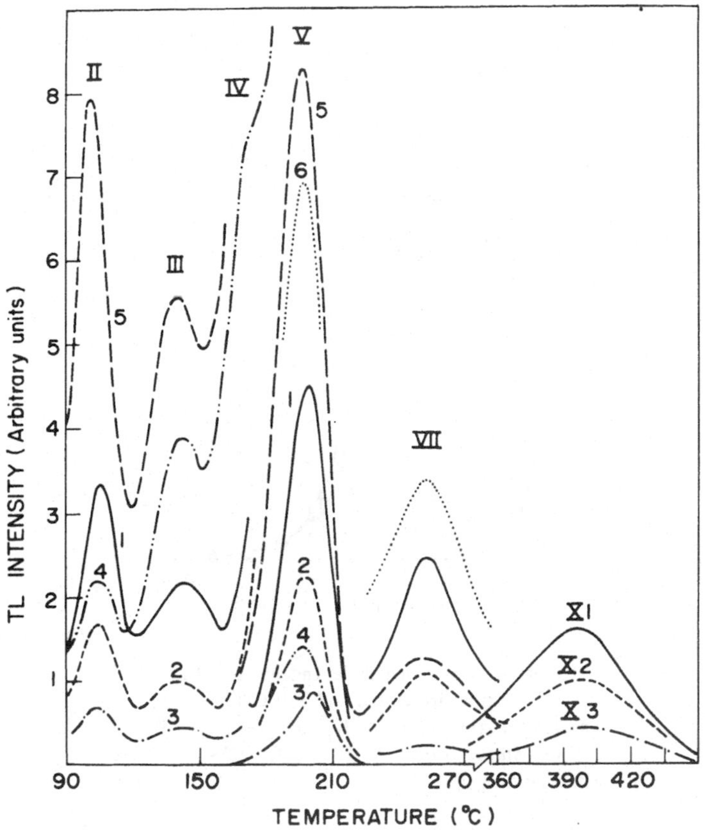

FIGURE 3. PTTL and changes in PTTL and sensitization in LiF due to repeated UV bleaching of residual peak X. Curve X1 after an exposure of 10^3 Gy and 15 min at 350°C. This is also sensitized powder. Curve 1 is due to 3 min of UV exposure to sensitized powder. The powder is then subjected to a cycle of 3 min UV and then 30 sec at 330°C. X2 and X3 are peak X remaining after 10 and 40 such cycles, and curves 2 and 3 are PTTL due to 3 min UV in these samples, respectively. Curves 4 to 6 are due to 3.4-Gy test exposure in unsensitized, sensitized, and after 40 cycles of bleaching in sensitized powder respectively. (From Jain, V. K., *Health Phys.*, 41, 363, 1981. With permission.)

linearity up to 6 m J cm^{-2} and independent of UV intensity between 0.1 to 50 $\mu W\ cm^{-2}$ within 10%.[30] They also noted significant (23%) increase in sensitivity if the temperature during PTTL increased from RT by even 15°C.

b. Mechanisms

Kos and Nink[50] have explained phototransferred thermoluminescence in LiF as follows: UV light releases electrons from F centers which not only refill electron traps but may also form additional Z_1 and Z_2 centers. These additional Z_2 centers, according to them, are formed due to conversion of Z_3 centers and are also created directly from F centers.[50] Jain has attributed the source of electrons to the Z_3 centers absorbing at 225 nm instead of the F centers.[51,52] The λ dependence of PTTL (Figure 4) shows that bleaching at 225 nm is favored. There are difficulties in accepting F centers as playing the central role in PTTL.[48,49] Optically destroying F centers with 250 nm light immediately begins destruction of TL but not of peak V and V_3 centers. There is no prominent glow peak corresponding to the thermal

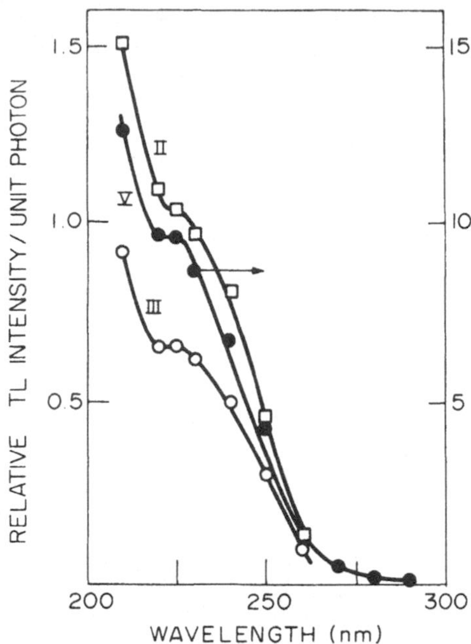

FIGURE 4. Wavelength dependence of PTTL in 10^3 Gy gamma-irradiated and 350°C/15 min annealed LiF. (From Jain, V. K. and Kathuria, S. P., *Phys. Status Solidi A*, 50, 329, 1978. With permission.)

release of F centers. But under different conditions of bleaching, F light destroys peak-V centers as also Z_3 centers. It has been argued that peak-V traps, which Kos and Nink attribute to Z_2 centers, cannot be formed from peak-X traps which constitute the Z_3 centers.[51-53] The TCLC model of Jain proposes that Z_3 centers are created as a result of breakup of complex TCLC centers.[27] TCLC are present in the 400°C-annealed LiF and produce the 137-nm absorption. On irradiation (and heating at about 100 to 150°C) the TCLC separate into trap TC centers and luminescence LC centers. (TC and LC recombine to form the complex TCLC on annealing around 400°C.) TCs are the Z_3 centers absorbing at 225 nm and give rise to peak X. Being created upon irradiation the response of this peak is supralinear from the beginning. This is so because the TL efficiency of peak X will be proportional to the product of the number of TCs present and the number of LCs giving to zero-order approximation a dependence on D^2. LCs are similar to the existing luminescence centers and when added in significant numbers increase the probability of recombination and emission causing the supralinearity of TL peaks including the dosimetry peak V. In the irradiated and partially annealed (350°C/15 min) LiF TLD only Z_3 centers (peak X) are present as RTL. Shining UV light releases electrons from these traps. After each gamma exposure there is change in the conditions of TL response. Since PTTL is obtained after gamma exposure and annealing, the PTTL reflects the conditions prevailing then with respect to the TL response. Strictly speaking the PTTL efficiency would be a convolution of the changing release of charge carriers from the supralinearly increasing peak X and the changing TL response. However, up to about 10 Gy, peak X is hardly discernible and peak-V response is linear; thereafter peak V grows highly supralinearly and peak-X response is also supralinear but less so. Therefore, the PTTL response is only slightly modified by the peak-X response and it appears to follow the gamma response.

Sagastibelza and Alvarez Rivas have recently proposed a new model for the TL processes in LiF.[54] According to this model F centers and interstitials are formed during irradiation. The interstitials are stabilized near impurities and dislocations. Thermoluminescence then arises from the recombination of F centers with mobile interstitial halogen atoms thermally released from traps such as impurities and dislocations. At the time these two types of lattice defects recombine, an electron-hole recombination also occurs and light is emitted. PTTL according to this model is caused by the migration, under illumination of F* centers, some of which undergo recombination with an interstitial at small interstitial aggregates. In addition to light, heat of about 6.3 eV is also released in this recombination. The latter causes the ejection of interstitials from the aggregate at which the mobile F* center has been annihilated. These released interstitials then regenerate the glow peak traps. Stability of traps under illumination with F light simultaneously with annealing at 330°C runs contrary to the above model. All centers formed as a result of interstitial aggregation as per this explanation should be destroyed by the above treatment. The highest temperature peak itself is completely bleached but the corresponding traps are not reduced as seen through subsequent test exposures.[51,52]

c. Reestimation of Dose

Gower et al.[41] observed that exposing gamma-irradiated LiF (TLD-100) to UV light (366 nm) after the dosimetry peak had been recorded enhanced the low temperature residual TL measured during a second readout, and this, they suggested, could be used to confirm the measurement of high doses of radiation. Mason and co-workers,[36,43,55] investigated the optimum conditions required for the reestimation of minimum possible dose in PTFE discs and extruded ribbons. The PTFE discs are annealed at 400°C for 1 hr and subjected to several cycles of normal readout procedure (heat in dry N_2 at 10°C sec^{-1} up to 290°C) and then annealed at 100°C for 1 hr. The extruded ribbons are just given the 400 and 100°C anneals for 1-hr each. The optimum time for UV exposure is 30 min of 254 nm light at ambient temperature. The threshold for reestimation is determined by the variation in the total intrinsic response to UV light, normal TL background in a second readout, tribo TL contribution, incandescence from the heater pan, etc. Defining the threshold of detection for reestimated dose as three times the standard deviation in the TL response of unirradiated dosimeters after exposure to UV, the threshold for PTFE discs was determined as 11 mGy, and 20 mGy could be measured with a precision of ± 2 mGy. For extruded ribbons the corresponding figures were 28 and 40 ± 4.5 mGy. This is to be compared with the normal readout threshold of 0.10 mGy and a precision of ± 0.3 mGy at 20 mGy. Increase in phosphor temperature during UV exposure augments the phototransferred signal. In PTFE discs PTTL at 80°C gave maximum 10% of the original signal which at ambient temperatures is only 3% at 200 mGy. This reduced the threshold to 2.5 mGy and a practical threshold of 7.5 $\pm 13\%$ mGy for a group of 10 dosimeters could be achieved.

The TL response is LET dependent. Therefore, PTTL is also LET dependent and so is the reestimation efficiency.[56] Since the LET dependence may vary from batch to batch,[57] it is necessary to determine the reestimation response of each batch.

Routine reassessment of absorbed dose by PTTL is carried out at the National Radiological Protection Board (U.K.) in their automated thermoluminescence dosimetry service which employs two LiF PTFE discs, 0.4- and 0.2-mm thick.[26,58] UV light irradiation at 254 nm is carried out while the dosimeters are held at 115°C. As described earlier and reproduced in Figure 5, the phototransfer process depends on time and temperature while shining UV light. The calibration curves for reassessment are shown in Figure 6. The reuse anneal temperature is 15 min at 300°C, and it is indicated that a 150-hr annealing at 300°C is sufficient to remove all memory of the reassessment. It is reported that 80% of the reassessed doses was within $\pm 30\%$ or 15 mGy of primary assessed doses. Reassessment is done for

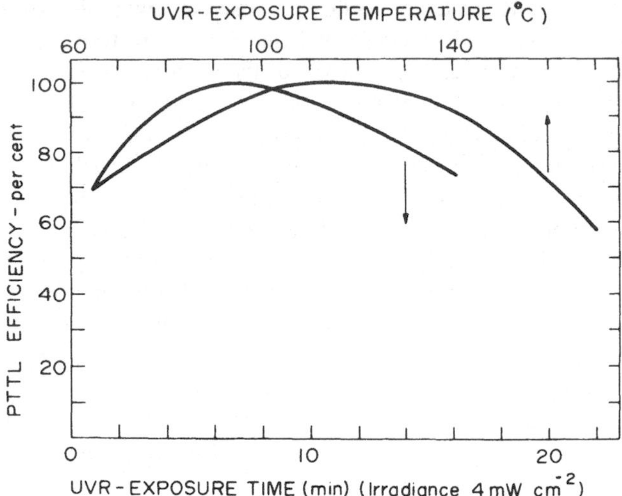

FIGURE 5. Dependence of PTTL efficiency on UV irradiation temperature and exposure time. (Adapted from McKinlay, A. F., Bartlett, D. T., and Smith, P. A., *Nucl. Instrum. Methods,* 175, 57, 1980. With permission.)

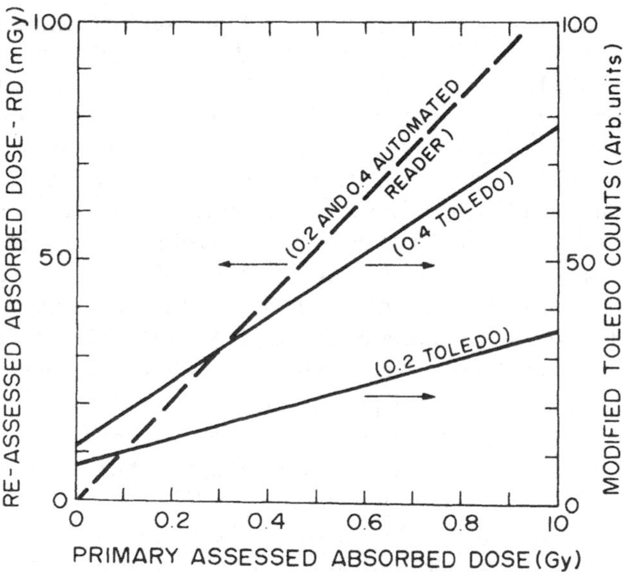

FIGURE 6. Calibration curves for PTTL reassessment of customer dosimeters read out on Toledo and Automated Reader. (Adapted from McKinlay, A. F., Bartlett, D. T., and Smith, P. A., *Nucl. Instrum. Methods,* 175, 57, 1980. With permission.)

three categories of dosimeters: (1) for assessed doses greater than 20 mGy on either disc; (2) if the primary assessment is lost, and (3) suspected by the user (prior to readout) to have had a high dose.[26]

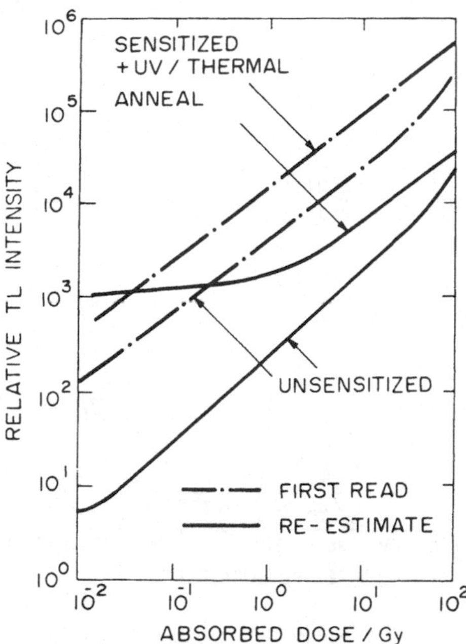

FIGURE 7. Simultaneous sensitization and reestimation in ^{7}LiF Harshaw chips: sensitized + simultaneous thermal/UV annealing data — 2 × 10^3 Gy, 300°C 6 mW cm^{-2} of 254 nm UV, 1 hr. Reestimation 5 mW cm^{-2}, 100°C/5 min. (Adapted from Charles, M. W. and Khan, Z. U., Proc. 5th Int. Congr. IRPA, Jerusalem, 1981. With permission.)

3. Sensitization and Reestimation of Dose

a. Practical Aspects

The sensitivity of the dosimetry peak for the measurement of small doses is increased by a factor of about 6 by preirradiation (to about 10^3 Gy or saturation dose for the dosimetry peak) and annealing (at 350°C for 15 min). However, the phosphor so sensitized poses problems when used for the measurement of low doses due to the presence of high temperature peaks. Mayhugh and Fullerton suggested the use of 254 nm UV light simultaneously with thermal annealing in the sensitization procedure (they termed it UV anneal) to make the latter useful.[59] The thermal energy gap between the optically excited state of the RTL peak and the free state has been determined to be 0.12 eV.[60] Bartlett and Sandford tested the UV anneal-sensitized LiF for reestimation of dose and found it impossible below 1 Gy.[61] Charles and Khan,[62,63] who also investigated the simultaneous sensitization and reestimation of dose, arrived at the following optimum conditions: 2 × 10^3 Gy gamma dose, simultaneous anneal and UV:300°C and 6 mW cm^{-2} of 254 nm UV for 1 hr, cooled to RT, and then 5 mW cm^{-2} of 254 nm UV at 100°C for 5 min. They found 0.2 Gy as the smallest dose (2σ) which could be reestimated in ^{7}LiF chips as compared to $\leq$0.01 Gy for unsensitized LiF chips.[63] But the sensitization factor also got reduced from 6 to 3. The increase in the minimum reestimable dose is due to the large background as seen in the response curves in sensitized and unsensitized ^{7}LiF chips for reestimation of dose (Figure 7).

b. Mechanism of Reestimation of Dose in Sensitized LiF

In sensitized LiF the normal sensitization procedure leaves a very large amount of residual

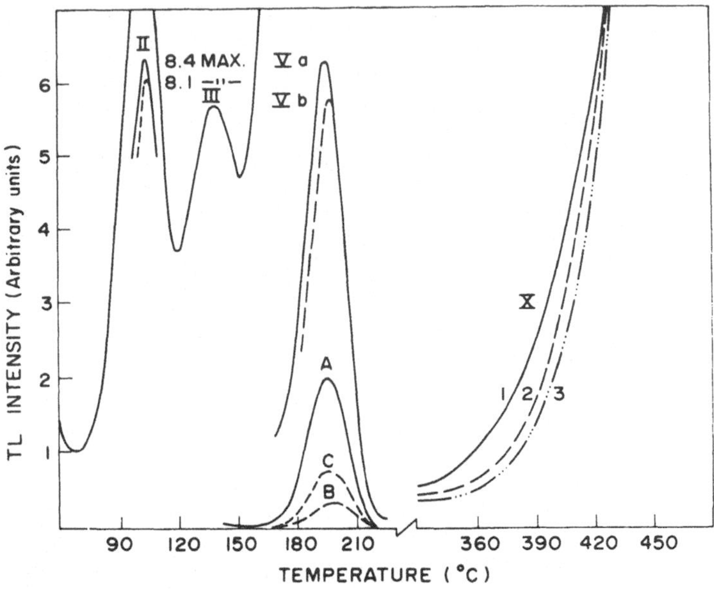

FIGURE 8. Effect of UV-thermal 330°C bleaching on sensitized TLD-100. Peak X: residual peak X is shown on the RHS by curves marked 1 and 2, respectively, for 15 and 45 min. Curve 3 is thermal background. Peak V: curves marked Va and Vb are due to 3.5-Gy test dose in the 15- and 45-min samples, respectively; curves A and B are 3 min PTTL in the 15- and 45-min samples, respectively; curve C is PTTL in the 45-min sample after the 3.5-Gy test dose. (From Jain, V. K., *Health Phys.*, 41, 363, 1981. With permission.)

TL corresponding to the sensitizing dose. This precludes the possibility of any reestimation because the proportion of the added dose to the large RTL already present will be negligible. However, the UV thermal-annealing method of sensitization has been shown by Jain to result in the emptying of the traps corresponding to the RTL peak, i.e., Z_3 centers.[27] As a result, peak X is not observed and sensitization is retained (Figure 8). In a successfully UV thermal-annealed sample phototransfer should not be observed. The success of the UV anneal and the minimum reestimable dose is determined by the phototransfer background signal. Charles and Khan also note the destruction of Z_3 centers but do not associate sensitization with these centers because their destruction results in only a small loss of sensitization. Instead they attribute sensitization to deep traps which can only be observed by absorption measurements below 200 nm.[63]

The TCLC model of Jain hypothesizes the destruction of Z_3 centers while retaining the sensitization. The emptying of Z_3 centers by UV light does not affect the luminescence centers and hence sensitization is retained. The reduction in sensitization factor noted after UV annealing at about 330°C is caused by the loss of LCs due to their recombination with TCs to form the complex TCLC.[27] Similarly, some background in the UV anneal-sensitized LiF is because of the difficulty in emptying all the Z_3 centers. The requirement is to empty Z_3 centers completely and quickly; keep the phosphor at the elevated temperature for as small a time as possible so that the TCLC formation and consequent reduction in the sensitization factor are minimal. This is achieved by carrying out UV annealing using a high intensity source, preferably 225 nm at about 330°C.

4. *Miscellaneous Effects*

PTTL in LiF has been employed by Kathuria and Sunta to determine the thermal quenching factor (TQF) in the high temperature peak whence PTTL is generated.[64] For this purpose 10 UV exposures each of 1-min duration are given, the PTTL is read out each time up to 300°C, and totaled. The amount of PTTL generated is measured to be 42.4 times the corresponding decrease in the RTL. Applying a retrapping factor of 2, the PTTL obtained is about 85 times the decrease in the RTL. This gives a TQF of 85. Commenting on the above work, Horowitz[65] has calculated that in the initial 1 min, the ratio of decrease in RTL to the PTTL produced is even greater, ~120, the TQF thus being 240. However, Jain has raised the issue of the effect of sensitization on PTTL which has been ignored.[65] If that is taken into account, i.e., a sensitization factor of 6, the PTTL would be only 7 times. Thus TQF amounts to only 14 and not 85, even if a retrapping factor of 2 is retained.

TL peaks at 100 and 140 K have been induced by a focused 30-ns pulse from ruby laser through internal fracture at LNT. Reddish coloration was noted at the damaged site which annealed at 550°C. It appears laser light can ionize electrons in LiF through a multiphoton process or through defect species.[66]

B. Calcium Fluoride

1. PITL

PSTL in calcium fluoride has been investigated with a large number of dopants but the more prominent ones are CaF_2:Mn, CaF_2:Dy, and CaF_2:nat. In a previously unexposed or fully annealed CaF_2 phosphor, there may not be much PITL.[67,68] However, Schayes et al.[69,70] have described a purely light-excited TL in a natural fluorite sample selected for dosimetry. The light-induced main TL peak is at a temperature higher than those of the dosimetry and PTTL peaks (Figure 9) and saturates quickly at <20 mGy equivalent. Repeated irradiation by UV light reduces this light sensitivity and the saturation occurs at <2 mGy. CaF_2:Mn, on the other hand, is reported to have little intrinsic sensitivity to light.[67] Similarly, the PITL in CaF_2:Dy is very small,[45] but high temperature heat treatment of CaF_2:Dy increases its intrinsic sensitivity to light.[71-73] Heating in an air flow furnace at 900°C for 15 min followed by cooling to RT results in high sensitivity to UV light. The sensitivity continues to increase up to several such cycles and then suddenly drops. The exact number of cycles for maximum sensitization depends on several factors like the rate of air flow, cooling rate, individual CaF_2:Dy chip, etc. Sensitization up to a factor of 500 has been reported.[71-74] At the same time the sensitivity to ionizing radiations decreases drastically.[75] The response of sensitized CaF_2:Dy to UV is limited to a narrow band around 254 nm (Figure 2)[73] and is insensitive to visible light, providing a UV dosimeter. The TL glow curve shapes for UV light and for gamma irradiation tend to be slightly different but the TL emission spectrum is the same. The increased sensitivity is retained even after the normal reuse anneal (at 400°C) and, therefore, can be used routinely. Fading, however, is very high, 30 to 40% in just 12 days.[71] Short-term fading is also high even when the low temperature peaks are annealed out (100°C/15 min) and the samples are kept in the dark.[73] Further, sensitivity variations from phosphor to phosphor demand strict selection from a large batch or calibration of individual discs if reasonable precision is required.

The increase in PITL sensitivity is possibly due to an oxidation reaction because if the heat treatment is carried out in vacuum or inert atmosphere increased PITL is not observed. The band gap of CaF_2 is greater than 10 eV and the formation of oxide with a much smaller band gap could make it UV sensitive. In Figure 2 the wavelength dependence of CaO is also given along with that of sensitized CaF_2:Dy. There is distinct similarity in the two responses though that of the former is shifted slightly towards longer wavelength.

2. PTTL

a. General

PTTL in fluorite has been quite extensively studied by various groups: Belgian, Brazilian,

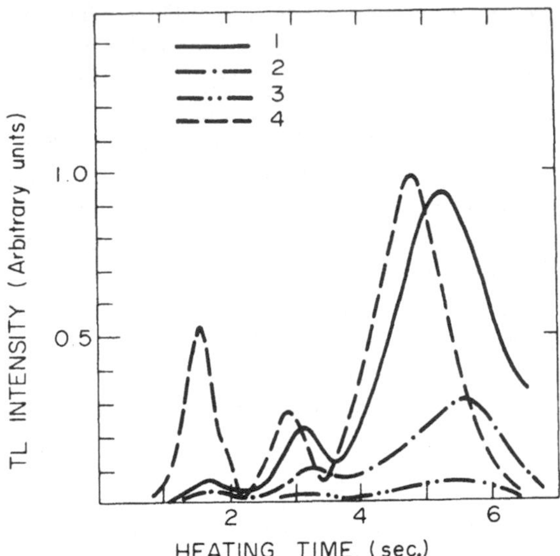

FIGURE 9. Glow curves of standard glass envelope MBLE dosimeter (CaF_2:nat). (1) Exposed to broad spectrum UV light. (2) New dosimeter emptied of all traps and exposed as above. (3) New dosimeter exposed repeatedly to UV light reducing the sensitivity to light of virgin dosimeter; peak III′ height reaches asymptotically ≃ 2 R equivalent. (4) Exposed to gamma rays. (Adapted from Brooke, C. and Schayes, R., Proc. Symp. Solid State and Chemical Radiation Dosimetry in Medicine and Biology, STI/PUB/138, International Atomic Energy Agency, Vienna, 1976, 31. With permission.)

Indian, and Russian. MBLE, Belgium developed natural fluorite personal dosimeters for large-scale use and worked out in detail the potential for the estimation of accumulated doses (memory effect/reestimation of dose). Not only MBLE dosimeters, but different natural fluorite samples and intentionally doped CaF_2, i.e., CaF_2:Mn, CaF_2:Dy, and other fluorites have also been investigated for the effect of exposures to ultraviolet radiation.

i. CaF_2:Mn

CaF_2:Mn has only one peak at 260°C (2.7°C sec^{-1} heating rate) and responds linearly to gamma rays up to 2×10^3 Gy. Nevertheless, it shows PTTL after the sample has been read out to 460°C and exposed to UV. The dependence on λ is shown in Figure 10.

The PTTL glow curve has its main peak at 285°C and a smaller one at 450°C which saturates rapidly. The PTTL response is found to be 3.5% of the original response in the range of 2 to 50 Gy and 5% in the range of 100 to 500 Gy (for 5.5 Ws cm^{-2} UV energy) thus showing a jump in PTTL output not observed in the gamma response. Shining UV light on gamma-irradiated samples causes bleaching of their TL. Also for high LET radiation, protons, and alpha particles, PTTL is increased. It has, therefore, been suggested that radiation damage may be creating the traps responsible for PTTL.[67]

Bernhardt and Herforth studied optically stimulated phosphorescence in CaF_2:Mn for radiation dosimetry.[76] Even in unexposed samples, optically excited phosphorescence is seen and the two effects are additively superposed. The response is linear up to 10^3 Gy, and the lower limit of measurement, which depends on the stimulating light intensity and the dosimeter temperature, is about 0.01 Gy. There is fading and interference due to phototransfer if dosimeters are reused after inadequate (450°C) annealing.

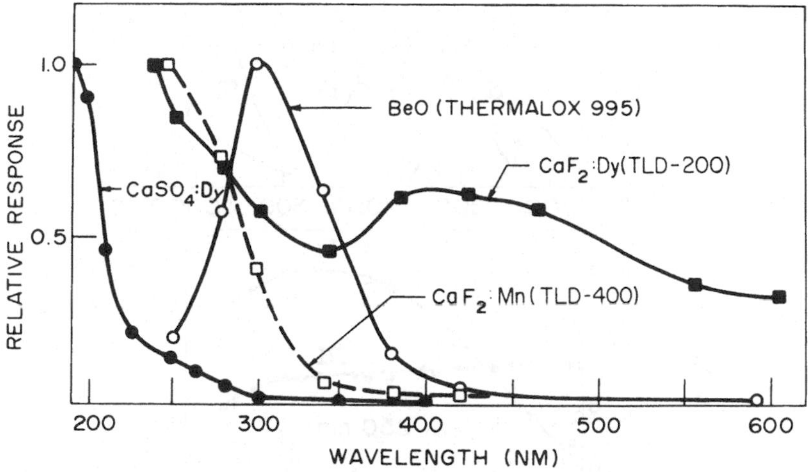

FIGURE 10. Transferred TL vs. wavelength.[72,96]

Bassi et al.[45,73] have reported on PTTL in CaF_2:Dy. The wave-length dependence, which is given in Figure 10, shows that it is sensitive to light over a wide range with a broad maximum occurring from 390 to 450 nm. The PTTL for a pregamma dose of 2 Gy could be used for UV dosimetry in the range of 10^{-1} to 10^1 of the maximum erythemal dose for skin.

ii. CaF_2:nat

In natural fluorite after normal readout or annealing at 400°C, peaks I to IV are erased and V and VI remain as residual. Shining light on such a sample regenerates peaks I to IV with one difference: instead of the dosimetry peak III, a new peak III′ at a slightly higher temperature develops most prominently.[69,70,77-81] This is also observed in a sample which is annealed at 600°C to drain off all natural TL, irradiated, and then submitted to the above procedure. Natural fluorite is found in many colors and there are sample-to-sample differences in the prominence of peaks.[82] Figure 11 shows the RTL and the PTTL induced by 250 and 365 nm light in blue natural fluorite heated to 600°C for 10 min, exposed to 7.0×10^3 Gy, and annealed for 90 min at 400°C.[78] The effect of the transferring light energy on the growth of peak III and the bleaching of residual TL are also shown in the figure. Sunta found the dependence of PTTL extends from <250 to ~450 nm for peak III, up to ~420 nm for peaks I and II, and contains maxima at 315, 367, and 410 nm and minima at 350 and 400 nm.[68] McCullough et al.[86] however, did not find such variations in the λ dependence of the natural fluorite samples (Figure 12). In a 600°C annealed and irradiated sample (50 Gy), if peaks I and II are read out by heating to 150°C and the sample exposed to light above 500 nm, peak III is bleached, I and II are regenerated, and IV and V are enhanced. This shows that light of $\lambda > 500$ nm selectively bleaches peak III and the charge carriers are liberated to be captured at all available trapping sites.[68]

b. Mechanism

The mechanism of transferred TL has been worked out to be due to the exponential decay of two centers probably corresponding to peaks V and VI.[78-80] The PTTL intensity is found to be directly proportional both to the RTL as well as the incident UV fluence. Sunta observed that the emission spectrum of the PTTL peaks and the RTL peaks was due to a Ce^{3+} ion locally compensated by O^{2-}.[83] The Ce^{3+} ion, present substitutionally (replacing Ca^{2+}) in the body center position and the O^{2-} ion, at one of the octahedral positions of the

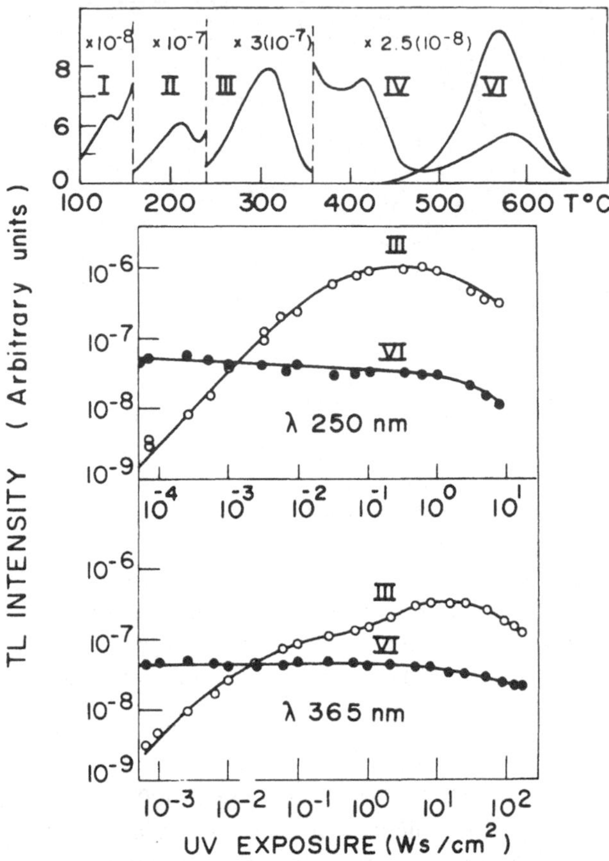

FIGURE 11. PTTL in CaF_2:nat. Blue fluorite treated as follows: 600°C/10 min + 6.7 × 10^3 Gy + 400°C, 90 min + 0.447 Wsec cm^{-2} of 250 nm light. For PTTL growth curves additional 400°C, 15 min annealing. (Adapted from Las, W. C. and Watanabe, S., Proc. 4th Int. Conf. Luminescence Dosimetry, Vol. 3, Niewiadomski, T., Ed., Institute of Nuclear Physics, Krakow, 1974, 1187. With permission.)

cube (replacing F^-) forms an "aggregate" center. Irradiation converts Ce^{3+} to Ce^{2+} by capture of an electron and O^{2-} to O^- by capture of a hole. The hole is freed by the UV light; it migrates and is trapped at one of the hole centers corresponding to the TL peaks. On subsequent heating for TL, these holes are rendered mobile and recombine at the only available recombination site, i.e., Ce^{2+}, thus giving the Ce^{3+} spectrum.

c. Reestimation of Dose

The MBLE, Belgium studied in detail the "memory effect" in their natural fluorite field type and low Z dosimeters and established the lower limits of dose which could be reestimated (Figure 13).[70,81,84,85] The wavelength of the light and the kind of glass covering the dosimeter are both important factors limiting the minimum dose which can be assessed by light transfer. An additional factor is the intrinsic light sensitivity which is minimized by repeated broad spectrum light exposures and heating.[84] The accumulated dose can be checked repeatedly, up to 10 times, with a precision of 15% or better for a total dose exceeding several hundred mGy. The lower limit depends critically on the experimental conditions. While several

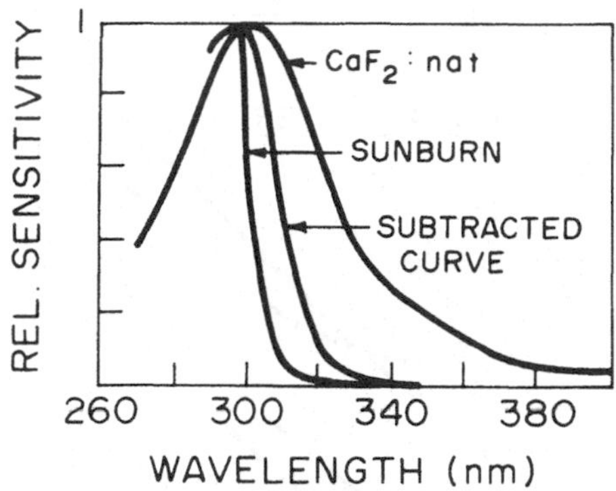

FIGURE 12. Spectral response for sunburn, PTTL in CaF_2:nat, and PTTL in CaF_2:nat through a Schott WG 320 2-mm thick filter subtracted from the PTTL without filter. (Adapted from McCullough, E. C., Fullerton, G. D., and Cameron, J. R., *J. Appl. Phys.*, 43, 77, 1972. With permission.)

workers found proportionality between the residual TL (corresponding to accumulated dose) and PTTL up to about 10^2 Gy, Okuno and Watanabe, working on a Brazilian green fluorite, found little effect on peak II and only above 30 Gy on peak III and other complications based on annealing and light conditions.[80]

d. UV Dosimetry

PTTL in natural fluorite also provides a means for the measurement of a wide range of UV exposures, depending on the residual TL.[77-79] McCullough et al.[86] carried out detailed investigations on the potential of PTTL in natural fluorite for the measurement of terrestrial solar ultraviolet radiation in the sunburn range (290 to 320 nm). To match the spectral sensitivity of the proposed dosimeter with the above range, a 2-mm thick sharp cutoff filter, Schott WG 320, is used. The TL from the filtered dosimeter is subtracted from that of the unfiltered TL dosimeter. Figure 12 shows the spectral response of: (1) CaF_2:nat., (2) subtracted TL, and (3) sunburn. Due to the higher sensitivity of the subtracted TL at most wavelengths, measurements carried out with calibration at any single wavelength overestimate the erythemally effective energy and consequently underestimate exposure time to produce a minimal erythema. The dosimeter after a priming dose of about 40 Gy of ^{137}Cs gamma rays and annealing at 400°C gave PTTL equivalent to about 0.4 Gy of gamma radiation for 12 mJ cm^{-2} of 300 nm radiation (mean erythmal dose). The PTTL is linear up to a priming dose of 100 Gy and each 12 mJ cm^{-2} results in a 1% drop-off in sensitivity due to transfer. Wilson et al.[77] suggested a slight variation on the direct PTTL measurement. The dosimeter is given a large gamma exposure and annealed at 400°C. It is then exposed to a predetermined short UV exposure and the PTTL is measured which "samples" the RTL. The dosimeter is now used for the unknown large UV exposure. The PTTL which would have been produced by the UV exposure is drained by heating to 400°C. It is then again given the short sampling UV exposure and PTTL is measured. The PTTL given by the two sampling exposures is then used to estimate the large UV exposure.

e. Other Studies

PSTL investigations of the nature of defect centers and luminescence process have been especially fruitful in fluorite. In Tm^{3+}-doped CaF_2 the PSTL spectrum was found to be similar to the emission spectrum of divalent Tm, indicating that electrons liberated by the

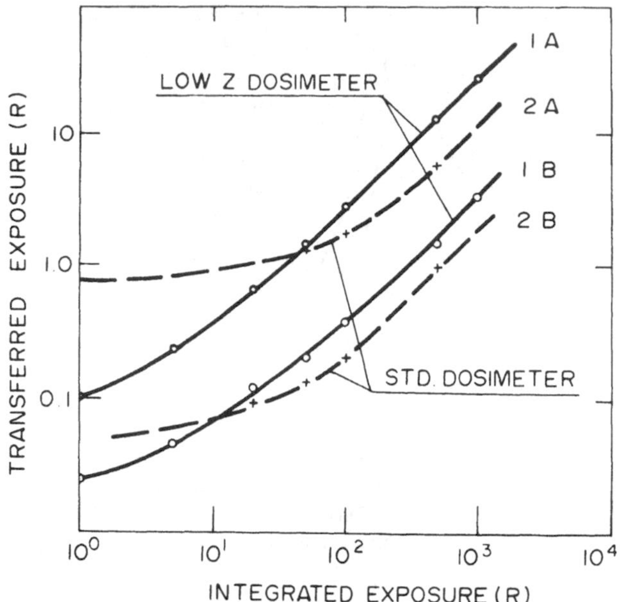

FIGURE 13. Light-induced TL reading (in dose equivalent) for standard and low Z-type dosimeters. The dosimeters were read and erased before exposure to light. Light sources: broad UV spectrum. (A) 1 hr, 6 Wsec cm^{-2}. (B) 5 min, 0.5 Wsec cm^{-2}. (Adapted from Brooke, C. and Schayes, R., Proc. Symp. Solid State and Chemical Radiation Dosimetry in Medicine and Biology, STI/PUB/138, International Atomic Energy Agency, Vienna, 1967, 31. With permission.)

divalent RE ions play the dominant role.[3] Kask et al.[4] working on Er- and Ho-doped crystals studied PSTL at 300, 77, and 4.2 K and managed to segregate the TL peaks into three families depending on the character of the elementary charge carrier. One of the elementary charge carrier centers was identified as F_2^-, consisting of a hole localized at two neighboring fluorine ions in lattice positions and different states of the hole related to it could be correlated to different TL peaks. Also for centers of the same family, existence of tunnel transitions from both the ground and the excited state between the capture centers was suggested. Liberated elementary defects appeared to undergo repeated localization and defects corresponding to high temperature TL peak pass through a state belonging to a low temperature peak. Kornienko et al.,[87] carrying out similar studies in fluorite crystals doped with 11 different RE ions, separated the peaks into two families. Their approach consisted of irradiating at 77 K, thermally bleaching peak I, cooling back to 77 K, and observing PSTL. Next, TL up to the second peak was bleached, the sample cooled, and PSTL observed. This process was repeated up to the last peak. Analyzing the position and intensities of the TL peaks in the two families, they concluded that the TL processes in doped fluorite depend on the type of RE impurity ion, that only TL up to 250 K could be related to the motion of holes, and that above 250 K the TL cannot be explained by the motion of one type of radiation defect.

C. Calcium Sulfate

1. PITL

$CaSO_4$:Mn, $CaSO_4$:Dy, and $CaSO_4$:Tm have been prominently studied for light-induced TL. $CaSO_4$:Mn is a material of historical interest having been prepared by Wiedemann and Schmidt in 1895 by mixing a few percent of $MnSO_4$ with $CaSO_4$.[2] When exposed to a

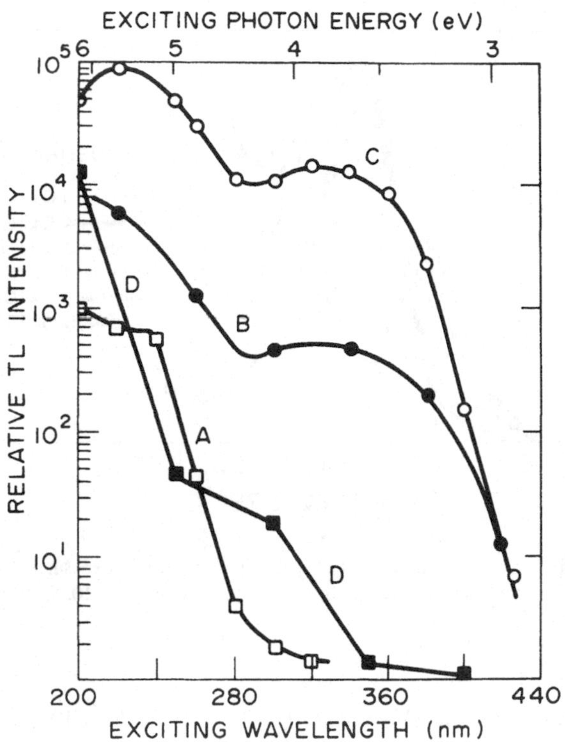

FIGURE 14. Excitation spectrum for Al_2O_3 and $CaSO_4$:Dy. (A) PTTL. (B) Sensitized by 500 J cm^{-2} UV. (C) Sensitized by 10^2-Gy gamma radiation. (D) $CaSO_4$:Dy.[91,112]

condensed spark in air at a distance of a few centimeters and then heated to about 200°C, it emitted green light. Later Lyman showed its spectral sensitivity to start abruptly at 130 nm and extend at least up to 14 nm. He also used $CaSO_4$:Mn to measure the transparency of air between 110 to 130 nm.[88] Watanabe found its spectral sensitivity for inducing TL to extend from 80 to 150 nm with the maximum at 103 nm, dropping by a factor of 10^{-3} at 142 nm.[89] Tousey et al. used it for the measurement of solar extreme ultraviolet in rocket flights.[90] $CaSO_4$:Mn has little sensitivity for light above $\lambda \sim 134$ nm. The TL peak occurs at 110°C which leads to considerable fading at normal temperatures, but the response is linear over 4 decades and independent of phosphor temperature from -20 to 70°C at the time of exposure.[89] Calcium sulfate doped with Dy or Tm also shows PITL.[91-93] The wavelength dependence of PITL in $CaSO_4$:Dy is such that in going from 200 to 350 nm it decreases in steps spanning a factor of 10^{-4} (Figure 14, curve D).[91] At 254 nm there is still considerable response and the glow curve has the same prominent peaks as for gamma irradiation but their relative heights are different. Peak temperatures shift back and forth with the amount of UV exposure in the range of 10^2 to 10^7 Jm^{-2}, as is also observed in gamma ray-excited glow curves (Figure 15). Above 10^3 Jm^{-2} the TL output becomes supralinear and a preexposure of 3.0×10^6 Jm^{-2} produces a sensitization factor of 5.2[92] Preheating $CaSO_4$:Dy at 1300°C for 2 hr causes an increase of factor 21 in the UV response, possibly due to the formation of CaO which has very high intrinsic sensitivity to UV.[94] Elevated temperatures during exposure also affect the TL output. The response to gamma rays decreases by 5% at 100°C and by 14% at 150°C, but increases by 20% if visible light is present during irradiation. The temperature dependence of UV response is seen to follow

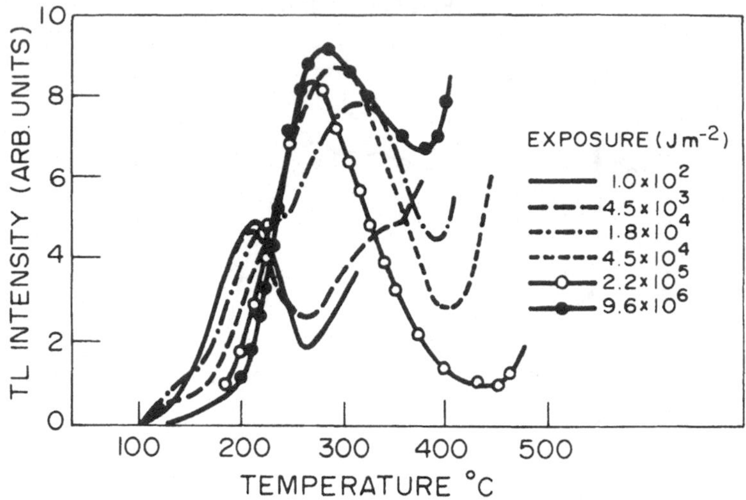

FIGURE 15. TL glow curves of $CaSO_4$:Dy at various UV exposures. (Adapted from Shastry, S. S. and Shinde, S. S., *Phys. Med. Biol.*, 24, 1033, 1979. With permission.)

an empirical relation: R = 0.1263t − 2.53 (28°C ≤ t ≤ 150°C) where R, the integrated TL normalized to 1 at 28°C, is equal to 18 at 150°C.[95]

2. PTTL

a. Response Characteristics

PTTL in $CaSO_4$:Dy or Tm decreases by a factor of 10 and becomes very small in going from 200 to 360 nm of UV light exposure (Figure 10). It also depends upon the temperature during UV exposure and on the postirradiation annealing temperature. Using 365 nm UV to minimize PITL, Nagpal and Rendurkar observed that (1) 400°C postirradiation annealing gave maximum PTTL and (2) PTTL increased with temperature during UV exposure up to 120°C. The activation energy Δ E to free electron from the optically excited state (by 365 nm) was determined to be Δ E = 0.097 eV from the slope $\Delta E/k$ of the plot of PTTL vs. 1/T.[95] A number of peculiarities are found in the reported response characteristics of $CaSO_4$. Although the main dosimetry peak near 200°C saturates at about 10^3 Gy, PTTL does not saturate up to 10^6 Gy. The work of Caldas and Mayhugh shows that the PTTL caused by 250 nm monochromatic light in samples (Harshaw $CaSO_4$:Dy) irradiated with 50 kV X-rays and annealed at 280°C for 15 min grows sublinearly as $R^{0.55}$ up to 10^6 Gy, where R is the dose.[97] Similarly, PTTL caused by the unfiltered UV light from a germicidal lamp in samples exposed to ^{60}Co gamma ray and annealed at 400°C for 30 min also grows as $R^{0.55}$ up to 10^6 Gy. However, the more recent results of Pradhan et al.[98] and Chandra et al.[99] on laboratory-prepared material are different and show the response to increase nonlinearly up to about 5 × 10^5 Gy and decrease thereafter.

b. UV Dosimetry, Reestimation of Dose, and Fading

The UV-induced TL in the peak near 200°C is proportional to the RTL peak near 450°C and is reported to be maximum for a pregamma exposure of 5 × 10^5 Gy. The response is observed to be independent of intensity (cf. Reference 10) and linear from 10^{-2} to 10 Jm^{-2} after which saturation sets in.[96]

Reestimation of exposure in the range of 3.0 × 10^{-4} to 1.3 × 10^{-1} C/kg^{-1} by means of PTTL has also been examined. After annealing at 380°C for 6 min, 20 min of stimulation by 365 nm UV at 95°C gives enough PTTL to reestimate 3.0×10^{-4} C/kg^{-1} within ±20%.[95]

The presence of low temperature TL peaks in $CaSO_4$:Dy gives an afterglow which decays to insignificant levels in several hours. When illuminated by UV light, fast decaying phosphorescence is observed which is up to 10^4 times the afterglow intensity. Integrating the phosphorescence for about 30 sec of stimulation after about 1 min can be correlated in a linear manner with the initial gamma dose without actually losing the signal.[100]

$CaSO_4$:Dy PTFE discs exhibit 10% fading in 1 day and 20% in 1 week when the irradiated samples were stored in darkness and the readings were taken in a preheat cycle to remove the low temperature peaks. Daylight and fluorescent light introduced a 10-μGy equivalent signal very quickly (1 to 2 min) and the maximum signal was no more than 20 μGy equivalent after several hours of exposure.[101,102] In PTFE discs, fading due to sunlight in 12 hr and 7 days, respectively, was as follows: when covered with metal filter—insignificant and 10%; under perspex or when wrapped in paper — 35 and 70%; when bare — 85 and 97%. Black PVC reduced the fading to 5% in 7 days. Bare discs when exposed to fluorescent light faded by 8% in 15 days and by 24% in 23 days.[103]

V. OXIDES

A. Aluminum Oxide

PSTL in Al_2O_3 has been investigated in a variety of samples: pure,[104-106] commercial,[107-109] impurity doped,[107,108] and in stones like ruby[109] and sapphire.[110] Wide variations in the number and temperature of glow peaks have been reported, but in the same sample, gamma- and UV-induced TL have generally been found to be similar. In spec. pure Al_2O_3, however, preferential trap filling is reported in the 200 to 280°C region. Buckman et al.[110] noted that only certain types of Al_2O_3 were sensitive to UV radiation. Ziniker et al.[106] found that neither the highly pure nor the highly impure samples showed detectable UV sensitivity, but samples having impurities closer to an equilibrium level due to the slow growth of the crystals showed UV sensitivity. Vishnevsky et al.[111] observed that trace amounts of Ti impurity were important for the TL of Al_2O_3. However, the attempts of Cooke et al.,[108] to determine the influence of Ti on its TL sensitivity were not successful. Mehta and Sengupta[112] purified commercial Al_2O_3 and heat-treated it in flame at about 2000°C. The material thus obtained is about five times as sensitive as TLD-100 for gamma rays and contains Si and Ti as the TL active impurities. The importance of Ti as an impurity was further verified by doping other samples. As opposed to most other Al_2O_3 samples in which only one peak at 170°C has been reported, this has a prominent peak at 280°C and also several others including one at about 640°C (heating rate 3°C/sec^{-1}) (Figure 16).[112] The UV excitation spectrum is shown in Figure 14. It falls off very rapidly at wavelengths longer than about 250 nm. The emission spectrum has its peak at 420 nm. The PITL does not show any fading in 4 weeks of storage at RT. The response to 254 nm light, however, is not linear initially, but becomes so after about 5×10^2 mJ cm^{-2} of exposure (Figure 17).

Cooke observed that repeated UV exposures altered the sensitivity of the sample and pregamma exposure and annealing at temperatures of less than 600°C increased the response sixfold.[107] Mehta and Sengupta obtained a sensitization factor of 250 for a pregamma dose of 10^2 Gy and a factor of 15 for 500 J cm^{-2} of 254 nm UV radiation (followed by annealing at about 525°C).[112] This enhancement in UV sensitivity is essentially due to PTTL. In the sensitized Al_2O_3 even the UV excitation spectrum changes (Figure 14). Excitation spectra and the response curve indicate that sensitized alumina could be used for broad spectrum UV dosimetry over a wide range of exposures: 10^{-2} to 5×10^4 mJ cm^{-2} (Figure 17).

The TL process in Al_2O_3 is generally described as one of hole release from traps recombining at the impurity center.[106,108,111] The intrinsic UV response is attributed to the release

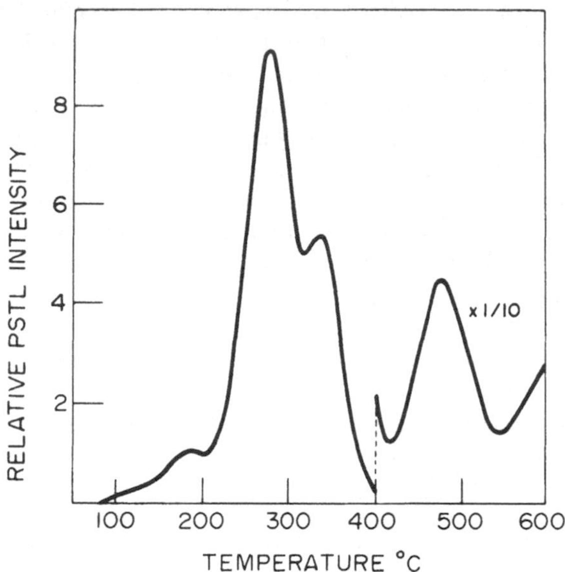

FIGURE 16. UV-induced glow curve in Al_2O_3 pellet; exposure 100 J cm^{-2}, 254 nm. (Adapted from Mehta, S. K. and Sengupta, S., *Phys. Med. Biol.*, 23, 471, 1978. With permission.)

of electrons from the 225-nm F centers and the phototransferred UV sensitivity due to electron release from the 325-nm, 625°C deep traps.[112]

B. Magnesium Oxide

Magnesium oxide possesses high intrinsic sensitivity to ultraviolet light. Many different commercial samples have been investigated for their TL response.[113-117] The positions and relative intensities of the glow peaks depend very much on the source of the samples (impurity content, heat treatments, etc.) and as such peak temperatures obtained by different workers do not agree. Ziniker et al.[113] recorded the glow curves by exciting at −150°C and found peaks at −40, 70, 90, 100, 150, and 200°C and suspected even higher temperature peaks. While most reports describe the main peak to be around 90°C, Dhar et al.[115] found a peak at 145°C as most prominent. Kirsh et al.[116] excited MgO by 110 to 170 nm light at −195°C and observed main peaks at −140 and 60°C and minor ones at −95, −65, and −35°C. Tl of gamma irradiated and 275°C-heated samples, when excited by VUV light at LNT, showed the above peaks shifted to lower temperatures by more than 20°C.

Magnesium oxide generally has Fe, Ca, and to a smaller extent Si and Al as common impurities. It has been observed that purer samples are more sensitive and could be used for dosimetry. Annealing at 2000°C for 2 hr in an argon atmosphere further increases its sensitivity.[115] Though the response of prominent peaks is linear over a considerable range of UV doses, fading of the main peak is quite fast: up to 20% in 48 hr. Takeuchi et al.[117] noted that in a sample irradiated to 450 mJ cm^{-2} and stored in the dark for 200 hr, the 90°C peak decayed by 20% while the 145 and 190°C peaks increased by 10%. Reproducibility of the TL readings is rather poor; to improve the reproducibility preirradiation of MgO to the UV saturation dose, heating up to 250°C and cooling to RT is suggested.[117] Samples left in fluorescent light after reading developed 0.5% of the original TL after 24 hr. A sensitization phenomenon in previously unirradiated crystals when subjected to the whole

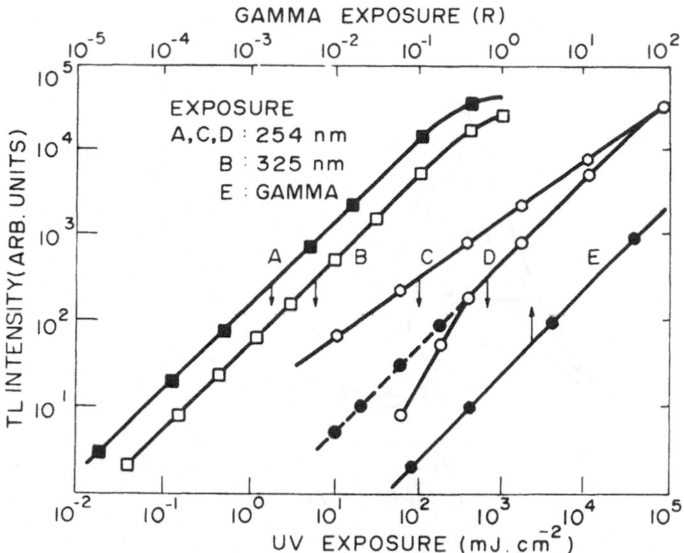

FIGURE 17. TL response in Al_2O_3. (A, B) PTTL, 10^2-Gy gamma sensitized. (C) 500 J cm^{-2}, 254 nm UV sensitized. (Broken line D) 3 J cm^{-2}, 254 nm UV sensitized. (Solid line D) PITL. (E) Gamma response. (Adapted from Mehta, S. K. and Sengupta, S., *Phys. Med. Biol.*, 23, 471, 1978. With permission.)

spectrum of the xenon source light before being exposed to the monochromatic UV radiation is also reported.[116] The amount of UV exposure is critical, an excess causing a drastic reduction. The enhanced sensitivity, however, decays with time and is not permanent.

The wavelength dependence of the TL response is also very sensitive to the impurity content (Figure 18). While Dhar et al.[115] found the response in their samples to be energy independent from 280 to 330 nm, Ziniker et al.[113,114] in the range of 250 to 333 nm and Takeuchi et al.[117] in the range of 220 to 280 nm found the response in their samples to be highly energy dependent. The quantum efficiency is stated to increase with decreasing λ, is several orders of magnitude larger at 220 nm, and includes peaks in the 260 to 290-nm region.[113,116] An interband transition-related increase in sensitivity and structure due to exciton spectrum is also observed upon excitation in the VUV region (110 to 170 nm).[116] However, no special feature is seen at 143 nm corresponding to the usually accepted value for the forbidden gap in MgO. It is believed that most of the peaks in MgO are generally hole release peaks, though the presence of some electron peaks has also been suggested.[113] UV light releases holes from the impurities and they form V centers. During TL reading, holes from V centers are released and recombine at the impurity giving light.[118,119]

Significant PTTL is not observed in MgO.

C. Beryllium Oxide

PITL in BeO is small and saturates at less than 1 mGy gamma-equivalent dose.[45,120,121] On exposure to visible light BeO exhibits strong afterglow[122,123] and has been examined for possible application to dosimetry.[124] Albrecht and Mandeville, investigating the nature of traps responsible for the storage of energy, X-rayed the crystals and kept them in the dark for long periods to allow the afterglow to decay completely.[122] They found that electrons from deep traps could be removed by photostimulation, which was wavelength dependent and had its maximum at 410 nm, which was later confirmed by other workers.[123] However,

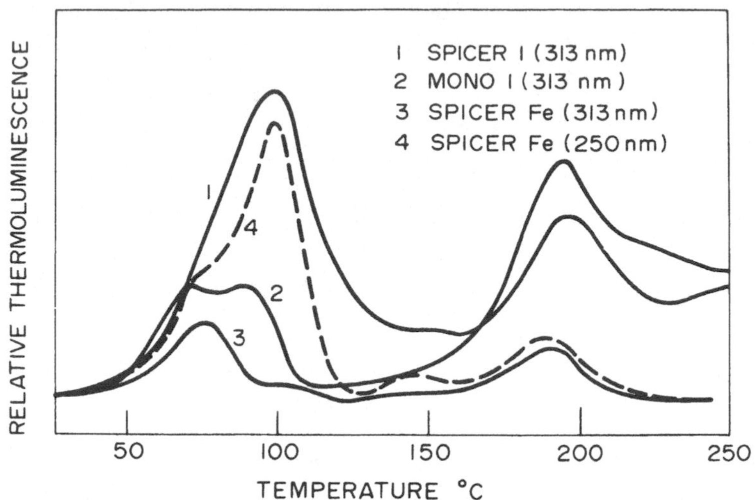

FIGURE 18. TL of MgO single crystals induced by 10 min of UV irradiation. Heating rate 20 K min^{-1}. (Adapted from Ziniker, W. M., Rusin, J. M., and Stoebe, T. G., *J. Mater. Sci.*, 8, 407, 1973. With permission.)

reports on Thermalox® 995 place the peak for PTTL efficiency at about 310 nm (Figure 10).[72]

The main peak in gamma-irradiated BeO occurs at 180°C (heating rates 1 to 5°Csec^{-1})[121,125,126] and in Thermalox 995 at 220°C (heating rate 20°Csec^{-1}), which shifts to higher temperatures with increasing dose.[120] A high temperature peak is reported to occur variously at 230,[125] 280,[121] or 350°C.[120] The maximum readout temperature, or the postirradiation annealing temperature for inducing PTTL, has generally been 350°C, but no TL peak has been observed beyond 350°C and no PTTL after annealing at 600°C. Although the dosimetry peak is generally also the PTTL main peak, Yasuno and Yamashita, working with alkali ion (Li, Na, or K)-doped BeO, found after exposure to 3000 lux fluorescent lamp for 1 hr that the PTTL peak occurred only at 230°C for gamma preexposure of 7.50 Gy and was supralinear in response.[125] Generally the gamma response is reported supralinear for doses >1 Gy, but the PTTL is less so and even becomes linear beyond 10 Gy.[120,123] As a result the ratio of the first readout after UV exposure decreases with increasing dose from 25 to 2.5%. Possibility of reassessment of measured low doses — 10 mGy — has been indicated. However, in beta-irradiated BeO, Hobzová found PTTL to increase from 1.7 to 3% at low doses to 7.5% at high doses. At 8 Gy it attained a maximum within 2 to 3 min and declined thereafter due to bleaching, but the 280°C peak remained constant[121] (Figure 19).

Fading in BeO even at RT is quite significant and becomes very serious under fluorescent light, decaying by 5% in the first 10 min and by over 90% in 1 day.[125-127] Crase and Gammage[128] and Gammage and Cheka[129] observed fading to be high both under fluorescent and incandescent lighting. Only under red light was there no fading and under yellow light the fading could be made tolerable.

D. Quartz

1. PITL

Natural or synthetic quartz annealed up to 800°C is not sensitive to UV light. However, after annealing the sample at 900°C for 2 hr all the gamma-induced peaks, viz. at −85, −65, −25, −3, 60, 110, and 165°C, which have been identified to originate from a defect

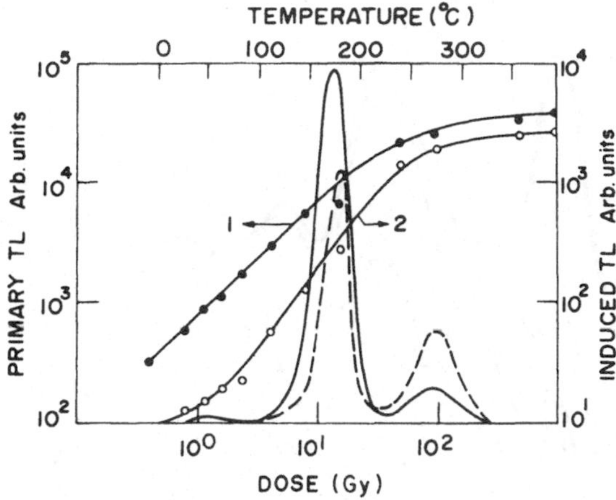

FIGURE 19. Phototransferred glow curve and response in BeO Thermalox 995. (1) Primary TL. (2) PTTL. Glow curves: /\ for 7-Gy beta dose and - - - for 5 min UV after read out scale × 40. (Adapted from Hobzová, L., Proc. 4th Int. Conf. Luminescence Dosimetry, Vol. 3, Niewiadomski, T., Ed., Institute of Nuclear Physics, Krakow, 1974, 1081. With permission.)

center are excited by UV but not the ones at −110 and 300°C, which are correlated to the presence of impurities, e.g., Ti.[130] The heat treatment does not appear to introduce any new absorption bands in the crystal, but UV stimulation of TL requires excitation in both the 185- and 254-nm regions. This suggests a two-step cascading to the conduction band following by capture at the trapping sites. Annealing may be providing a mechanism for this process.

2. *PTTL*

In irradiated quartz annealing at 250°C leaves a peak of 310°C and possibly others at higher temperatures. Shining UV light (254 nm) induces PTTL which reaches saturation quickly and then starts decaying. The PTTL response is quite different if it is observed by thermally removing the first peak and keeping the rest as RTL, then draining the first two peaks and keeping the rest as RTL and so on. From the PTTL so recorded it appears that the phototransfer takes place mainly from the peak next to the one bleached and not the higher temperature ones (Figure 20). Similarly, photobleaching of peaks in a sample irradiated to saturation leads to the peaks bleaching one by one, first peak followed by the second and so on. But in an unsaturated sample, the lowest and the highest temperature peaks decay fastest and the others fall in between.[131] These rather unusual effects in the interaction of UV-irradiated natural quartz remain unexplained. It is possible that the different traps are not independent and there exists some sort of link among them.

E. Zirconia and Zircon

The band gap of zirconia is 4.9 eV making band-to-band transitions possible on exposure to 254 nm light. Much less effective is 365 nm light. In zircon, 365 nm gives good PTTL and little PITL. The glow curves are similar in both for ionizing radiation and UV light.[132] The sensitivity of zirconia, which has several peaks, increases with high temperature treatment (up to about 1200°C) which transforms ZrO_2 from tetragonal to a monoclinic form. A peak at −30°C has been associated with tetragonal ZrO_2. Iaconni et al.[133] have demonstrated that the trapping centers in ZrO_2 are constituted by Zr coordination polyhedron and the

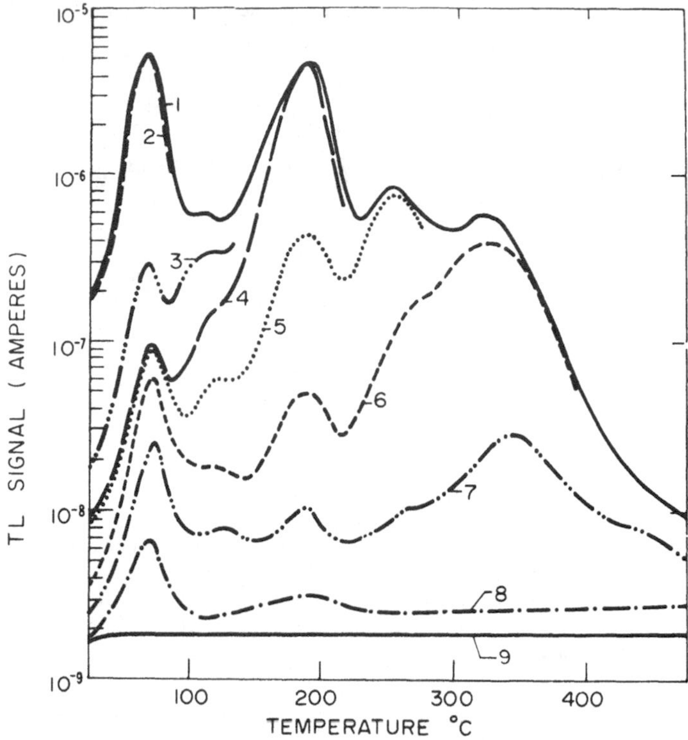

FIGURE 20. Generation of TL peaks on UV irradiation in pink quartz. (1) Full glow curve after a gamma dose of 2.5 × 10³Gy. (2) Sample 1. TL recorded up to 80°C. (3) Sample 2 + 3 min UV and TL recorded up to 130°C. (4) Sample 3 + 3 min UV and TL recorded up to 220°C. (5) Sample 4 + 3 min UV and TL recorded up to 275°C. (6) Sample 5 + 3 min UV and TL recorded up to 375°C. (7) Sample 6 + 3 min UV and TL recorded up to 425°C. (8) Sample 7 + 3 min UV and TL recorded up to 550°C. (9) Sample 8 + 3 min UV and TL recorded. (Adapted from David, M. and Sunta, C. M., *Ind. J. Pure Appl. Phys.*, 19, 1041, 1981. With permission.)

emission centers are Ti^{3+} activators in a Zr^{4+} site surrounded by six oxygens. In zircons prepared by the hydrothermal method, UV (254 nm) excitation produces two peaks at 60°C and 120°C in common with those found in ZrO_2. On the basis of this and detailed study of thermal effects and TL emission spectra Caruba et al.[134] concluded that small quantities of ZrO_2 are included in the zircon lattice. In zircon grains separated from beach sands, Jain[135] found TL sensitivity to increase with heat treatment up to 1500°C and six peaks at 100, 180, 230, 275, 365, and 400°C are obtained. The prominence of different peaks varies with the temperature of the heat treatment. Annealing at about 300°C after gamma irradiation leaves residual TL and considerable PTTL is produced by exposure to 365 nm light. For heating at 1500°C and above, presence of ZrO_2 is noted. The emission in the zircon sand was found to be due to Tb^{3+}.[135] Fluorescence studies in zircons have shown the presence of RE ions.

VI. OTHER POPULAR TL MATERIALS

A. Magnesium Orthosilicate

The ultraviolet response characteristics of Mg_2SiO_4:Tb are complex and depend on the

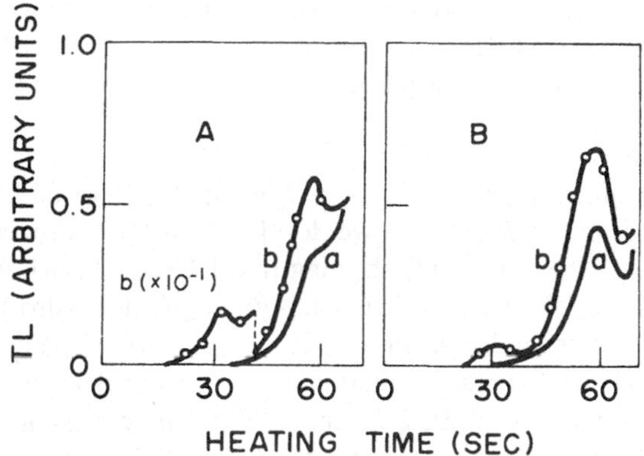

FIGURE 21. Glow curves in $Li_2B_4O_7$ with (0.1% by mol Mn) A and with (0.3% by mol Mn and 0.25% by wt Si) B. (A) 10^5 Gy ^{60}Co-gamma. (B) Postannealed at 300°C/1 hr and exposed to 3 J cm^{-2} of UV, 254 nm. (Adapted from Pradhan, A. S., Bhatt, R. C., and Supe, S. J., *Int. J. Appl. Radiat. Isot.*, 31, 671, 1980.)

thermal and radiation history of the phosphor. The preparation temperature is important and the γ/UV response ratio increases with increasing preparation temperature.[136] There is no sensitivity to white light from a tungsten lamp and to UV of 365 nm, but at 254 nm the intrinsic and transferred TL response of this material is very high.[137-139] In the gamma-irradiated samples there is UV-stimulated luminescence depending directly on the gamma exposure if the latter exceeds 1 C kg^{-1}, which consequently may be used for high level gamma dosimetry. Lakshmanan et al.,[138,140] using Mg_2SiO_4:Tb manufactured by the Dai Nippon Co., found its TL properties to be encouraging and different from the earlier reports. The main peak appearing around 200°C does shift in temperature for both UV and gamma exposures and the UV response is supralinear in the range of 10 to 10^4 Jm^{-2}, but there is no fading for at least 3 weeks and no effect of fluorescent light. Gamma exposure, 1.3 × 10^3 C kg^{-1}, followed by 300°C/1 hr annealing leaves a peak around 450°C as residual and leads to sensitization by a factor of 55 for a UV test exposure of 2.5 × 10^2 Jm^{-2}. Strangely, UV response in the sensitized phosphor is sublinear. The sensitization is presumed to be due to phototransfer because the residual peak decreases.

Mg_2SiO_4:Tb is a high sensitivity phosphor which has not been studied sufficiently thoroughly.

B. Lithium Borate and Cadmium Borate

The effect of UV light on lithium borate and cadmium borate is very different and interesting. PTTL is not observed in either. In $Li_2B_4O_7$:Mn and $Li_2B_4O_7$:Mn,Si two TL peaks at 210 and 370°C (heating rate 6°C sec^{-1}) occur but neither sensitization nor PTTL is observed after gamma irradiation and 300°C/1 hr annealing. Shining 254 nm UV light results, instead, in an increase of the residual 370°C peak although a weak 210°C peak also appears (Figure 21). In the absence of the 370°C peak, UV has no effect. The authors assume the creation of new UV-sensitive centers associated with high temperature peaks which are destroyed by the removal of the 370°C peak.[98]

Manganese-activated $Cd_2B_4O_7$, on the other hand, is intrinsically sensitive to UV in the range of 248 to 312 nm and has four peaks at −165, −60, 95, and 130°C (heating rate 1.5°C sec^{-1}). All the peaks have the same emission spectrum: a broad band with maximum at 630 nm. An interesting phenomenon of coupled traps is observed. The traps corresponding to 130°C peaks are filled only via the traps for the −165°C peak; if 1 μm light is shined simultaneously with UV excitation both the peaks are affected, whereas after the UV excitation only the − 165°C peak is affected. Furthermore, for the latter no thermally stimulated conductivity is observed indicating that the released electron goes directly to the Mn activator.[141]

C. Calcium Carbonate

Natural calcite containing large quantities of Mg (up to 10%) and Mn (up to 1%) as impurities and another sample containing lower levels (up to 1% Mg) were observed to have glow peaks at 100, 220, 320, and 350°C (heating rate 1°Csec^{-1}) on UV irradiation from a low pressure mercury lamp.[142,143] Synthetic crystals grown by the hydrothermal method with Pb and Mn impurities and several other impurities (Si, Al, Mg, Na) did not show any TL for excitation by λ >300 nm, but after excitation by 254 nm, peaks were observed at − 170, − 135, − 105, 25, 65 to 90, 140, 160, 220, and 320°C. The emission spectra consisted of wide bands at 450 and 620 nm and, except for the 65 to 90°C peak, the 450-nm band was predominant.[144] Natural crystals after annealing at 430°C showed only 65 to 90°C peaks. Peak temperatures for gamma and UV excitation may,[144] or may not,[143] coincide. The temperature shift (higher side) and the TL intensity, in the UV-excited sample compared to the gamma-irradiated one, decrease with increasing temperature. UV effects are wavelength dependent and resemble the optical absorption spectra; 290 and 250 nm light has the greatest effect on the intensity of peaks. UV bleaches peaks present in the natural sample and builds up the decayed or the thermally drained peaks. Some of the observations on UV effects have been explained by postulating the presence of an optically active defect center which is capable of being ionized upon illumination and provide free electrons and positively charged nonradiative recombination centers. The electrons populate traps responsible for peaks in the region of 100°C, and the radiationless recombination centers affect intensities of other glow peaks.

D. Bone Mineral

Hydrazine deproteinated bone mineral develops sensitivity to UV light (from Xe lamp) if heat treated at 400°C. The UV-induced peak is observed at 85°C and attains a maximum after 60 min of annealing. The heat treatment is seen to introduce an ESR signal and to cause an increase in overall TL intensity analogous to growth to saturation. Gamma and X-rays do not produce any TL. Both the TL intensity and the ESR signal decay exponentially if the sample is subjected to a continuous cycle of UV exposure and TL readout. These results indicate that UV transfers charge carriers from the deep ESR-sensitive traps to the 85°C TL traps.[145] Hydroxyapatite and amorphous calcium phosphate also show somewhat similar TL properties.[146]

VII. ORGANIC SYSTEMS

A. Green Plants

1. Introduction

This section reviews very briefly the application of PSTL as a tool to probe the internal structure of large organic molecules and the process of photosynthesis in green plants. The photosynthetic activity of green plants is accompanied by fluorescence and delayed fluorescence. The component of delayed fluorescence, under certain experimental conditions, is studied as thermoluminescence. Changes in the yield pattern of PSTL have been used very effectively to investigate the oxygen-evolving photosynthetic photochemical reactions and the energy storage states of the photosynthetic membranes.

The photosynthetic electron transport in higher plants and algal cells is driven by two photosystems (PS I and PS II) functioning in series. The two photosystems are connected by a thermochemical bridge. Most of the light absorbed by photosynthetic pigments upon illumination is used up in photosynthesis but a few percent of this light is reemitted as fluorescence. The major portions of the pigment systems are composed of various spectral forms of chlorophyll a and other accessory pigments. An assemblage of about 300 to 400 molecules comprises a photosynthetic unit with its own reaction center molecule. Light energy absorbed in any one of the accessory pigments leads to excitation energy transfer to chlorophyll a molecules. Since the primary photochemical reaction of photosynthesis is extremely efficient, this transfer process has to be very efficient too. Thus the light energy absorbed by various pigments including chlorophyll a leads to the trapping of energy by reaction center chlorophyll (the trap). The fluorescence yield of PS I is weak and is less affected by its photochemistry. On the other hand, the photochemistry of PS II is in competition with chlorophyll a fluorescence intensity. PS II also shows delayed fluorescence.[147] Delayed fluorescence is of extremely low intensity and is believed to be due to the instability at room temperature of some negative and positive charges in the reaction center which recombine to release chemical energy as delayed light emission.

2. PSTL

a. Glow Curve

The pigment-containing chloroplasts when cooled while being irradiated, or irradiated at low temperature, have some of the metastable charges frozen stabilized and require energy to be activated. On warming such frozen chloroplasts, the stabilized positive and negative charges overcome the activation energy barrier and recombine to emit light from the chlorophyll molecule which is excited by release of their recombination energy. The phenomenon of thermoluminescence in dried chloroplasts was first observed by Arnold and Sherwood in 1957.[148] Six glow peaks have been observed in the glow curve of intact spinach, between liquid nitrogen temperature and 60°C as shown in Figure 22.[149,150] As expected, the glow peak temperatures observed by different investigators vary but agree with each other within the range of experimental variability. The glow peaks are designated as Z (− 160°C), Z_v (variable), A (− 10°C), B_1 (25°C), B_2 (40°C), and C (55°C).

b. Origin of Glow Peaks

The Z band is not directly related to the photosynthetic reaction because it is present in the glow curve of a leaf boiled for the inactivation of photochemical reaction and because it is excited much more efficiently by blue light than by red light.[146] However, the finding that the Z band could also be excited by red light or gamma radiation and that its emission maximum occurs at 740 nm led Sane et al.[149] to consider the Z band as arising from a triplet state of chlorophyll.

The Z_v band is extremely weak and shows variation in its emission temperature depending on the excitation temperature. The exact origin of this band is not settled but appears to arise from charges stabilized near the PS II reaction center. Desai et al.[151] suggested that the Z_v band is due to the Z band and peak I of Desai et al.,[152] depending upon the temperature of excitation.

The A, B_1, and B_2 bands are most closely related to the photochemical reactions in photosynthesis. It has been shown that preheating of spinach leaves of up to 90°C eliminates the A, B, and C bands and that species lacking in PS II activity do not emit them.[153] These bands are, therefore, believed to arise from recombination of the charges produced by photosynthetic photochemical reaction and stabilized at low temperature on various charge carriers of the electron transport chain. The effect of various chemicals, of importance in the study of photosynthesis, like DCMU,* CCCP, NH_2OH, TRIS, has also been investigated to correlate thermoluminescence with detailed photosynthetic reactions.

* DCMU: 3-(3′,4′-dichlorophenyl)-1, 1-dimethyl urea; CCCP: carbonyl cyanide m-chlorophenylhydrazone; TRIS: Tris (hydroxymethyl) aminomethane.

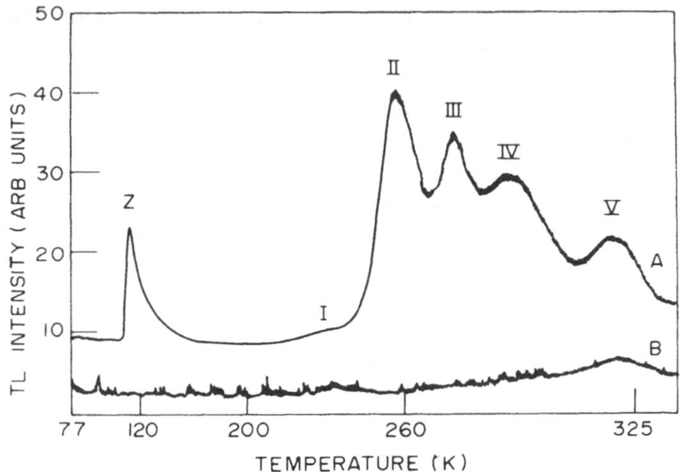

FIGURE 22. Glow curves of spinach leaves frozen to 77 K. (A) Leaf frozen in medium intensity of white light and exposed for 2 min. (B) Relaxed leaf frozen in dark. (Adapted from Desai, T. S., Sane, P. V., and Tatake, V. G., *Photochem. Photobiol.*, 21, 345, 1975. With permission.)

c. PSTL and the Oxygen-Evolving System

By choosing mutant species specifically lacking in PS I or PS II activity, thermoluminescence has been correlated with PS II activity. Furthermore, chloroplasts in which PS I and PS II are functionally developed but the oxygen-evolving system is latent show Z_v band and a very weak C band but no A, B_1, and B_2 bands. When the oxygen-evolving system is activated (by continuous light), the three bands simultaneously appear. Similarly it has been demonstrated that in Mn-deficient cells (Mn enzyme is involved during photoactivation of PS II) the glow curve is devoid of B band and on addition of Mn^{2+} to the deficient cells followed by shortly spaced flash illumination the B band appears rapidly with concomitant reactivation of the oxygen-evolving system. From such studies it has been concluded that the positive charges for the three glow peaks A, B_1, and B_2 are those stabilized in the oxygen-evolving system most probably as oxidized Mn atoms in the "Mn-enzyme-protein".

d. Further Detailed Relationship of TL and Oxygen Evolution

In photosynthetic oxygen evolution, four positive charges stored stepwise with flashes on an imaginary charge accumulator, S, cooperate to oxidize two molecules of water. The charge accumulator undergoes a quadruple cyclic change to evolve oxygen at every four flashes. Since the thermoluminescence bands A, B_1, and B_2 are found to be closely linked to the oxygen-evolving activity, it is inferred that the positive charges responsible for the three bands are stored on the S system so that the three bands may undergo quadruple oscillation when excited by saturating short flashes.[147] This has indeed been observed for the three bands. As a result of detailed investigations the three bands have been related to the different states of the oxygen clock. The various nuances of these investigations have recently been reviewed.[154]

The relationship between electron transport and glow peaks has been studied using differential excitation of PS I and PS II and by fractionation of the chloroplast membranes. These studies have shown that peaks appearing at −40 and −10°C arise in PS II, the peaks appearing at 10 and 25°C are related to electron transport between the two photosystems, whereas the peak appearing at 45°C arises from PS I excitation.[152,155] An analysis of the glow peaks shows that the higher temperature ones do not follow the Randall-Wilkins theory.[156]

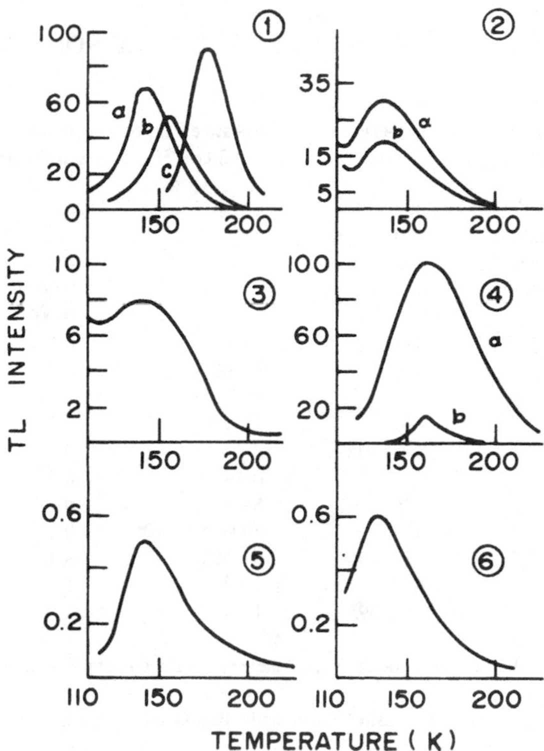

FIGURE 23. TL glow peaks in textile fibers. (1a) Terital. (1b) Vertan. (1c) Dacron. (2a) Lilion. (2b) Nylon. (3) Movil. (4a) Orlon. (4b) Velicren. (5) Koplon. (6) Wool. (Adapted from Aramu, F., Maxia, V., and Rucci, A., *J. Luminescence*, 3, 438, 1971. With permission.)

B. PSTL of Textile Fibers

Textile fibers both man-made and natural show UV-excited TL. Interestingly in each fiber the glow peak is found to occur at a different temperature (Figure 23). Aramu et al.[157] recorded glow peaks after irradiation with unfiltered UV light and heating at the rate of 1.2°C sec^{-1}. The peak temperatures were as follows: terital (−131 ± 2°C), vestan (−118 ± 1°C), dacron (−98 ± 1°C), nylon (−139 ± 4°C), lilion (−137 ± 1°C), movil (−134 ± 2°C), orlon (−113 ± 3°C), velicren (−113 ± 1°C), koplon (−130 ± 1°C), and wool (−141 ± 2°C). Polyester fibers — terital, vestan, and dacron — in spite of their similar composition, give glow peaks with different temperature maxima and have high yield. Polyammidic fibers — nylon and lilion — also have strong TL yields, but exhibit little difference between the 66 and 6 chains compounds. Polyvinylcyanide fibers — orlon and velicren — give well-developed glow peaks of different areas, but the same temperature maximum. In the case of movil polyvinylchloride fiber, the emission starts as an afterglow at LNT and the peak occurs at a temperature lower than in the case of other cyanide fibers. The emission of koplon cellulosic fiber and of natural wool are very weak. This may be due to the small concentration of defects present in these natural materials. The activation energies of the TL peaks in these fibers have been calculated to range from 0.066 to 45 eV and to be of kinetic order one.

Table 1
COMPARATIVE PITL RESPONSE PARAMETERS OF SOME MATERIALS[a]

Material	UV range[b] (nm)	Main peak temperature (K)	Exposure range (mJ/cm^{-2})	Relative sensitivity	Remarks
TLD-100	254	363	—	—	2 mGy[c]
CaF_2-nat	Broad spectrum	530	—	—	5—20 mGy[c]
BeO	Broad spectrum	450—490	—	—	1 mGy[c]
CaF_2:Mn	Broad spectrum	425, 450	—	0.0001	—
$Li_2B_4O_7$	254	395, 485	—	0.0005	—
$CaSO_4$:Mn	<150	395	1—10^4	0.0001	Linear
$CaSO_4$:Dy	<400	490	4—400	0.001	Linear
			400—5 × 10^3		Supralinear
$CaSO_4$:Tm	<400	490	1—100	0.03	Linear
CaF_2:Dy	(254)	450—500	0.2—3	0.1	Linear
			0.4—15	0.5	Linear
ZrO_2	(254)	375	Small	0.1	Fast decay
Al_2O_3	< 320	550	500—5 × 10^4	0.1	Linear
MgO	<350	420, 460	5—500	<1	Linear
	280—330	420	1—500	1	Linear
Mg_2SiO_4:Tb	<360	450, 595	1—10^4	1	Supralinear

[a] The figures given in this table are only indicative, as the exact values depend on a number of factors and experimental conditions.

[b] The response usually increases with decreasing wavelength. Bracketed figures indicate peak value.

[c] Gamma-equivalent saturation dose.

Table 2
COMPARATIVE PTTL RESPONSE CHARACTERISTICS OF SOME TL MATERIALS[a]

Material	UV range[b] (nm)	Main peak temperature (K)	Sensitizing exposure (Gy)	Postannealing temperature (K)	Useful range ($mJ\ cm^{-2}$)
LiF	<300	470	10^3	575—625	10^{-4}—6
CaF_2:nat	<360	535	10^4	675	1—3 × 10^4
CaF_2:Mn	<360	530	10^3	735	1—5.5 × 10^3
$CaSO_4$:Tm or Dy	<360	475	5 × 10^5	675	10^{-3}—1
Al_2O_3:Si,Ti	<425	550	10^2	800	10^{-2}—100
	<320	550	3J cm^{-2}	800	10—5 × 10^4
BeO	Broad spectrum (312)	460, 550	10^3	673	1.5—120
CaF_2:Dy	Broad spectrum (390—450)	420, 475	10^3	575	—

[a] The figures given in this table are only indicative, as the exact values depend on a number of factors and experimental conditions.

[b] The response usually increases with decreasing wavelength. Bracketed figures indicate peak value.

VIII. CONCLUSIONS

The PITL and PTTL characteristics of some TL materials are summarized in Tables 1 and 2, respectively. Photostimulated thermoluminescence studies constitute an essential component in the complete investigation of TL characteristics of any material. PSTL is found to be a useful tool in understanding the mechanisms of thermoluminescence. It is also observed to be of great help in isolating peaks in a multipeak glow curve.[158] It provides means for the measurement of ultraviolet radiation over a wide range of energies and intensities. PSTL is successfully used in the reestimation of dose once measured and thus erased.

ACKNOWLEDGMENTS

The author acknowledges with gratitude the support and the facilities provided by the BARC authorities, in particular Mr. S. D. Soman, Head, Health Physics Division. Thanks are also due to Dr. P. V. Sane for suggestions on the section dealing with TL of organic systems.

REFERENCES

1. **Stoddard, A. E.,** Effects of illumination upon sodium chloride thermoluminescence, *Phys. Rev.*, 120, 114, 1960.
2. **Wiedeman, E. and Schmidt, G. C.,** Ueber Luminescenz, *Ann. Phys. Chem. N. F.*, 54, 604, 1895.
3. **Kiss, Z. J. and Staebler, D. L.,** Dynamics of oxidation-reduction processes in rare-earth doped CaF_2, *Phys. Rev. Lett.*, 14, 691, 1965.
4. **Kask, N. E., Kornienko, L. S., Lozhnikov, A. A., and Chernov, P. V.,** Photostimulated thermoluminescence of fluorite crystals with admixed erbium and holmium ions, *Sov. Phys. Solid State Phys.*, 12, 2792, 1971.
5. **Batsanova, S. S., Korobeinikova, V. N., Kazakov, V. P., and Kobets, L. F.,** Thermoluminescence of CaF_2-Tb^{3+} crystals, *Sov. Phys. Opt. Spectrosc.*, 30, 265, 1971.
6. **Kristianpoller, N. and Israeli, M.,** On glow curves obtained by ionizing and non-ionizing radiation, *Phys. Status Solidi B*, 47, 487, 1971.
7. **Kirsh, Y. and Kristianpoller, N.,** UV induced processes in pure and doped SrF_2, *J. Luminescence*, 15, 35, 1977.
8. **Staebler, D. L. and Schnatterly, S. E.,** Optical studies of a photochromic color centre in rare-earth-doped CaF_2, *Phys. Rev. B*, 3, 516, 1971.
9. **Becker, M., Kiessling, J., and Scharmann, A.,** Thermoluminescence of CaF_2:Sm, *Phys. Status Solidi A*, 15, 515, 1973.
10. **Nagpal, J. S.,** Ultraviolet induced processes in thermoluminescent phosphors, *Phys. Status Solidi A*, 57, K63, 1980.
11. **Tournon, J. and Berge, P.,** Thermoluminescence à basse température de cristaux de LiF, *Phys. Status Solidi A*, 5, 117, 1964.
12. **Braner, A. and Israeli, M.,** Effects of illumination on the thermoluminescence of alkali halides, *Phys. Rev.*, 132, 2501, 1963.
13. **Bossachi, B., Fieschi, R., and Scaramelli, P.,** Photostimulated thermoluminescence in potassium chloride single crystals, *Phys. Rev.*, 138, A1760, 1965.
14. **Fieschi, R. and Scaramelli, P.,** Evidence for L bands in NaCl, *Phys. Rev.*, 145, 622, 1966.
15. **Fieschi, R., Golo, M., and Scaramelli, P.,** Photostimulated thermoluminescence in alkali halide crystals, in Proc. Int. Conf. Luminescence, Szigeti, G., Ed., Hungarian Academy of Sciences, Budapest, 1966, 86.

16. **Scaramelli, P.,** F′ and M′ traps in alkali halides studied by means of photostimulated thermoluminescence, *Lett. Nuovo Cimento,* 45B, 119, 1967.
17. **Crippa, P. R. and Paracchini, C.,** Photostimulated thermoluminescence in KBr single crystals, *J. Phys. Soc. Jpn.,* 24, 92, 1968.
18. **Fieschi, R. and Scaramelli, P.,** Photostimulated thermoluminescence in alkali halide crystals, in *Thermoluminescence of Geological Materials,* McDougall, D. J., Ed., Academic Press, London, 1968, 291.
19. **Fieschi, R. and Paracchini, C.,** Photostimulated thermoluminescence in additively coloured KCl, *Phys. Rev.,* 182, 815, 1969.
20. **Diffey, B. L.,** Ultraviolet radiation physics and the skin, *Phys. Med. Biol.,* 25, 405, 1980.
21. **Pitts, D. G. and Tredici, T. J.,** The effects of ultraviolet on the eye, *Am. Ind. Hyg. Assoc. J.,* 32, 235, 1971.
22. **Sliney, D. H.,** Non-ionizing radiation, in *Industrial Environmental Health,* Academic Press, New York, 1972, 117.
23. American Conference of Governmental Industrial Hygienists, Threshold Limit Values for Chemical Substances and Physical Agents in the Workroom Environment. Ultraviolet Radiation, Cincinnati, 1973.
24. National Institute for Occupational Safety and Health, Criteria for a Recommended Standard — Occupational Exposure to Ultraviolet Radiation, U.S. Department of Health, Education and Welfare, Washington, D.C., 1972.
25. **Brooke, C.,** Properties of different phosphors as used in packaged TLD systems, multiple readings of dose by UV light transfer, in *Solid State Dosimetry,* Amelinckx, S., Batz, B., and Straumane, R., Eds., Gordon & Breach, New York, 1969.
26. **McKinlay, A. F., Bartlett, D. T., and Smith, P. A.,** The development of phototransferred thermoluminescence (PTTL) technique and its application to the routine re-assessment of absorbed dose in the NRPB automated personal dosimetry system, *Nucl. Instrum. Methods,* 175, 57, 1980.
27. **Jain, V. K.,** Phototransfer, sensitization and re-estimation of dose in lithium fluoride TLD, *Health Phys.,* 41, 363, 1981.
28. **Cameron, J. R., Suntharalingam, N., and Kenney, G. N.,** *Thermoluminescence Dosimetry,* University of Wisconsin Press, Madison, 1968.
29. **Mayhugh, M. R. and Fullerton, G. D.,** Thermoluminescence in LiF: Useful Sensitization by Pre-irradiation, Rep. No. COO-1105-209, U.S. A.E.C., NTIS, Springfield, Va.
30. **Buckman, W. G. and Payne, M. R.,** Photostimulated thermoluminescence of lithium fluoride as an ultraviolet radiation dosimeter, *Health Phys.,* 31, 501, 1976.
31. **Lippert, J. and Mejdahl, V.,** Thermoluminescence readout instrument for measurement of small doses, in Luminescence Dosimetry, U.S.A.E.C. CONF-650637, NTIS, Springfield, Va., 1967, 204.
32. **Bjärngard, B. E., McCall, R. C., and Bernstein, I. A.,** Lithium fluoride-teflon thermoluminescence dosimeter, in Luminescence Dosimetry, U.S.A.E.C. CONF-650637, NTIS, Springfield, Va., 1967, 308.
33. **Bjärngard, B. E. and Jones, D.,** Experience with a new thermoluminescence method for finger and hand dosimetry employing lithium fluoride-teflon dosimetry, in *Proc. 1st Int. Congr. Radiat. Prot.,* Snyder, W. S., Ed., Pergamon Press, New York, 1968, 473.
34. **Freeswick, D. C. and Shambon, A.,** Light sensitivity of LiF thermoluminescent dosimeters, *Health Phys.,* 19, 65, 1970.
35. **Marshall, T. O., Shaw, K. B., and Mason, E. W.,** The consistency of the dosimetric properties of ^{7}LiF in teflon discs over repeated cycles of use, in Proc. 3rd Int. Conf. Luminescence Dosimetry, Mejdahl, V., Ed., Rep. No. 249, AEC/IAEA, Riso, Denmark, 1971, 530.
36. **Mason, E. W.,** Thermoluminescence response of ^{7}LiF to ultraviolet light, *Phys. Med. Biol.,* 16, 303, 1971.
37. **Webb, G. A. M. and Phykitt, H. P.,** Possible elimination of the annealing cycle for thermoluminescent LiF, in Proc. 3rd Int. Conf. Luminescence Dosimetry, Mejdahl, V., Ed., Rep. No. 249, AEC/IAEA, Risö, Denmark, 1971, 185.
38. **Townsend, P. D., Clark, C. D., and Levy, P. W.,** Thermoluminescence in lithium fluoride, *Phys. Rev.,* 155, 908, 1967.
39. **Pearson, D. and Cameron, J. R.,** Repopulation of TL Traps in LiF, Rep. No. COO-1105-140, U.S. A.E.C., NTIS, Springfield, Va., 1968.
40. **Mayhugh, M. R.,** TL in LiF (TLD-100) and $CaSO_4$:Dy, Rep. No. COO-1105-214, U.S. A.E.C., NTIS, Springfield, Va., 1974.
41. **Gower, R. G., Hendee, W. R., and Ibbott, G. S.,** Ultraviolet-induced changes in residual thermoluminescence from gamma-irradiated lithium fluoride, *Health Phys.,* 17, 607, 1979.
42. **Sunta, C. M., Bapat, V. N., and Kathuria, S. P.,** Effects of deep traps on supralinearity, sensitization and optical thermoluminescence in LiF TLD, in Proc. 3rd Int. Conf. Luminescence Dosimetry, Majdahl, V., Ed., Rep. No. 249, AEC/IAEA, Risö, Denmark, 1971, 146.
43. **Mason, E. W. and Linsely, G. S.,** Properties of some deep traps in lithium fluoride, in Proc. 3rd Int. Conf. Luminescence Dosimetry, Mejdahl, V., Ed., Rep. No. 249, AEC/IAEA, Risö, Denmark, 1971, 164.

44. **Nakajima, T.,** On the causes of changes in sensitivity due to re-use of LiF thermoluminescence dosimeters, *Health Phys.*, 16, 509, 1969.
45. **Bassi, P., Busuoli, G., Lembo, L., and Rimondi, O.,** Transferred and intrinsic thermoluminescence for UV dosimetry, in Proc. 4th Int. Conf. Luminescence Dosimetry, Niewiadomski, T., Ed., Institute of Nuclear Physics, Krakow, 1971, 1073.
46. **Jain, V. K. and Ganguly, A. K.,** Some aspects of thermal radiation and LET effects on the thermoluminescence of lithium fluoride, Rep. B.A.R.C. I-466, AEC, Bhabha Atomic Research Center, Bombay, 1977.
47. **Pearson, D. W. and Cameron, J. R.,** Wavelength Dependence of Photostimulated TL in LiF, Rep. No. COO-1105-149, U.S. Atomic Energy Commission, Washington, D.C., 1969.
48. **Jain, V. K. and Kathuria, S. P.,** Z_3 centre thermoluminescence in LiF TLD phosphor, *Phys. Status Solidi A*, 50, 329, 1978.
49. **Pearson, D. W. and Cameron, J. R.,** Supralinearity of UV Repopulated TL from TLD-100, Rep. No. COO-1105-156, U.S. A.E.C., NTIS, Springfield, Va., 1970.
50. **Kos, J. H. and Nink, R.,** Mechanism of restored thermoluminescence in lithium fluoride, *Phys. Status Solidi A*, 44, 505, 1977.
51. **Jain, V. K.,** Radiation-induced sensitization and UV effects in lithium fluoride TLD phosphor, *Phys. Status Solidi A*, 60, 351, 1980.
52. **Jain, V. K.,** Z_3 centre thermoluminescence and sensitization mechanism in lithium fluoride, *Phys. Status Solidi A*, 66, 341, 1981.
53. **Horowitz, Y. S.,** Criticism of the Z-centre model in LiF-TLD, *Phys. Status Solidi A*, 69, K29, 1982.
54. **Sagastibelza, F. and Alvarez Rivas, J. L.,** Thermoluminescence in LiF (TLD-100) and LiF crystals irradiated at room temperature, *J. Phys. C*, 14, 1837, 1981.
55. **Mason, E. W., McKinlay, A. F., and Saunders, D.,** The re-estimation of absorbed doses of less than 1 rad measured with LiF TLD, *Phys. Med. Biol.*, 22, 29, 1977.
56. **Douglas, J. A., Baker, D. M., Marshall, M., and Budd, T.,** The effect of LET on the efficiency of dose re-estimation in LiF using UV photo-transfer, *Nucl. Instrum. Methods*, 175, 54, 1980.
57. **Horowitz, Y. S., Fraier, I., Kalefezra, J., Pinto, H., and Goldbart, Z.,** Non-universality of the TL-LET response in thermoluminescent LiF: the effect of batch composition, *Phys. Med. Biol.*, 24, 1268, 1979.
58. **Bartlett, D. T., McKinlay, A. F., and Smith, P. A.,** The routine re-assessment of absorbed dose in LiF/polytetrafluoroethylene thermoluminescent dosimeter elements, *Nucl. Technol.*, 49, 504, 1980.
59. **Mayhugh, M. R. and Fullerton, G. D.,** Thermoluminescence in LiF: sensitization useful at low exposures, *Health Phys.*, 28, 297, 1975.
60. **Sunta, C. M. and Watanabe, S.,** Thermoluminescence of LiF TLD-100 by phototransfer, *J. Phys. D*, 9, 1271, 1976.
61. **Bartlett, D. T. and Sandford, D. J.,** Incompatibility of sensitization and re-estimation of dose in lithium fluoride thermoluminescent phosphor, *Phys. Med. Biol.*, 23, 332, 1978.
62. **Charles, M. W. and Khan, Z. U.,** The development of ultra-thin dosimeters for skin dose assessment, IAEA Symp. Adv. Radiat. Prot. Monitoring, Paper IAEA-SM-229/24, Stockholm, 1978.
63. **Charles, M. W. and Khan, Z. U.,** Simultaneous sensitization and re-estimation in thermoluminescent LiF, in Proc. 5th Int. Congr. IRPA, Jerusalem, 1981.
64. **Kathuria, S. P. and Sunta, C. M.,** Thermal quenching by phototransfer in LiF TLD 100, *Phys. Med. Biol.*, 26, 707, 1981.
65. **Horowitz, Y. S.,** Comment on "Thermal quenching by phototransfer in LiF TLD-100", *Phys. Med. Biol.*, 27, 604 and 1190, 1982; Jain, V. K., *Phys. Med. Biol.*, 27, 1525, 1982.
66. **Frechette, V. D. and Cline, C.,** Stimulation of TL in LiF by ruby laser light, *Appl. Phys. Lett. (U.S.A.)*, 10, 39, 1967.
67. **Puite, K. J.,** The effect of ultraviolet exposure on thermoluminescent CaF_2:Mn, *Int. J. Appl. Radiat. Isot.*, 19, 397, 1968.
68. **Sunta, C. M.,** Optically produced thermoluminescence glow peaks in fluorite, *Phys. Status Solidi A*, 37, K81, 1970.
69. **Schayes, R., Brooke, C., Kozlowitz, I., and Lheureux, M.,** Thermoluminescent properties of natural calcium fluoride, in Luminescence Dosimetry, U.S.A.E.C. CONF-650637, Attix, F. H., Ed., NTIS, Springfield, Va., 1967, 138.
70. **Brooke, C. and Schayes, R.,** Recent developments in thermoluminescent dosimetry; extensions in the range of applications, in Proc. Symp. Solid State and Chemical Radiation Dosimetry in Medicine and Biology, STI/PUB/138, IAEA, Vienna, 1967, 31.
71. **Bassi, P., Busuoli, G., and Rimondi, O.,** UV dosimetry by intrinsic TL of CaF_2:Dy, *Health Phys.*, 31, 179, 1976.

72. **Bassi, P., Busuoli, G., and Rimondi, O.,** Thermoluminescent dosimeters for UV radiations, in Proc. 4th Int. Cong. Radiat. Prot., Vol. 2, IRPA, Paris, 1977, 321.
73. **Bassi, P., Busuoli, G., and Rimondi, O.,** A practical dosimeter for UV light, *Nucl. Instrum. Methods,* 143, 195, 1977.
74. **Pradhan, A. S., Kher, R. K., and Bhatt, R. C.,** Effect of temperature treatment on sensitivity of CaF_2:Dy to ultraviolet and ionizing radiations, *Int. J. Appl. Radiat. Isot.,* 29, 329, 1978.
75. **Pradhan, A. S. and Bhatt, R. C.,** Ultraviolet and gamma-ray induced TL emission spectrum of CaF_2:Dy. Effect of temperature treatment, *Int. J. Appl. Radiat. Isot.,* 32, 179, 1981.
76. **Bernhardt, R. and Herforth, L.,** Radiation dosimetry by optically stimulated phosphorescence of CaF_2:Mn, in Proc. 4th Int. Conf. Luminescence Dosimetry, Niewiadomski, T., Ed., Institute of Nuclear Physics, Krakow, Poland, 1974, 1091.
77. **Wilson, C. R., Lin, F. M., and Cameron, J. R.,** Preliminary Investigation in the use of Thermoluminescence for Ultraviolet Radiation, Rep. No. C00-1105-136, U.S. A.E.C., NTIS, Springfield, Va., 1967.
78. **Las, W. C. and Watanabe, S.,** Mechanism of UV induced thermoluminescence in natural calcium fluorite, in Proc. 4th Int. Conf. Luminescence Dosimetry, Vol. 3, Niewiadomski, T., Ed., Institute of Nuclear Physics, Krakow, 1974, 1187.
79. **Sunta, C. M.,** Thermoluminescence of natural CaF_2 and its applications, in Proc. 3rd Int. Conf. Luminescence Dosimetry, Rep. No. 249, Danish AEC/IAEA, Risö, Denmark, 1971, 392.
80. **Okuno, E. and Watanabe, S.,** UV induced TL in natural CaF_2, *Health Phys.,* 23, 377, 1972.
81. **Schayes, R., Brooke, C., Kozlowitz, I., and Lheureux, M.,** New developments in thermoluminescent dosimetry, *Health Phys.,* 14, 251, 1968.
82. **Jain, V. K. and Mitra, S.,** Effect of thermal treatment on the thermoluminescence of yellow and colourless fluorites from Amba Dongar, Gujrat (India), *Thermochim. Acta,* 35, 349, 1980.
83. **Sunta, C. M.,** Mechanism of phototransfer of thermoluminescence peaks in natural CaF_2, *Phys. Status Solidi A,* 53, 127, 1979.
84. **Van Espen, E.,** Memory effects with M.B.L.E. CaF_2 thermoluminescent dosimeters, in *Radiation Protection,* Proc. 1st Int. Congr. IRPA, Vol. 1, Snyder, W. S., Ed., Pergamon Press, Oxford, 1966, 449.
85. **Harvey, J. R.,** Some applications of low dose thermoluminescence measurements with calcium fluoride, in Luminescence Dosimetry, U.S.A.E.C. CONF-650637, Attix, F. H., Ed., NTIS, Springfield, Va., 1967, 331.
86. **McCullough, E. C., Fullerton, G. D., and Cameron, J. R.,** Thermoluminescence in natural calcium fluoride as a dosimeter for terrestrial solar ultraviolet radiation, *J. Appl. Phys.,* 43, 77, 1972.
87. **Kornienko, L. S., Lozhnikov, A. A., Nazarov, V. I., and Chernov, P. V.,** *Opt. Spectrosc.,* 35, 651, 1973.
88. **Lyman, T.,** The transparency of air between 1100 and 1300 A, *Phys. Rev.,* 48, 149, 1935.
89. **Watanabe, K.,** Properties of a $CaSO_4$:Mn phosphor under vacuum ultraviolet excitation, *Phys. Rev.,* 83, 785, 1951.
90. **Tousey, R., Watanabe, K., and Purcell, J. D.,** Measurements of solar extreme ultraviolet and X-rays from rockets by means of a $CaSO_4$:Mn phosphor, *Phys. Rev.,* 83, 792, 1951.
91. **Chandra, B., Ayyangar, K., and Lakshmanan, A. R.,** Ultraviolet response of $CaSO_4$:Dy, *Phys. Med. Biol.,* 21, 67, 1976.
92. **Shastry, S. S. and Shinde, S. S.,** High-exposure ultraviolet response and sensitization of $CaSO_4$:Dy, *Phys. Med. Biol.,* 24, 1033, 1979.
93. **Nagpal, J. S.,** Phototransferred thermoluminescence studies in $CaSO_4$:Dy, *Phys. Med. Biol.,* 25, 549, 1980.
94. **Pradhan, A. S., Kher, R. K., and Bhatt, R. C.,** TL sensitivity of $CaSO_4$:Dy. Effect of temperature treatment, *Int. J. Appl. Radiat. Isot.,* 30, 127, 1979.
95. **Nagpal, J. S. and Pendurkar, H. K.,** Effect of temperature during UV and gamma irradiation of TL phosphor, *Nucl. Instrum. Methods,* 159, 521, 1979.
96. **Nambi, K. S. V. and Higashimura, T.,** Tm and Dy activated $CaSO_4$ for UV dosimetry, in Proc. 3rd Int. Conf. Luminescence Dosimetry, Mejdahl, V., Ed., Rep. No. 249, AEC/IAEA, Risö, Denmark, 1971, 1107.
97. **Caldas, L. V. E. and Mayhugh, M. R.,** Photo-TL in $CaSO_4$:Dy. High exposure dosimetry, *Health Phys.,* 31, 451, 1976.
98. **Pradhan, A. S., Bhatt, R. C., and Supe, S. J.,** Photoluminescence and the behaviour of high-temperature glow peaks in $Li_2B_4O_7$: Mn and $CaSO_4$:Dy phosphors, *Int. J. Appl. Radiat. Isot.,* 31, 671, 1980.
99. **Chandra, B., Bhatt, R. C., and Supe, S. J.,** High level gamma-dosimetry using $CaSO_4$:Dy phosphor with high Dy-concentration, *Int. J. Appl. Radiat. Isot.,* 32, 553, 1981.
100. **Pradhan, A. S. and Bhatt, R. C.,** Photostimulated luminescence and thermoluminescence in $CaSO_4$:Dy, *Phys. Status Solidi A,* 68, 405, 1981.

101. **Webb, G. A. M., Dauch, J. E., and Dodin, G.,** Operational evaluation of a new high sensitivity thermoluminescent dosimeter, *Health Phys.*, 23, 89, 1972.
102. **Nakajima, T.,** Optical and thermal effects on TL response of Mg_2SiO_4(Tb) and $CaSO_4$(Tm) phosphors, *Health Phys.*, 23, 133, 1979.
103. **Adtani, M. M., Sawant, R. V., Shetty, B., and Supe, S. J.,** Light induced fading in $CaSO_4$:Dy teflon thermoluminescence discs, *Radiat. Prot. Dosimetry*, 2, 119, 1982.
104. **Buckman, W. G.,** Aluminum oxide thermoluminescence properties for detecting radiation, *Health Phys.*, 22, 402, 1972.
105. **Basu, A. S. and Ganguly, A. K.,** Thermoluminescence properties of pure Al_2O_3, in Proc. Natl. Symp. Thermoluminescence and its Applications, Kalpakkam, Madras, February 12 to 15, 1975, 345.
106. **Ziniker, W. M., Rusin, J. M., and Stoebe, T. G.,** Thermoluminescence and activation energies in Al_2O_3, MgO and LiF (TLD-100), *J. Mater. Sci.*, 8, 407, 1973.
107. **Cooke, D. W.,** Enhancement of the ultraviolet sensitivity of aluminium oxide by gamma radiation, *Health Phys.*, 27, 130, 1974.
108. **Cooke, D. W., Roberts, H. E., and Alexander, Jr., C.,** Thermoluminescene and emission spectrum of UV-grade Al_2O_3 from 90—500 K, *J. Appl. Phys.*, 49, 3451, 1978.
109. **Buckman, W. G., Philbrick, C. R., and Underwood, N.,** The characteristics of ruby as a thermoluminescent dosimeter, Proc. 2nd Int. Conf. Luminescence Dosimetry, U.S.A.E.C. CONF-680920, Auxier, J. A., Becker, K., and Robinson, E. M., Eds., NTIS, Springfield, Va., 1968, 82.
110. **Buckman, W. G., Sutherland, D. C., and Cooke, D. W.,** The detection of ultraviolet radiation using the thermoluminescence of sapphire, Proc. 4th Annu. Midyear Topical Symp. Natl. Health Phys. Soc., 1970, 407.
111. **Vishnevsky, V. N., Gnyp, R. G., Pidzyrailo, N. S., and Tolchinska, R. M.,** X-ray and thermal luminescence of pure and doped corundum single crystals, *Ukr. Phys. (U.S.S.R.)*, X1 (9), 991, 1966.
112. **Mehta, S. K. and Sengupta, S.,** Photostimulated thermoluminescence of Al_2O_3 (Si, Ti) and its application to ultraviolet radiation dosimetry, *Phys. Med. Biol.*, 23, 471, 1978.
113. **Ziniker, W. M., Merrow, J. K., and Mueller, J. I.,** Thermally stimulated luminescence in MgO, *J. Phys. Chem. Solids*, 33, 1619, 1972.
114. **Ziniker, W. M., Rusin, J. M., and Stoebe, T. G.,** Thermoluminescence and activation energies in Al_2O_3, MgO and LiF (TLD-100), *J. Mater. Sci.*, 8, 407, 1973.
115. **Dhar, A., DeWerd, L. A., and Stoebe, T. G.,** Direct response ultraviolet thermoluminescent dosimeter, *Med. Phys.*, 3, 415, 1976.
116. **Kirsh, Y., Kristianpoller, N., and Chen, R.,** Vacuum ultraviolet induced thermoluminescence in gamma irradiated MgO powder, *Philos. Mag.*, 35, 653, 1977.
117. **Tekeuchi, N., Inabe, K., Yamashita, J., and Nakamura, S.,** Thermoluminescence of MgO single crystals for UV dosimetry, *Health Phys.*, 31, 519, 1976.
118. **Las, W. L., Mathews, R. J., and Stoebe, T. G.,** Mechanism for thermoluminescence in MgO and $CaSO_4$, *Nucl. Instrum. Methods*, 175, 1, 1980.
119. **Mollenkopf, H. C., Halliburton, L. E., and Kohnke, E. E.,** Initiation exoelectron emission by thermally released holes in single crystal MgO, *Phys. Status Solidi A*, 19, 243, 1973.
120. **Scarpa, G., Benincasa, G., and Ceravolo, L.,** Further studies on the dosimetric use of BeO as a thermoluminiscent material, in Proc. 3rd Int. Conf. Luminescence Dosimetry, Rep. No. 249, Mejdahl, V., Ed., AEC/IAEA, Risö, Denmark, 1971, 427.
121. **Hobzová, L.,** UV-induced thermoluminescence in BeO, in Proc. 4th Int. Conf. Luminescence Dosimetry, Vol. 3, Niewiadomski, T., Ed., Institute of Nuclear Physics, Krakow, 1974, 1081.
122. **Albrecht, H. O. and Mandeville, C. E.,** Storage of energy in beryllium oxide, *Phys. Rev.*, 101, 1250, 1956.
123. **Tochilin, E., Goldstein, N., and Miller, W. G.,** Beryllium oxide as a thermoluminescent dosimeter, *Health Phys.*, 16, 1, 1969.
124. **Rhyner, C. R. and Miller, W. G.,** Radiation dosimetry by optically stimulated luminescence of BeO, *Health Phys.*, 18, 681, 1970.
125. **Yasuno, Y. and Yamashita, T.,** Thermoluminescent phosphors based on beryllium oxide, in Proc. 3rd Int. Conf. Luminescence Dosimetry, Rep. No. 249, Mejdahl, V., Ed., AEC/IAEA, Risö, Denmark, 1971, 290.
126. **Yamashita, T., Yasuno, Y., and Ikeda, M.,** Beryllium oxide doped with lithium or sodium for thermoluminescence dosimetry, *Health Phys.*, 27, 201, 1974.
127. **Scarpa, G.,** The dosimetric use of beryllium oxide as a thermoluminescent material: a preliminary study, *Phys. Med. Biol.*, 15, 667, 1970.
128. **Crase, K. W. and Gammage, R. B.,** Improvements in the use of ceramic BeO for TLD, *Health Phys.*, 29, 739, 1975.

129. **Gammage, R. B. and Cheka, J. S.,** Further characteristics important in the operation of ceramic BeO TLD, *Health Phys.*, 32, 189, 1977.
130. **Medlin, W. L.,** Thermoluminescence in quartz, *J. Chem. Phys.*, 38, 1132, 1963.
131. **David, M. and Sunta, C. M.,** Thermoluminescence of quartz. VI. Effect of ultraviolet rays, *Ind. J. Pure Appl. Phys.*, 19, 1041, 1981.
132. **Jain, V. K.,** Some aspects of the thermoluminescence of zircon (sand) and zirconia, *Ind. J. Pure Appl. Phys.*, 15, 601, 1977.
133. **Iaconni, P., Lapraz, D., and Caruba, R.,** Traps and emission centres in thermoluminescent ZrO_2, *Phys. Status Solidi A*, 50, 275, 1978.
134. **Caruba, R., Turco, G., Iaconni, P., and Lapraz, D.,** La thermoluminescence de zircons de synthèse, *Bull, Soc. Fr. Minéral. Cristallogr.*, 98, 85, 1975.
135. **Jain, V. K.,** Thermoluminescence glow curve and spectrum of zircon (sand), *Bull. Mineral.*, 101, 358, 1978.
136. **Toryu, T., Sakamoto, H., Hitomi, T., Kotera, N., and Yamada, H.,** Composition dependency of thermoluminescence of new phosphors for radiation dosimetry, in *Luminescence of Crystals, Molecules and Solutions*, Proc. Int. Conf. Luminescence, William, F., Ed., Plenum Press, New York, 1973, 685.
137. **Bhasin, B. D., Sasidharan, R. S., and Sunta, C. M.,** Preparation and thermoluminescent characteristics of terbium doped magnesium orthosilicate phosphor, *Health Phys.*, 30, 139, 1976.
138. **Lakshmanan, A. R., Shinde, S. S., and Bhatt, R. C.,** Ultraviolet-induced thermoluminescence and phosphorescence in Mg_2SiO_4:Tb, *Phys. Med. Biol.*, 23, 952, 1978.
139. **Jun, J. S. and Becker, K.,** Thermoluminescent dosimetry with terbium activated magnesium orthosilicate, *Health Phys.*, 28, 459, 1975.
140. **Lakshmanan, A. R., Pendurkar, H. K., and Vora, K. G.,** Gamma-radiation-induced sensitization and phototransfer in Mg_2SiO_4:Tb TLD phosphor, *Nucl. Instrum. Methods*, 159, 585, 1979.
141. **Dittmann, R., Hahn, D., and Muller, U.,** On the luminescence and thermoluminescence of manganese activated cadmium borate, *J. Luminescence*, 3, 230, 1970.
142. **Vaz, J. E. and Zeller, E. J.,** Thermoluminescence of calcite from high gamma radiation doses, *Am. Mineral.*, 51, 1156, 1966.
143. **Vaz, J. E., Kemmey, P. J., and Levy, P. W.,** The effects of ultraviolet light illumination on the thermoluminescence of calcite, in *Thermoluminescence of Geological Materials*, McDougall, D. J., Ed., Academic Press, New York, 1966, 111.
144. **Lapraz, D. and Iaconni, P.,** On some luminescent and optical properties of synthetic calcite single crystals, *Phys. Status Solidi A*, 36, 603, 1976.
145. **Chapman, M. R., Miller, A. G., Burnell, J. M., and Stoebe, T. G.,** Thermoluminescence in the bone mineral system, in Proc. 5th Int. Conf. Luminescence Dosimetry, Scharmann, A., Ed., Sao Paolo, Physikalisches Institut, Giessen, 1977, 342.
146. **Chapman, M. R., Miller, A. G., and Stoebe, T. G.,** Thermoluminescence in hydroxyapatite, *Med. Phys.*, 6, 494, 1979.
147. **Arnold, W. and Azzi, J.,** The mechanism of delayed light production by photosynthetic organisms and a new effect of electric fields on chloroplasts, *Photochem. Photobiol.*, 14, 233, 1971.
148. **Arnold, W. and Sherwood, H. K.,** Are chloroplasts semiconductors? *Proc. Natl. Acad. Sci. U.S.A.*, 43, 105, 1957.
149. **Sane, P. V., Tatake, V. G., and Desai, T. S.,** Detection of the triplet states of chlorophyllis in vivo, *FEBS Lett.*, 45, 290, 1974.
150. **Ichikawa, T., Inoue, Y., and Shibata, K.,** Characteristics of thermoluminescence bands of intact leaves and isolated chloroplasts in relation to the water-splitting activity in photosynthesis, *Biochim. Biophys. Acta*, 408, 228, 1975.
151. **Desai, T. S., Tatake, V. G., and Sane, P. V.,** Characterization of the low temperature thermoluminescence band Z_v in leaf: an explanation for its variable nature, *Biochim. Biophys. Acta*, 462, 775, 1977.
152. **Desai, T. S., Sane, P. V., and Tatake, V. G.,** Thermoluminescence studies on spinach leaves and euglena, *Photochem. Photobiol.*, 21, 345, 1975.
153. **Arnold, W. and Azzi, J.,** Chlorophyll energy levels and electron flow in photosynthesis, *Proc. Natl. Acad. Sci. U.S.A.*, 61, 29, 1968.
154. **Inoue, Y. and Shibata, K.,** Thermoluminescence from photosynthetic apparatus of green plants in photosynthesis, in *Energy Conversions by Plants and Bacteria*, Govindjee, Ed., Academic Press, New York, 1982.
155. **Sane, P. V., Desai, T. S., Tatake, V. G., and Govindjee,** On the origin of glow peaks in euglena cells, spinach chloroplasts and sub-chloroplasts fragments enriched in system I or II, *Photochem. Photobiol.*, 26, 33, 1977.

156. **Tatake, V. G., Desai, T. S., Govindjee, and Sane, P. V.,** Energy storage states of photosynthetic membranes: activation energies and lifetime of electrons in the trapstates by thermoluminescence method, *Photochem. Photobiol.*, 33, 243, 1981.
157. **Aramu, F., Maxia, V., and Rucci, A.,** On the thermoluminescence of textile fibres, *J. Luminescence,* 3, 438, 1971.
158. **Ratnam, V. V. and Gartia, R. K.,** Photostimulated glow curve method of isolating thermoluminescence peaks, *Phys. Status Solidi A,* 27, 627, 1975.

INDEX

A

B

C

D

E

F

G

H

I

K

L

M

N

Q

R

S

T

U

V

W

X

Z